Geographic
Information
Systems
in
Water Resources
Engineering

Geographic Information Systems in Water Resources Engineering

Lynn E. Johnson

CRC Press
Taylor & Francis Group
Boca Raton   London   New York

CRC Press is an imprint of the
Taylor & Francis Group, an **informa** business

CRC Press
Taylor & Francis Group
6000 Broken Sound Parkway NW, Suite 300
Boca Raton, FL 33487-2742

First issued in paperback 2020

© 2009 by Taylor & Francis Group, LLC
CRC Press is an imprint of Taylor & Francis Group, an Informa business

No claim to original U.S. Government works

ISBN-13: 978-0-367-57742-1 (pbk)
ISBN-13: 978-1-4200-6913-6 (hbk)

Library of Congress Cataloging-in-Publication Data

Johnson, Lynn E.
    Geographic information systems in water resources engineering / author, Lynn E. Johnson.
        p. cm.
    "A CRC title."
    Includes bibliographical references and index.
    ISBN 978-1-4200-6913-6 (alk. paper)
    1. Water resources development--Geographic information systems. 2. Water resources development--Systems engineering. I. Title.

TC409.J66 2008
628.1028--dc22   2008044085

Visit the Taylor & Francis Web site at
http://www.taylorandfrancis.com

and the CRC Press Web site at
http://www.crcpress.com

Contents

Preface

Geographic information systems (GISs) are strongly impacting the fields of water resources engineering, environmental science, and related disciplines. GIS tools for spatial data management and analysis are now considered state of the art, and application of these tools can lead to improved analyses and designs. Familiarity with this burgeoning technology may be a prerequisite for success in our efforts to create reliable infrastructure and sustain our environment.

GIS is rapidly changing the ways that engineering planning, design, and management of water resources are conducted. Advances in data-collection technologies—using microprocessor-based data-collection platforms and remote sensing—provide new ways of characterizing the water environment and our built facilities. Spatial databases containing attribute data and imagery over time provide reliable and standardized archival and retrieval functions, and they allow sharing of data across the Internet. GIS analysis functions and linked mathematical models provide extensive capabilities to examine alternative plans and designs. Map-oriented visualizations in color, three-dimensional, and animation formats help communicate complex information to a wide range of participants and interest groups. Moreover, interactive GIS database and modeling capabilities permit stakeholders to participate in modeling activities to support decision making. GIS is an all-encompassing set of concepts and tools that provides a medium for integrating all phases of water resources engineering planning and design.

This book provides relevant background on GIS that is useful in understanding its advanced applications in water resources engineering. The book has been developed with two primary sections. For the first part of the book (Chapters 1–4), the emphasis is on developing an understanding of the nature of GIS, recognizing how a GIS is used to develop and analyze geographic data, differentiating between the various types of geographic data and GISs, and summarizing data development and database concepts. Primary field-data collection and methods of interpretation and analysis are also introduced. The second part of the book (Chapters 5–12) focuses on the various subdomains of water resources engineering, the data involved, linkage of GIS data with water resource analysis models, and management applications. Applications include watershed hydrologic and groundwater modeling, water and wastewater demand forecasting, pipe network modeling, nonpoint sources of water pollution, floodplain delineation, facilities management, water resources monitoring and forecasting, and river-basin management decision-support systems. The applications include descriptions of GIS database development, analysis background theory, and model integration with the GIS.

The chapter titles in the book are as follows:

1. Introduction
2. Introduction to Geographic Information Systems
3. GIS Data and Databases
4. GIS Analysis Functions and Operations
5. GIS for Surface-Water Hydrology
6. GIS for Groundwater Hydrology
7. GIS for Water-Supply and Irrigation Systems
8. GIS for Wastewater and Stormwater Systems
9. GIS for Floodplain Management
10. GIS for Water Quality
11. GIS for Water Resources Monitoring and Forecasting
12. GIS for River Basin Planning and Management

At the end of each chapter there is a list of references related to the specific topic covered in that chapter. The GIS literature is large and growing rapidly, and relevant works are found in a diversity of sources. Some are found in refereed journals of civil engineering, water resources, and planning. Other works are found in government agency publications, academic programs, and on Web sites for both. Although the collected references and Web links are considered a valuable resource, I make no claim that it is comprehensive. For many readers, the references listed will be sufficient; for others wishing to go farther, they will serve only as a beginning.

Acknowledgments

A number of people have assisted me, directly or indirectly, in the preparation of this book. Some years ago D. P. Loucks at Cornell University provided an opportunity for me to work on his project on Interactive Computer Graphics for Water Resources Planning. That work provided a stepping-off point for continuing growth in the use of GIS for data management and modeling. John Labadie at Colorado State University has generously provided selected materials for inclusion in the book and has provided helpful reviews. David Maidment's program at the University of Texas Center for Research in Water Resources is exemplary in its breadth of endeavor on water resources GIS and the availability of its publications. Certainly, thanks are due to the many authors, organizations, and publication outlets from which materials for this book are drawn. While every effort has been made to trace the owners of copyrighted material, in a few cases there may be inadvertent omissions; for these I offer apologies and request that you provide corrections to me by e-mail at Lynn.E.Johnson@ucdenver.edu

Author

Lynn Johnson has been interested in rivers and water systems since he began his career gauging streams for the U.S. Geological Survey. His professional practice and research on water resources systems have involved extensive use of maps of various kinds and led to computerized versions of these and then to GIS. After receiving undergraduate degrees in geology and civil engineering at SUNY Buffalo and a master's degree in water resources management at the University of Wisconsin at Madison, he conducted various hydrologic and water resources investigations for private and public agency clients. He completed the Ph.D. at Cornell University in water resources systems and developed an early interactive GIS application for river-basin modeling. He has successfully conducted funded research activities for the National Science Foundation, National Oceanic and Atmospheric Administration, National Weather Service, U.S. Environmental Protection Agency, U.S. Army Corps of Engineers, U.S. Agency for International Development, and a variety of other water-management agencies. He is currently professor of civil engineering (water resources and geographic information systems) at the University of Colorado Denver. He teaches graduate and undergraduate courses in geographic information systems (GIS), water resources systems modeling and planning, hydrology, and environmental engineering. He has also led the use of on-line learning techniques in the Master of Engineering–GIS Program at CU Denver. The opportunity for interactions with students having various disciplinary backgrounds through the use of GIS continues to be a source of joy and satisfaction.

Audience

This book is directed to water resources and environmental engineers, scientists, and hydrologists who are interested in GIS applications for hydrological and water resource systems modeling, urban facilities management, and river-basin decision support. Given the interdisciplinary character of GIS, the book will be of interest to civil engineers, geologists and geographers, water-use planners, environmentalists, and public works officials. The book can serve as a graduate-level text in engineering and environmental science programs, as well as a reference for engineers, environmentalists, and managers seeking to enhance the linkage between GIS data sets and water resources systems models.

Audience

This book is directed to water resources and environmental engineers, scientists, and hydrologists who are interested in GIS applications for hydrological and water resource systems modeling, water facilities management, and river basin decision support. Given the interdisciplinary character of GIS, the book will be of interest to civil engineers, geologists and geographers, water planners, environmentalists, and public works officials. The book can serve as a graduate-level text in engineering and environmental science programs, as well as a reference for engineers, environmentalists, and managers seeking to enhance the linkage between GIS databases and water resource system models.

Selected Acronyms

ABR	Average Basin Rainfall
ABRFC	Arkansas-Red River Forecast Center (NWS)
AD	Average Day Demand
AMBER	Areal Mean Basin Effective Rainfall
ANC	Auto-NowCast
ANN	Artificial Neural Networks
APAPS	Automated Data Processing System (USGS)
API	Application Program Interface
ALERT	Automated Local Evaluation in Real Time
AML	Arc Macro Language
AMSR	Advanced Microwave Scanning Radiometer
AOP	Annual Operating Plan
ASCE	American Society of Civil Engineers
ASCII	American Standard Code for Information Interchange
AWWA	American Water Works Association
BASINS	Better Assessment Science Integrating Point and Nonpoint Sources
BOD	Biochemical Oxygen Demand
BMP	Best Management Practice
BRA	Basin Rate of Accumulation
BRM	Big River Model
CAD	Computer-Aided Design
CADSWES	Center for Advanced Decision Support for Water and Environmental Systems
CAPPI	Constant Altitude Plan Position Indicator
CASE	Computer-Aided Software Engineering
CCTV	Closed-Circuit Television
CDSS	Colorado Decision-Support System
CERL	Construction Engineering Research Lab (U.S. Army Corps of Engineers)
CFS	Climate Forecast System
CIS	Customer Information System
CMI	Crop Moisture Index
COM	Component Object Model
CPC	Climate Prediction Center (NOAA)
CRPAB	Colorado River Policy Advisory Board
CRSM	Colorado River Simulation Model
CRSS	Colorado River Simulation System
CRWCD	Colorado River Water Conservation District
CRWR	Center for Research in Water Resources (Univ. Texas, Austin)
CU	Consumptive Use
CUAHSI	Consortium of Universities for the Advancement of Hydrologic Science
CWCB	Colorado Water Conservation Board
CWNS	Clean Watersheds Needs Survey
DBMS	Data Base Management System
DCIA	Directly Connected Impervious Area
DCP	Data Collection Platform
DCS	Data Capture Standards (FEMA)

DEM	Digital Elevation Model
DFIRM	Digital Flood Insurance Rate Map
DLG	Digital Line Graph
DMI	Data Management Interface
DMS	Document Management System
DNR	Department of Natural Resources (State of Colorado)
DO	Dissolved Oxygen
DOQQ	Digital Orthoimagery Quarter Quadrangles
DPA	Digital Precipitation Array
DSS	Decision Support System
DTM	Digital Terrain Model
DWR	Department of Water Resources (State of Colorado)
EDNA	Elevation Derivatives for National Applications
EM	EnviroMapper
EOS	Earth Observation Satellite
EPA	Environmental Protection Agency
ESRI	Environmental Systems Research Institute, Inc.
ET	Evapo-Transpiration
ETM+	Enhanced Thematic Mapper Plus
F2D	Flood 2-Dimensional Rainfall-Runoff Model
FDA	Flood Damage Analysis
FEMA	Federal Emergency Management Agency
FFG	Flash Flood Guidance
FIRM	Flood Insurance Rate Map
FIS	Flood Insurance Studies
FWPP	Flood Warning and Preparedness Program
Geo-MODSIM	GIS-based MODSIM (Modular Simulation program)
GeoRAS	Geospatial River Analysis System
GIS	Geographical Information System
GLEAMS	Groundwater Loading Effects of Agricultural Management Systems
GMS	Groundwater Modeling System
GMIS	Groundwater Modeling Interface System
GNIS	Geographic Names Information System
GOES	Geostationary Operational Environmental Satellite
GPS	Global Positioning System
GRASS	Geographic Resources Analysis Support System
GUI	Graphical User Interface
HAS	Hydrologic Analysis and Support
HEC	Hydrologic Engineering Center (U.S. Army Corps of Engineers)
HEC-RAS	HEC River Analysis System
HIS	Hydrologic Information System
HL-RMS	Hydrology Lab-Research Modeling System (NWS)
HMS	Hydrologic Modeling System (HEC)
HMT	Hydrometeorological Testbed (NOAA)
HRAP	Hydrologic Rainfall Analysis Project
HTML	Hypertext Markup Language
HTTP	Hypertext Transfer Protocol
HUC	Hydrologic Unit Code
I/I	Infiltration/Inflow
ICPA	Interstate Compact Policy Analysis
IDF	Intensity-Duration-Frequency

IID	Imperial Irrigation District
IMAP	Information Management Annual Plan
IMC	Information Management Committee
IMS	Infrastructure Management System
IPET	Interagency Performance Evaluation Task Force Team (New Orleans)
IR	InfraRed
IS	Impervious Surface
LAI	Leaf Area Index
LAPS	Local Area Prediction System (NOAA)
LCR	Lower Colorado River
LFWS	Local Flood Warning System
LIDAR	LIght-Detection And Ranging
LP	Linear Programming
LPG	Linear Programming Gradient
LSM	Land Surface Model
LULC	Land Use–Land Cover
MAF	Million Acre Feet
MAP	Mean Areal Precipitation
MCE	Multiple Criteria Evaluation
MD	Maximum Day demand
MH	Maximum Hour demand
MMS	Materials Management System
MODFLOW	Modular Finite-Difference Groundwater Flow Model
MODSIM	Modular Simulation program
MPE	Multisensor Precipitation Estimator
MRLC	Multi-Resolution Land Characteristics Consortium
MSS	Multispectral Scanner
NAD	National Assessment Database
NAIP	National Agricultural Imagery Program
NASA	National Aeronautics and Space Administration (U.S.)
NASIS	National Soil Information System
NCAR	National Center of Atmospheric Research
NCWCD	Northern Colorado Water Conservancy District
NDVI	Normalized Difference Vegetation Index
NED	National Elevation Dataset
NFIP	National Flood Insurance Program
NLP	Non-Linear Programming
NESDIS	National Environmental Satellite Data Information Service
NEXRAD	Next Generation Weather Radar
NHD	National Hydrography Dataset
NLCD	National Land Cover Dataset
NLDAS	North American Land Data Assimilation System
NOAA	National Oceanic and Atmospheric Administration
NOHRSC	National Operational Hydrologic Remote Sensing Center
NPDES	National Pollutant Discharge Elimination System
NRCS	Natural Resources Conservation Service
NRC	National Research Council
NSA	National Snow Analyses
NWS	National Weather Service
NWIS	National Water Information System (USGS)
O-O	Object-Oriented (database)

OGC	Open Geospatial Consortium
OHP	One-Hour Precipitation
OSD	Official Soil Series Description
PDSI	Palmer Drought Severity Index
PMF	Probable Maximum Flood
PPS	Precipitation Processing System (radar)
PRISM	Parameter-elevation Regressions on Slope Model
PVA	Property Valuation Administration
QPE	Quantitative Precipitation Estimate
QPF	Quantitative Precipitation Forecast
RAD	Reach Address Database
RDBMS	Relational Data Base Management System
RDBMS	Relational Database Management System
REMF	Real Estate Master File
ResSim	Reservoir Simulation model (HEC)
RF	Representative Fraction
RFC	River Forecast Center (NWS)
RGDSS	Rio Grande Decision Support System
RIT	Reach Indexing Tools
RT	Regression Tree
SAC-SMA	Sacramento Soil-Moisture Accounting
SCADA	Supervisory Control and Data Acquisition
SDCWA	San Diego County Water Authority
SDF	Stream Depletion Factor
SDMS	Spatial Data Management System
SDSS	Spatial Decision Support Systems
SEO	State Engineer's Office
SFHA	Special Flood Hazard Area
SLAR	Side-Looking Airborne Radar
SMA	Soil Moisture Accounting
SQL	Structured Query Language
SSM/I	Special Sensor Microwave/Imager
SSURGO	Soil Survey Geographical (database)
STATSCO	State Soil Geographic (database)
STORET	STOrage and RETrieval
STP	Storm Total Precipitation
SVG	Scalable Vector Graphics
SWBMS	Soil Water Balancing Model System
SWE	Sensory Web Enablement
SWMM	Storm Water Management Model
TAC	Technical Advisory Committee
TAZ	Traffic Analysis Zone
TDH	Total Dynamic Head
THP	Three-Hour Precipitation
TIGER	Topologically Integrated Geographic Encoding and Referencing
TIN	Triangulated Irregular Network
TITAN	Thunderstorm Identification, Tracking and Analysis system (NCAR)
TM	Thematic Mapper
TMDL	Total Maximum Daily Load
TOPAZ	Topographic Parameterization model
TSS	Total Suspended Solids

UDFCD	Urban Drainage and Flood Control District (Denver, CO)
UGA	Urban Growth Area
UH	Unit Hydrograph
UML	Universal Modeling Language
USBR	United States Bureau of Reclamation
USDA	U.S. Department of Agriculture
USGS	United States Geologic Survey
UZFWM	Upper-Zone Free Water Maximum
VAA	Value-Added Attribute
VDB	Visual Data Browser
VOC	Volatile Organic Compound
WADISO	Water Distribution System Analysis and Optimization
WADSOP	Water Distribution System Optimization
WAM	Watershed Assessment Model
WASP	Water quality Analysis Simulation Program
WATERS	Watershed Assessment, Tracking and Environmental Results
WBD	Watershed Boundary Dataset
WDAD	Watershed Monitoring and Analysis Database
WEAP	Water Evaluation And Planning System
WMS	Work Management System
WPCA	Water Pollution Control Act
WQDM	Water Quality Data Model
WQM	Water Quality Model
WQSDB	Water Quality Standards Database
XML	eXtensible Markup Language

UDFCD	Urban Drainage and Flood Control District, Denver, CO
UGA	Urban Growth Area
UH	Unit Hydrograph
UML	Unified Modeling Language
USBR	United States Bureau of Reclamation
USDA	U.S. Department of Agriculture
USGS	United States Geological Survey
UZRWM	Upper Zone Real Free Water Maximum
VAA	Value-added Attribute
VDB	Visual Data Browser
VOC	Volatile Organic Compound
WASAO	Water Distribution System Analysis and Optimization
WaterGEMS	Water Distribution System Optimization
WAM	Watershed Assessment Model
WASP	Water Quality Analysis Simulation Program
WATERS	Watershed Assessment, Tracking and Environmental Results
WBD	Watershed Boundary Dataset
WMAD	Watershed Monitoring and Analysis Database
WEAP	Water Evaluation And Planning System
WMS	Watershed Management System
WPCA	Water Pollution Control Act
WQDM	Water Quality Data Model
WQM	Water Quality Module
WQSDB	Water Quality Standards Database
XML	Extensible Markup Language

1 Introduction

1.1 OVERVIEW

Geographic information system (GIS) concepts and technologies are being used extensively in water resources engineering planning and design, and are changing the way these activities are conducted. We are in an age when natural resources are increasingly scarce and the effects of human activity are pervasive. In this situation, the best tools available must be used to characterize the environment, predict impacts, and develop plans to minimize impacts and enhance sustainability. GIS technologies, tools, and procedures have substantial benefits for resource inventories, modeling, and choice communication to involved agencies and concerned citizens.

This chapter introduces GIS and the water resources systems to which it is applied. A general overview of water resources and GIS is presented, including how maps have historically been used to support water resources development. The scope and character of water resources systems are then described in more detail, leading to an overview of GIS applications. The chapter concludes with a brief review of topics covered in the book.

1.2 WATER RESOURCES AND GIS

Information about water resources and the environment is inherently geographic. When surveyor John Wesley Powell explored the Colorado River and the Grand Canyon in 1869, part of his contribution was to make a map of the region (Figure 1.1). In doing so, he provided a cartographic basis for others to gain an understanding of the region and to formulate plans for further exploration and development. Later on, Powell initiated efforts to assess the water supply and to acknowledge natural limits to settling the land (Worster 2001; NPR 2002). Powell established a river gauge station along a stretch of the Rio Grande in New Mexico in 1889, the first of its kind in the nation, setting in motion programs for resource inventories in the western United States. Powell and his colleagues are credited with the terms *runoff* and *acre-foot* as part of these early efforts to figure out how to assess how much water was available (deBuys 2001). He advocated setting up government by watersheds and resisted the poorly planned expansion of settlements in the West that did not acknowledge water supply limitations. And he believed it was the role of government to hire and train the experts who could come to the West, inventory the resources, and set up the laws and the framework within which sustainable settlement could take place. Powell's legacy was the founding of resource accounting and planning processes that continued for the following century.

Maps, whether on paper or in digital GIS formats, continue to be the medium for the expression of engineering plans and designs. We are concerned with the spatial distribution and character of the land and its waters. Weather patterns, rainfall and other precipitation, and resultant water runoff are primary driving forces for land development, water supplies, and environmental impacts and pollution. Our water resources systems comprise dams and reservoirs, irrigated lands and canals, water supply collection and distribution systems, sewers and stormwater systems, and floodplains. These systems are tailored in response to a complex mix of topography and drainage patterns, population and land use, sources of water, and related environmental factors.

The planning and engineering design processes used in the development and management of water resources involve different levels of data abstraction. Data are collected and used to characterize the environment at some level of detail, or scale. In seeking to make decisions about plans and

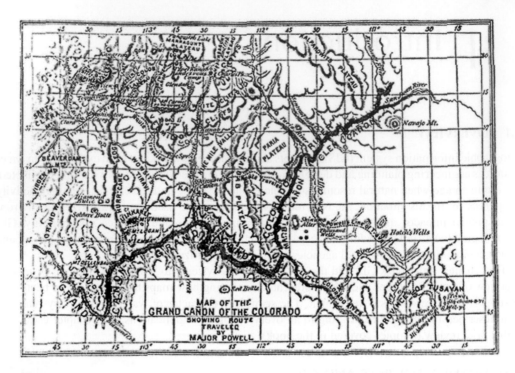

FIGURE 1.1 Map of the Grand Canyon of the Colorado River showing the route traveled by Powell, 1875. (Image courtesy of Edwin J. Foscue Library, Southern Methodist University.)

FIGURE 1.2 Digital elevation model (DEM) for Grand Canyon, Arizona. (*Source:* USGS 2007.)

designs, data must be collected to describe the resource, and procedures or models must be developed to predict the resultant changes. These data and models help us understand the real world, and this understanding guides our decision making. An example of new mapping technology is satellite imagery from which detailed terrain maps can be created; Figure 1.2 shows a digital elevation model of the Grand Canyon that Powell's group traversed.

In contrast with Powell's exploratory efforts at mapping the Colorado River, there are now extensive and sophisticated digital renderings of the river basin that are the foundation for modern water management and decision support. For example, the National Weather Service (NWS), the U.S.

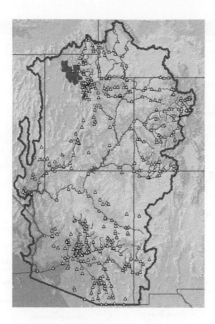

FIGURE 1.3 Colorado River basin-water management infrastructure: NWS river forecast points where predictions of future flows are made. (*Source:* NWS 2007.)

Geological Survey (USGS), the U.S. Bureau of Reclamation (USBR), and other federal and state agencies have deployed a large number of monitoring gauges for stream flow, rainfall, and weather data throughout the basin (Figure 1.3).

The USBR coordinates operation of the system of reservoirs and diversions in the Colorado River basin. Computer models of the river system simulate snowmelt and rainfall runoff as well as reservoir operations throughout the river network; these models are linked to GIS databases on snowmelt; related hydrological processes; and water demands for domestic, industrial, and agricultural uses. Figure 1.4 illustrates the river basin computer model RiverWare® interface, showing the reservoirs, diversions, and related processes as an integrated collection of intelligent "objects"; this model is described in more detail in Chapter 12 (GIS for River Basin Planning and Management). Powell's original mapping has now evolved to a complete digital depiction of the land and its hydrological and water-management infrastructure.

GIS presents information in the form of maps and feature symbols, and is integrated with databases containing attribute data on the features. Looking at a map gives knowledge of where things are, what they are, and how they are related. A GIS can also provide tabular reports on the map features; create a list of all things connected in a network; and support simulations of river flows, travel time, or dispersal of pollutants. A GIS is a computer-based information system that supports capture, modeling, manipulation, retrieval, analysis, and presentation of spatial data. This is a standard definition that does not highlight the uses of GIS as an integrator of data-management operations and decision support in an organization. A more expansive view is that the purpose of a GIS is to provide a framework to support decisions for the intelligent use of Earth's resources and to manage the built and natural environments. The purposes and concepts of GIS are key to the understanding and successful application of this technology, and are discussed in more detail in Chapter 2 (Introduction to Geographic Information Systems).

1.3 WATER RESOURCES ENGINEERING

Water resources engineering is concerned with the analysis and design of systems to manage the quantity, quality, timing, and distribution of water to meet the needs of human societies and the

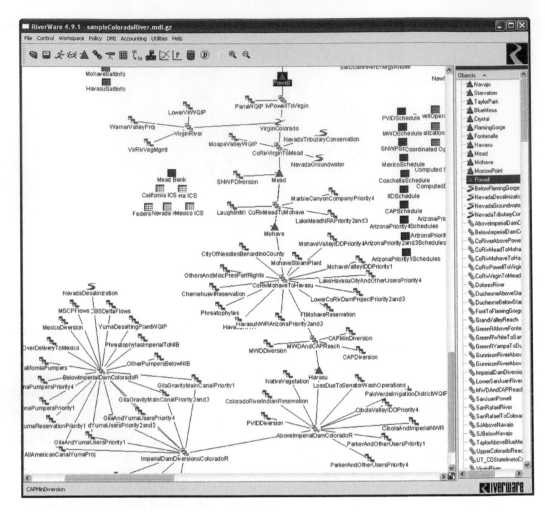

FIGURE 1.4 Computer model of the Lower Colorado River is used to simulate reservoir and diversion operations. (*Source:* CADSWES 2007. With permission.)

natural environment (Chin 2006). Water resources are of critical importance to society because these systems sustain our livelihood and the ecosystems on which we depend. However, there may be too little or too much water; and what there is may not be located where we need it, or it may be too polluted or too expensive. There is a growing worldwide water crisis, which is likely to further expand as a result of population growth, land-use changes, urbanization and migration from rural to urban areas, and global climate changes. All of these factors emphasize the need for wise development and management of our water resources. Facilities for water supply and wastewater disposal, collection and control of flood runoff, and maintenance of habitat are examples of the relevant applications of water resources engineering.

In this book, the emphasis is on water and the water-related environment. Collection and archiving of basic data on water flows, terrain, soils, and related environmental resources are essential to the rational use and protection of these resources. Beyond the physical features, there are the economic, social, and political dimensions of water systems. And, historically, the existence and expansion of civilizations have been controlled to a great degree by the abundance or shortage of water. Because of this, the field of water resources has a distinctly engineering orientation directed to the design of facilities, which blends with a more scientific direction seeking to better define the resources.

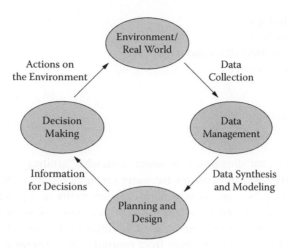

FIGURE 1.5 Water resources planning and design processes begin with data collection on the real world and proceed through data management and modeling to generate information for decision making on alternative plans and designs.

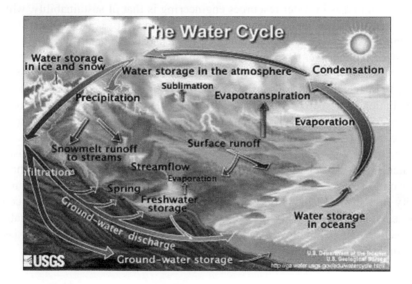

FIGURE 1.6 The water cycle. (*Source:* USGS 2008.)

Water resources infrastructure development occurs as a long process involving information gathering and interpretation, plan development, decision making and financing, construction, and operation. Powell's efforts were just a first step in that process for the Colorado River. Engineering planning and design processes involve a variety of procedures for data collection and management, data synthesis and system modeling, and development of information for decision making. The schematic diagram of Figure 1.5 illustrates the process of setting objectives, data collection and synthesis, planning and design, gathering of information for decision making, and taking action. It begins and ends with the real world, a world that is inherently spatial.

Water resources engineering builds on the core science of hydrology, which deals with the occurrence, distribution, movement, and properties of water on the Earth's surface and underground (Figure 1.6). A spectrum of domains for the application of GIS to water resources engineering are addressed in this book, including:

1. Surfacewater hydrology
2. Groundwater hydrology
3. Water supply for municipalities and irrigation
4. Wastewater and stormwater
5. Floodplains
6. Water quality
7. Monitoring and warning
8. River basins

The distinctions between domains may seem somewhat arbitrary, as hydrologic concepts are inherent in each. However, procedures for water resources analyses and design within each domain are distinguished by some commonality of geography and practice and by the institutional responsibilities governing motivations and requirements. Outcomes from analyses and designs may involve construction and operation of facilities for (a) control of the spatial and temporal distributions of surface water resulting from rainfall events, (b) water use systems designed for water supply and wastewater collection, or (c) environmental restoration systems directed to sustaining functions of the natural environment. The analyses may also lead to information products that guide responses to hazardous weather events or operations of reservoirs for multiple objectives.

An emerging emphasis in water resources engineering is that of sustainability, which refers to the conduct of our activities, with a view toward long-term effects, consideration of externalities, and assessment of risks and uncertainties. Long-term effects accrue in our water resources systems due to the fundamental changes that we make to the natural systems. Externalities, or neighborhood effects, occur when there are unanticipated effects that were not considered during the planning and design of built facilities. These impacts are typically not compensated for and may result in unequal distribution of costs to underrepresented groups, including future generations. Risks and uncertainties are inherent in all of our complex systems due to our inability to understand and account for the range of possible outcomes of our interactions with the environment. For example, global climate change may result in basic changes to the hydrologic cycle and the distribution and timing of water fluxes. This emphasis on sustainability is driven by the knowledge that the volume of freshwater on Earth is practically constant, that we are using the water resources to their maximum capacity in many areas, and that increased human activities are increasingly contaminating our environment. While it is not clear at what point our activities become unsustainable, there are enough signals that population growth and associated increases in demands are serious issues that need to be addressed.

1.4   APPLICATIONS OF GIS IN WATER RESOURCES ENGINEERING

GIS provides an integrating data and modeling environment for the conduct of these activities. A GIS provides a means to collect and archive data on the environment. Measurements of location, distance, and flow by various devices are typically handled in digital formats and quickly integrated into a spatial database. Data processing, synthesis, and modeling activities can draw on these data using the GIS, and analysis results can be archived as well. The GIS spatial and attribute database can then be used to generate reports and maps, often interactively, to support decision making on which design alternatives are best and the impacts of these. Further, maps are a powerful communication medium; thus this information can be presented in public forums so that citizens concerned with planning and design choices can better understand and be more involved.

Planning and design in water resources engineering typically involve the use of maps at various scales and the development of documents in map formats. For example, in a river basin study, the map scale often covers a portion of a state and includes several counties and other jurisdictions. The river drains a certain geography having topographic, geologic (including types of soils), vegetative, and hydrologic characteristics. Cities and human-built facilities are located along the river and across the basin, and transportation and pipeline networks link these together. All of these data sets must be established in a common georeference framework so that overlays of themes can be made and the coincidence of features can be identified in the planning and design phase.

The GIS is applied to manage all of these data. It provides a comprehensive means for handling the data that could not be accomplished manually. The large amount of data involved requires a GIS, as there may be many thousands of features having a location, associated attributes, and relationships with other features. The GIS provides a means of capturing and archiving these data, and of browsing and reviewing the data in color-coded map formats. This data-review capability supports quality control, as errors can be more readily identified. Also, through visualization, the user can gain a better understanding of patterns and trends in the data in a manner not possible if the data were only in tabular format. The GIS provides an analysis capability as well. The database can be accessed by computer software and used as input to various modeling procedures to generate derived products.

In a river basin there are many applications of GIS, for example:

- Defining the watershed and its hydrologic and hydraulic characteristics so that models of rainfall-runoff processes can be applied to examine the impacts of land-use changes
- Mapping land-use and population demographics in support of water and wastewater demand estimation procedures
- Interpolating groundwater contaminant concentrations given sampled data at observation wells spaced throughout an aquifer, or estimating snowpack amounts at ungauged locations based on data obtained at gauged locations guided by factors of elevation and exposure
- Managing public infrastructure, such as scheduling maintenance on a sewage collection system, notifying residents of water-pipe rehabilitation work, or identifying areas of potential low pressure during fire-response planning scenarios
- Finding the coincidence of factors, such as erosion-prone areas having a certain combination of soil type, land cover, and slope
- Monitoring the occurrences and intensities of severe thunderstorms and providing tools for warning threatened populations of impending hazardous flood conditions
- Providing the logical network structure for coordinating simulation and optimization models that schedule the interactions between basin water supplies, reservoirs, diversions, and demands

In addition to the physical scope of engineering planning and design activities, the organizational context within which the GIS exists is important. Whether it is a large federal agency seeking to establish water supplies for a region or a small municipality trying to keep up with rapid development, the GIS requires the establishment of procedures and standards. Often, the GIS will require a change in the way an agency's work is done. Advances in data collection and engineering measurement technologies, changes in data formats and report-generation capabilities, and requirements for data sharing across jurisdictions can be different from established historical practices. All of these factors can lead to improved practice, but they can also cause stress by requiring training and change.

1.5 OVERVIEW OF BOOK

In this book, I attempt to summarize the state-of-the-art use of GIS in water resources engineering. To accomplish this, there are three chapters that address the foundational concepts of GIS:

- Chapter 2 (Introduction to GIS): This chapter presents an overview of GIS terminology and concepts.

- Chapter 3 (GIS Data and Databases): This chapter reviews (a) the methods and principles for map data collections and conversions and (b) the various attribute and spatial database models and their utility for management of geographic data.
- Chapter 4 (GIS Analysis Functions and Operations): This chapter addresses the broad scope of the kinds of analyses that can be accomplished with GIS.

Following these introductory materials, there are chapters that address GIS concepts and applications to the various domains of water resources engineering:

- Chapter 5 (GIS for Surface-Water Hydrology): This chapter reviews (a) the character of surface-water modeling data and parameterizations, (b) methods for developing these data sets using GIS, and (c) procedures for rainfall-runoff modeling.
- Chapter 6 (GIS for Groundwater Hydrology): This chapter addresses the groundwater domain; it reviews (a) groundwater systems and modeling state-of-the-art systems for quantity and quality and (b) the use of GIS techniques to support groundwater assessments and modeling.
- Chapter 7 (GIS for Water Supply and Irrigation Systems): This chapter considers the water supply domain for urban and irrigation services, including (a) water supply data and system design concepts and (b) GIS procedures and applications for accomplishing these designs.
- Chapter 8 (GIS for Wastewater and Stormwater Systems): This chapter reviews (a) urban wastewater and stormwater data and models and (b) GIS methods for the design and management of these systems.
- Chapter 9 (GIS for Floodplain Management): This chapter addresses (a) data and models used for floodplain modeling and management and (b) GIS procedures for data management and modeling of floodplains.
- Chapter 10 (GIS for Water Quality): This chapter considers (a) concepts and methods for assessing water quality of surface-water bodies and (b) GIS methods for data management and modeling of these systems.
- Chapter 11 (GIS for Water Resources Monitoring and Forecasting): This chapter reviews (a) real-time monitoring, forecasting, and warning data-collection and modeling systems and (b) GIS methods employed to enable these systems.
- Chapter 12 (GIS for River Basin Planning and Management): This chapter addresses river basin planning and management models and their integration with GIS, including highly integrated systems for decision support.

REFERENCES

Center for Advanced Decision Support for Water and Environmental Systems (CADSWES). 2007. http://cadswes.colorado.edu/.

Chin, D. A. 2006. *Water resources engineering*. New York: Pearson Prentice Hall.

deBuys, W., ed. 2001. *Seeing things whole: The essential John Wesley Powell*. Washington, D.C.: Island Press.

National Public Radio (NPR). 2002. The true legacy of John Wesley Powell: The explorer sounded early warnings about water in the West. http://www.npr.org/programs/atc/features/2002/sept/powell/.

National Weather Service (NWS). 2007. Colorado Basin River Forecast Center. http://www.cbrfc.noaa.gov/.

U.S. Geological Survey (USGS). 2007. National Elevation Dataset (NED) 1/3 arc-second DEM. http://ned.usgs.gov/.

U.S. Geological Survey (USGS). 2008. The water cycle. http://ga.water.usgs.gov/edu/watercycle.html.

2 Introduction to Geographic Information Systems

2.1 OVERVIEW

Geographic information systems (GIS) have become an increasingly important means for understanding and dealing with the pressing problems of water and related resources management in our world. GIS concepts and technologies help us collect and organize the data about such problems and understand their spatial relationships. GIS analysis capabilities provide ways for modeling and synthesizing information that contribute to supporting decisions for resource management across a wide range of scales, from local to global. A GIS also provides a means for visualizing resource characteristics, thereby enhancing understanding in support of decision making.

This chapter presents an overview of GIS. Several definitions of GIS are offered to introduce the concepts and technologies that comprise GIS. A general overview of GIS involves the technologies for data capture and conversion, data management, and analysis. Also, there is a need to be aware of the management dimensions of GIS, as implementation of GIS can require basic changes in the way engineering planning and design are accomplished. The chapter concludes with a brief review of popular GIS software.

2.2 GIS BASICS

2.2.1 DEFINITIONS

Various definitions have been offered that reinforce the major dimensions of GIS. Several of these definitions are listed below. Elements of a GIS include the data and information technology (i.e., computers, software, and networks) to support it. Spatial data include any data that have a geographic location. This "toolbox" definition focuses on the hardware and software components of a GIS. In its totality, a GIS can be viewed as a data-management system that permits access to and manipulation of spatial data and visual portrayal of data and analysis results. There are also the human and organizational aspects. For example, standards must be agreed upon to facilitate database integrity and sharing across the organization. There are also the people with GIS expertise who understand and can carry out the procedures and build and maintain the GIS. Finally, there is the organizational setting—the technical, political, and financial operating environments created by the interaction among stakeholders—in which the GIS is to function.

- GIS is a computerized system that is used to capture, store, retrieve, analyze, and display spatial data (Clarke 1995).
- GIS is "an information system that is designed to work with data referenced by spatial or geographical coordinates" (Star and Estes 1990).
- GIS "manipulates data about points, lines, and areas to retrieve data for ad hoc queries and analyses" (Duecker 1987).
- GIS consists of five basic elements: "data, hardware, software, procedure and people" (Dangermond 1988).

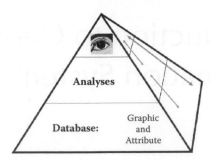

FIGURE 2.1 GIS pyramid illustrates that analyses and visualizations draw from the database that forms the foundation. Analyses link directly to the database, and visual interactions can occur between the viewer and the database and analyses.

- GIS comprises "four basic elements which operate in an institutional context: hardware, software, data and liveware" (Maguire 1991).
- GIS is "an institutional entity, reflecting an organizational structure that integrates technology with a database, expertise and continuing financial support over time" (Carter 1989).

A GIS pyramid (Figure 2.1) illustrates that the GIS is built on a foundation of spatial and attribute data, and that users can access the database to conduct analyses and generate visualizations of data and analyses. Practically, the volume of the pyramid devoted to the database is indicative of the time and effort required to build a successful GIS.

GIS concepts and technologies arise from a wide variety of fields, and GIS has become a generic term referring to all automated systems used primarily for the management of maps and geographic data. Alternative terms associated with GIS include:

- Automated mapping and facilities management (AM/FM): Used by public and private utility organizations to manage information on facilities (e.g., water, wastewater, telecommunications, electricity distribution); enables real-time inventory of facilities and production of maps for use in the field and for the creation of a map library.
- Computer-aided drafting/design (CAD): CAD is used to design, develop, and optimize products; used by engineering organizations to portray plans and specifications for constructed works: architecture, engineering, and construction (AEC), mechanical and electrical.
- Computer-assisted or computer-aided mapping (CAM): Interactive graphic systems for geocoded databases. Associated with surveyors, aerial photogrammetry, and airborne and satellite remote sensing.
- Spatial database-management systems (DBMS): Software for managing attribute data on mapped features; founded on relational DBMS and related technologies that facilitate enterprise management of an organization's data to allow many different users to share data and process resources while ensuring security and integrity.
- Land information system (LIS): Used by assessors and land management organizations for land ownership information on quantity, value, and ownership of land parcels.
- Multipurpose cadastre: Refers to an integrated LIS containing legal (e.g., property ownership or cadastre), physical (e.g., topography, human-made features), and cultural (e.g., land use, demographics) information in a common and accurate reference framework.

The development of GIS has relied on innovations made in many disciplines: geography, civil engineering, photogrammetry, remote sensing, surveying, geodesy, statistics, computer science, operations research, demography, and many other branches of engineering and the natural and social sciences. Indeed, an outstanding characteristic of GIS is its interdisciplinary character in its development as a collection of tools as well as the wide variety of applications. GIS cartographic concepts originated with the maps created by early explorers and have been extended by modern geographers to portray locations on and characteristics of the Earth. Engineering measurement theories and practices of surveyors and geodesists provided the means to describe property boundaries and locate Earth features accurately. Civil engineers have migrated to digital formats for land-development plans, including parcel boundaries as well as elements for water and sewer pipes, roads and streets, and other infrastructure. Satellite and airborne remote-sensing technologies have advanced to become a primary data source for high-resolution mapping of land characteristics; these apply for base mapping, in real time, and for assessing changes over time.

A GIS is sometimes distinguished from other computer-based systems that use geo-referenced information. What makes a GIS different is that it provides a more comprehensive environment for data integration and analysis. While these other systems can generate computer-stored maps, and perhaps can make database retrievals, the GIS integrates data for multiple themes, provides tools for analysis across themes (e.g., overlay), and can be integrated with other analysis routines to obtain a modeling and decision-support system. Another way that GIS is often distinguished from the companion technologies is that it has served an important role as an integrating technology. Rather than being completely new, GIS has evolved by linking a number of separate technologies into a single coordinated information system. However, distinctions between the various types of GISs may seem somewhat arbitrary. AM/FM systems have found extensive application in the utilities fields, where large databases accessible across the organization are evident. Public utilities play an important role for water resources, where water supply and sewer (sanitary and storm) are major themes for municipal utilities management.

2.2.2 GIS DATA AND DATABASES

It has become common to think of GIS databases as a series of map layers that are geographically referenced and registered to a common projection. Most GISs organize data by layers, each of which contains a theme of map information that is logically related by its location (Figure 2.2). Each of these separate thematic maps is referred to as a layer, coverage, or level. And each layer is precisely overlaid on the others so that every location is matched to its corresponding locations on all the

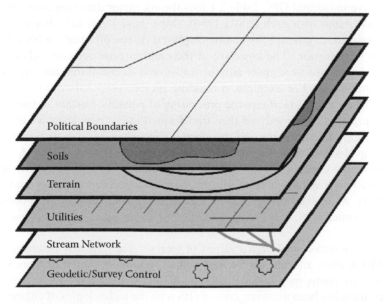

FIGURE 2.2 Map layers address multiple themes.

TABLE 2.1
Municipal GIS Data Sets

Data Category	Example Map Layers
Base-map data	Control points
	Topographic contours
	Building footprints
	Major locational references
Area data	Land-use areas
	Demographic areas
	Tax-rate areas
	Emergency-service areas
Environmental data	Soils maps
	Streams and water bodies
	Floodplain map
	Land cover
Network facilities data	Water system
	Sewer system
	Telecommunications
	Electrical network
Transportation network data	Street centerlines
	Road intersections
	Railroad lines
Land records data	Parcel boundaries
	Land-parcel boundaries
	Easements and rights-of-way

other maps. The bottom layer of this diagram is quite important; it represents the location reference system to which all the maps have been accurately registered.

The layer idea is central to the concept of the multipurpose casdastre that is invoked as part of a comprehensive municipal GIS. Table 2.1 lists the six major data categories and the types of thematic data associated with each (ESRI 1986). Once these maps have been registered within a standardized reference system, information displayed on the different layers can be compared and analyzed in combination. The objective of the multipurpose cadastre scheme is to provide a fully integrated database to support administrative and decision-making functions across all levels of the municipality. For example, a building project proposal could be readily checked against the floodplain map to facilitate the processing of permits. Further, information from two or more layers might be combined and then transformed into a new layer for use in subsequent analyses. This process of combining and transforming information from different layers is sometimes called "map algebra," as it involves adding and subtracting map values. If, for example, we wanted to consider the impacts of near-stream development, we could buffer the stream riparian zone to produce a new map, and overlay this new map on layers representing land use to identify potential conflicts. Also, even though the multipurpose cadastre concept has been presented in the context of a municipality, it is readily extended to larger areas such as a state, country, and even globally.

Input or capture of data comes from a variety of sources. The data may be converted from existing paper (or Mylar) plans and records, as well as data residing in digital databases (e.g., property records). These conversions may involve tablet digitizing and scanning to images. Over the past several decades, there has been a convergence of GIS with the technologies of engineering measurement that record field data in digital formats and can be ported directly into a GIS spatial database (e.g., surveying total station, global positioning systems [GPS]). Data-capture technologies include as well remote sensing by satellites and airborne platforms (photogrammetry). Satellite imagery is received in various wavelengths so that particular aspects of the land surface can be characterized through image-processing procedures. Imagery from airplane overflights is most often of the photographic type, particularly for the development of high-resolution topographic maps of urban areas and identification of urban features such as building footprints, street centerlines, manholes, and water distribution valves. Increasingly, light detection and ranging (LIDAR) is being used to provide the high-resolution topographic mapping required for detailed site planning and floodplain hydraulic studies. Regardless of the source, there is a requirement that spatial data be identified in some coordinate reference system.

GIS databases incorporate two distinct branches, the spatial database and the associated attribute database. Many GIS software packages maintain this distinction. The spatial data are characterized as having a "vector" structure composed of features represented as points, lines, and polygons. Other GIS spatial data are handled as images, or "rasters," having simple row-and-column formats. Attribute data are handled in relational database software composed of records and fields, and the power of the relational model is applied to these data. These feature data are "tagged" to the spatial database to facilitate tabular data retrievals. Details on GIS databases are presented in Chapter 3.

The pyramid structure (Figure 2.1) also suggests how a GIS is built. Since everything depends on the database, it must be developed first, or at least major portions must be developed so that the desired analyses and displays can be accomplished. This, in turn, suggests that the approach to building a GIS should begin with database design and development. The approach for building a GIS is important, as time and effort, and the corresponding expenses, must be invested before products can be produced. This suggests that a GIS must evolve from an inventory tool to an analysis tool and, ultimately, to a management tool; this view indicates some progression of acceptance into organizational decision making.

2.2.3 GIS ANALYSES

GIS analysis capabilities are specifically keyed to the spatial realm. An analysis function unique to GIS is the overlay operation, whereby multiple data themes can be overlain and the incidence of line and polygon intersections can be derived. This graphical and logical procedure is used in many ways to identify the correspondence between multiple data layers. Other GIS functions include networks and connectivity operations, terrain analyses, statistical interpolation, and other neighborhood procedures, as well as functions for spatial database development and maintenance. GIS analysis functions are described in more detail in Chapter 4.

One common GIS application is suitability analysis. An early example of suitability analysis is that of Ian McHarg (1969) in his seminal work *Design with Nature*. Suitability analysis uses classification operations in a process of scoring and overlaying to derive characterizations of the land for some purpose. For example, GIS coverages for soils, vegetative cover, and slope can be classified, scored, and combined in a manner to assess the potential of the land for erosion, or to identify sensitive lands not appropriate for development. In the classification process, judgments of "goodness" are made based on technical and prescriptive factors to assign the ratings. The results are used to guide management authorities in land-use allocation decisions.

The land-use example just described illustrates an additional dimension of the GIS definition—that of visualization and decision support. Here, the capacity of GIS to produce derived maps that portray the decision-relevant information in color-coded formats permits the communication of complex resource-management decisions to managers and interested citizens. Maps have high communication value, and most people can understand map displays that are properly prepared and include a legend. A planner's view can be that a primary purpose of GIS is to help decision makers make sensible decisions on the management of resources.

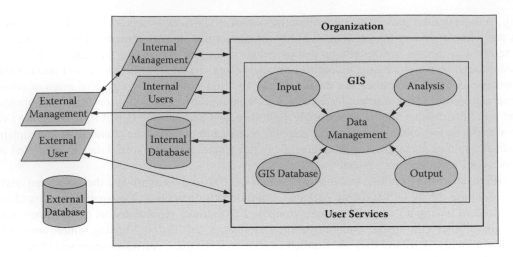

FIGURE 2.3 GIS in an organizational context involves internal and external information exchanges. (Adapted from Aronoff 1991.)

2.2.4 GIS MANAGEMENT

Implementation and management of a GIS are made more difficult, given that there are more than just technical issues involved. The key word is *information*, which is often the primary product of an organization, so this brings in organizational factors and concerns. A GIS that generates information useful for decision making exists within an institutional context (Figure 2.3). The information is important and therefore can be sensitive. As described by Aronoff (1991), the GIS is operated by staff who report to management. That management has a mandate to use the GIS to serve some user community, either internal or external to the organization, or both. Ultimately, the purpose and justification for the GIS is to assist the users in accomplishing the goals of their respective organizations.

2.3   MAPS AND MAP DATA CHARACTERISTICS

2.3.1   MAP FUNCTIONS

Maps have been used throughout history to portray the Earth's surface, location of features, and relations between features. Traditionally, maps were exclusively hand-drawn or drafted documents. The practice of cartography paralleled the exploration of the world as navigators established location reference schemes, classifications of features, labeling, and other annotations. Many of the symbols developed are retained in modern maps, such as blue lines for streams, double-line symbols for roads, and contour lines for topography.

As noted in Chapter 1, of the many contributions John Wesley Powell made in his explorations of the western United States, his creation of a map of the previously unknown Colorado River region had perhaps the greatest impact. The map destroyed the mystery of the canyon, showed where it flowed from and to, and showed the relative elevations along the way. That information provided following explorers, engineers, and land developers with a logistical perspective on how to get from here to there, and the land relief to be expected. It was the beginning of a systematic mapping of the West that led to the development of transportation routes, settlements, and irrigation and reservoir projects.

A map can accomplish many things in many ways. When you read a map, you observe the shapes and position of features, some attribute information about a feature, and the spatial relationships between features (Zeiler 1999). Some things that maps accomplish include:

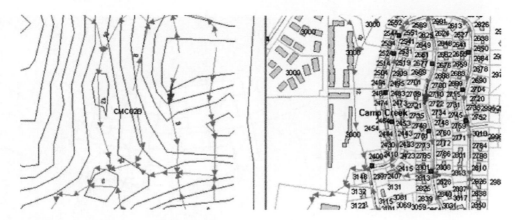

FIGURE 2.4 Portion of a sanitary sewer design plan showing (a) terrain contours and (b) connected services. (From Brown and Toomer 2003. With permission.)

- Identify what is at a location through placement of a feature's symbol in a reference frame
- Portray the relationship between features as connecting, adjacent, contained within, intersecting, in proximity, or higher/lower
- Display multiple attributes of an area
- Allow portrayal of and discernment between distributions, relationships, and trends
- Show classifications of feature attributes and graphic portrayals as thematic maps
- Visually encode feature attributes as text, values, or identifiers
- Detect changes over time using maps prepared at different times
- Integrate data from diverse sources into a common geographic reference, thereby allowing comparison

In environmental and water resources engineering, maps and plans are a basic medium for design. Infrastructure designs are portrayed in a map format to communicate the exact nature of the project in terms of specific locations and relationships over an area. For example, a sanitary sewer system is shown in Figure 2.4. The layout shows the location and flow path of the collection sewer system. Topographic contours describe the lay of the land. Slope values can be derived and act as input for pipe alignment and diameter computations. Pipe flows are derived from the specific properties and streets in this plan.

2.3.2 Coordinate Systems and Geocoding

If a map feature is to be comparable in space to other features, it must have a location. Spatial data compiled from various sources must be assembled into a consistent reference frame. All points on the Earth's surface can be defined in geographic coordinates as latitude, longitude, and elevation above mean sea level. A map projection is a mathematical transformation by which the latitude and longitude of each point on the Earth's curved surface is converted into corresponding (x,y) or (easting, northing) projected coordinates in a flat-map reference frame (Snyder 1987). Figure 2.5 illustrates the concept of map projection for the equatorial case. If data are available in one map projection and required in another, then specialized GIS software can perform the transformation into the new projected reference frame.

Knowledge of the map scale is needed to properly understand a map's accuracy. The map scale describes the relationship between the mapped size and the actual size. It is expressed as the ratio (or representative fraction) of the linear distances on the map and corresponding ground distances. Large-scale maps ($\approx$1:1000) cover small areas, but can include a high level of detail. Large-scale maps are most often used for municipal facilities plans, and these maps must be developed using

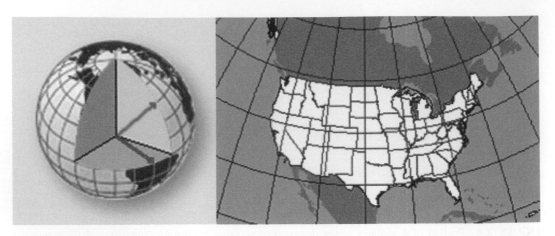

FIGURE 2.5   Geographic coordinates expressed as degrees latitude and longitude represent angular degrees calculated from the center of the earth. (*Source:* www-atlas.usgs.gov/articles/mapping/a_latlong.html)

photogrammetric techniques. Small-scale maps (≈1:250,000) depict larger areas with less detail. For example, the U.S. Geological Survey Digital Line Graph (DLG) series is issued as three primary types: (a) large-scale (7.5-min of latitude and longitude) DLGs correspond to the USGS 1:20,000-, 1:24,000-, and 1:25,000-scale topographic quadrangle maps; (b) intermediate-scale (1:100,000-scale) DLGs; and (c) small-scale (1:1,000,000-scale) DLGs for the National Atlas. The 1:24,000 scale is most often used for watershed studies. The 1:100,000 scale is used for national coverage of the U.S. stream network. Details on U.S. map standards can be found at http://nationalmap.gov/gio/standards/.

USGS maps adhere to the National Map Accuracy Standards (USGS 1999). As applied to the USGS 7.5-min quadrangle topographic map, the horizontal accuracy standard requires that the positions of 90% of all points tested must be accurate within 1/50th of an inch (0.05 cm) on the map. At 1:24,000 scale, 1/50th of an inch is 40 ft (12.2 m). The vertical accuracy standard requires that the elevation of 90% of all points tested must be correct within half of the contour interval. On a map with a contour interval of 10 ft, the map must correctly show 90% of all points tested within 5 ft (1.5 m) of the actual elevation.

2.3.3   DATA REPRESENTATIONS AND DATA MODELS

The nature of the data representation has a strong influence on the analysis that can be applied. Spatial data in GIS are most often organized into vector and raster (or surface) data structures (Figure 2.6). In the vector structure, geographic features or objects are represented by points, lines, and polygons that are precisely positioned in a continuous map space, similar to traditional hard-copy maps that identify landmarks, buildings, roads, streams, water bodies, and other features by points, lines, and shaded areas. In addition, each object in the vector structure includes topologic information that describes its spatial relation to neighboring objects, in particular its connectivity and adjacency. This explicit and unambiguous definition of and linkage between objects makes vector structures attractive and allows for the automated analysis and interpretation of spatial data in GIS environments (Meijerink et al. 1994).

On the other hand, surface, or raster (from display technology), data structures divide space into a two-dimensional (2-D) grid of cells, where each cell contains a value representing the attribute being mapped. A raster is an x,y matrix of spatially ordered numbers. Each grid cell is referenced by a row and column number, with the boundary of the grid being registered in space to known coordinates. Raster structures arise from imaging sources such as satellite imagery and assume that the geographical space can be treated as though it were a flat Cartesian surface (Burrough 1986). A point is represented by a single grid cell, a line by a string of connected cells, and an area by a group of adjacent cells. When different attributes are considered, such as soil and land use, each

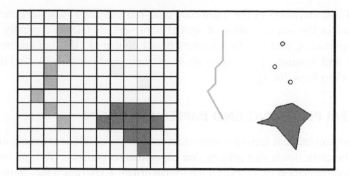

FIGURE 2.6 Raster (grid) and vector data structures provide complementary means for representing location and character of map features.

are represented by separate raster layers. Operations on multiple layers involve the retrieval and processing of the data from corresponding cell positions in the different layers. This overlay concept is like stacking layers (2-D grids) and then analyzing each cell location (Meijerink et al. 1994). The simplicity of data processing in raster structures has contributed to its popularity. Both vector and raster structures are valid representations of spatial data. The complementary characteristics of both structures have long been recognized, and modern GIS can process both structures, including conversions between structures and overlays of both structures. Additional details on GIS data structures are presented in Chapter 3.

Of primary interest for water resources, especially surface-hydrology applications, are representations of topography. *Digital elevation model* (DEM) is the general term used for topographic data models. DEMs are generally stored in one of three data structures: (a) raster or grid structures, (b) triangulated irregular network (TIN) structures, and (c) contour-based structures. Grid structures consist of a square grid matrix, with the elevation of each grid square (called a pixel) stored in a matrix node. Location is implicit from the row-and-column location within the matrix, given known boundary coordinates. In TIN structures, a continuous surface is generated from interconnected triangles with known elevation values at the vertices of the triangles. For each triangle, the location (x,y) and elevation (z) of the vertices are stored, as well as topological information identifying adjacent triangles. Triangles vary in size, with smaller triangles clustered in areas of rapidly changing topography and larger triangles in areas of relatively smooth topography. Contour-based structures consist of digitized contour lines defined by a collection of x,y coordinate pairs for contours of specified elevation. DEM data sources are described in Chapter 3.

2.4 USER INTERFACES AND INTERACTION MODES

A primary attraction of modern GIS is the user friendliness of the computer system interface provided by the various software vendors. Efficient retrieval of data depends not only on properly structured data in the database and the speed of retrieval, but also on well-designed interfaces and query languages. The human–computer interface provides the environment that enhances human interaction with the GIS. It makes it easy for the user to access data and analysis results, and to display these data in understandable formats. Most traditional information systems provide limited presentation formats, usually as text, tables, and graphs. While these formats are still useful, the spatial character of geodata allows additional possibilities, including map formats and visualization techniques.

GIS software usability has progressed from command-line modes to menu and forms modes to today's graphical user interfaces (GUIs) (Martin et al. 2005). A GUI enables a user to interact with the computer system by pointing to pictorial representations (icons) and lists of menu items on the screen using a mouse. The use of icons facilitates the functionality of data selection, data presentation, and data manipulation.

Visualization is an extension of the traditional data-retrieval and -display concepts. It includes techniques that aid in the interpretation of spatial data sets. A GIS is typically concerned with analysis and interpretation, and it is the graphics-based nature of GIS that enables the perception of spatial patterns and features of the information, extraction of parameters, and discrimination of classes of objects (Worboys 2004).

2.5 GIS SYSTEM PLANNING AND IMPLEMENTATION

Consideration of organizational factors is important to successful GIS implementation and management because of the critical role that information plays in an organization's function. In most cases, the design and implementation of a GIS is a long-term effort. Experience has shown that, as important as technical issues of software, hardware, and database design are, it is the people problems arising from access to the information and its use that determine whether a GIS will succeed or fail (Aronoff 1991).

During the planning of a major GIS acquisition or development, it is important to consider certain organizational attributes that will impact the chosen approach. These attributes, generally addressed in a needs assessment, should be evaluated in the broadest possible sense. In this way, the goals, equipment, costs, etc., of all impacted departments will be included in the implementation planning. Only after careful consideration of these attributes can the best possible implementation strategy be chosen. The following are some attributes to consider:

- Overall organizational function and goals
- Sources of data available as input to the GIS system
- GIS hardware/software/databases and products that are currently and planned to be utilized
- Management approaches that will and have previously guided the GIS program
- Costs of implementation, both historic and planned
- Benefits of implementation, both tangible and intangible
- Procedure to be used in evaluating and comparing the costs and benefits
- Generation procedures—review internal, external, current, and potential procedures
- Quality assurance/quality control procedures (QA/QC) and any applicable data standards
- End-user interactions/training—consider how the GIS group will communicate with its "clients"
- Evaluation/assessment procedures to be used to review the GIS implementation
- Legal issues pertaining to data distribution and ownership

Although it is difficult to quantify many of these attributes, it is a useful exercise to at least estimate the worth of each one. For instance, many organizations consider a formal cost/benefit analysis to be based on highly speculative information, although it is possible to measure the "goodness" of intangible benefits on a relative scale. Further, as these types of organizational issues are discussed during planning, a broader and more realistic picture of the resulting GIS implementation emerges.

2.6 GIS SOFTWARE

A large number of GIS software options are available as open-source or commercial products. Following is a brief summary of some of the more popular GIS packages. A large listing of GIS software can be found at http://en.wikipedia.org/wiki/GIS_software.

2.6.1 PROPRIETARY GIS

- ArcGIS®: ArcGIS is the name of a suite of GIS software product lines produced by ESRI (http://www.esri.com/). At the desktop GIS level, ArcGIS can include: ArcReader, which allows one to view and query maps created with the other Arc products; ArcView®, which allows one to view spatial data, create maps, and perform basic spatial analysis; ArcEditor, which includes all the functionality of ArcView as well as more-advanced tools for manipulation of shape files and geodatabases; or ArcInfo®, the most advanced version of ArcGIS, which includes added capabilities for data manipulation, editing, and analysis. There are also server-based ArcGIS products as well as ArcGIS products for personal digital assistants (PDAs). Extensions can be purchased separately to increase the functionality of ArcGIS.

- AutoCAD®: AutoCAD is a popular engineering CAD software produced by AutoDesk (http://usa.autodesk.com/). AutoCAD Map 3D software is a leading engineering platform that bridges the gap between CAD and GIS. When combined with Autodesk MapGuide® technology, AutoCAD Map 3D provides a way to publish data to the Web or an intranet.
- Cadcorp®: Cadcorp (http://www.cadcorp.com/) is the developer of GIS software and OpenGIS standard (e.g., Read/Write Open-Source PostGIS database). Products include a Spatial Information System (SIS), which runs on Microsoft Windows and encompasses desktop GIS modules, ActiveX- and COM-based developer kits, Web-based GIS software (GeognoSIS), and a mobile data-capture solution (mSIS).
- ERDAS IMAGINE®: ERDAS IMAGINE is a raster graphics editor and remote-sensing application designed by ERDAS, Inc. (http://www.erdas.com/). It is aimed primarily at geospatial raster data processing that allows the user to display and enhance digital images. It is a toolbox allowing the user to perform numerous operations on an image and generate an answer to specific geographical questions.
- IDRISI®: GIS developed by Clark Labs (http://www.clarklabs.org/products/) at Clark University, Massachusetts. IDRISI Andes is an integrated GIS and image-processing software providing over 250 modules for spatial analysis and display. Originally developed under United Nations sponsorship, the IDRISI is widely used worldwide.
- Intergraph®: Intergraph (http://www.intergraph.com/) provides software and services for infrastructure management for the electric, gas, water, pipeline, utility, and communications industries. Products include GeoMedia, GeoMedia Professional, GeoMedia WebMap, and add-on products for industry sectors, as well as photogrammetry.
- MapInfo®: MapInfo (http://www.mapinfo.com/) GIS software products include the desktop GIS software, MapInfo Professional, MapXtreme 2005, and MapXtreme Java for Web-based and desktop client mapping, as well as developer tools such as MapBasic.
- MicroStation®: MicroStation is a suite of CAD/GIS software products for 2-D and 3-D design and drafting, developed and sold by Bentley Systems (http://www.bentley.com/). It is used by engineering designers for transportation and for water and wastewater utilities. Bentley also offers GIS-based water resources modeling software for water, sewer, and stormwater systems (SewerCAD, WaterCAD, StormCAD).

2.6.2 Open-Source GIS

- GRASS: GRASS (Geographic Resource Analysis Support System) is a public-domain open-source raster GIS developed as a general-purpose spatial modeling and analysis package (Neteler and Mitasova 2008). GRASS is a raster/vector GIS, image processing system, and graphics production system. GRASS contains over 350 programs and tools to render maps and images on monitor and paper; manipulate raster, vector, and sites data; process multispectral image data; and create, manage, and store spatial data. GRASS uses both an intuitive Windows interface as well as command-line syntax for ease of operation.

REFERENCES

Aronoff, S. 1991. *GIS: A management perspective*. Ottawa: WDL Publications.
Brown, C., and K. Toomer. 2003. GIS in public works: Atlanta—save our sewers initiative. In *Proc. ESRI Annual Users Conference*. San Diego, Calif. http://gis.esri.com/library/userconf/proc03/abstracts/a0464.pdf.

Burrough, P. A. 1986. *Principles of geographical information systems for land resources assessment.* New York: Oxford Univ. Press.

Clarke, K. C. 1995. *Analytical and computer cartography.* 2nd ed. Englewood Cliffs, N.J.: Prentice-Hall.

Dangermond, J. 1988. Introduction and overview of GIS. Paper presented at Geographic Information Systems Seminar: Data sharing—Myth or reality. Ontario: Ministry of Natural Resources.

Duecker, K. J. 1987. Geographic information systems and computer-aided mapping. *J. Am. Plann. Assoc.* 53: 383–390.

ESRI (Environmental Systems Research Institute). 1986. *San Diego Regional Urban Information System Conceptual Design Study: System concept and implementation program,* vol. 1. Redlands, Calif.: ESRI Press.

Maguire, D. J. 1991. An overview and definition of GIS. In *Geographical information systems principles and applications,* ed. D. J. Maguire, M. F. Goodchild, and D. W. Rhind, 9–20. New York: Longman Scientific and Technical, John Wiley and Sons.

Martin, P. H., E. J. LeBoeuf, J. P. Dobbins, E. B. Daniel, and M. D. Abkowitz. 2005. Interfacing GIS with water resource models: A state-of-the-art review. *JAWRA* 41 (6): 1471.

McHarg, J. L. 1969. *Design with nature.* Garden City, N.J.: Doubleday.

Meijerink, A. M. J., H. A. M. Brouwer, C. M. Mannaerts, and C. R. Valenzuela. 1994. Introduction to the use of geographic information systems for practical hydrology. UNESCO International Hydrological Programme, Publication No. 23. Venice: UNESCO.

Neteler, M., and H. Mitasova. 2008. *Open source GIS: A GRASS GIS approach.* 3rd ed. Vol. 773 of The International Series in Engineering and Computer Science. New York: Springer. http://grass.itc.it/.

Snyder, J. P. 1987. Map projections: A working manual. U.S. Geol. Surv. Prof. Pap. 1395. Washington, D.C.: USGS.

Star, J., and J. Estes. 1990. *Geographic information systems: An introduction.* Englewood Cliffs, N.J.: Prentice Hall.

USGS. 1999. Map accuracy standards fact sheet FS-171-99. http://nationalmap.gov/gio/standards/.

Worboys, M. F., and M. Duckham. 2004. *GIS: A computing perspective.* 2nd ed. Boca Raton, Fla.: CRC Press.

Zeiler, M. 1999. *Modeling our world: The ESRI guide to geodatabase design.* Redlands, Calif.: ESRI Press.

3 GIS Data and Databases

3.1 OVERVIEW

Data and databases provide the foundation for water resources GIS development and management. Without an adequate database having accurate and complete entries, there can be no reporting and modeling to support the various decisions required. Sources of spatial data are myriad and include conversion of existing data from archived records and plans (maps) and creation of new data from field measurements. Technologies for field measurements continue to advance for on-ground, aerial, and satellite surveys. Consideration must be given to the inherent accuracy of the data per the intended uses. Given the existence of data in digital formats, it is then required to archive and manage those data for the purposes intended. This requires careful planning and design of spatial databases, which are the foundation of a successful GIS.

This chapter presents a review of (a) principles and methods for GIS data development and maintenance, (b) sources for GIS data relevant to the water resources field, and (c) various attribute and spatial database models and their utility for management of geographic data.

3.2 GIS DATA DEVELOPMENT AND MAINTENANCE

GIS functions for spatial data capture include the numerous technologies for data capture as well as the many ways for conversion of source data into GIS-compatible formats. These functions include:

- Tablet digitizing
- Scanning
- Format conversion
- Surveying and COGO (coordinate geometry)
- GPS (global positioning system)
- Photogrammetric data development
- Image processing
- Geometric transformations
- Projection conversions
- Attribute entry and editing
- Metadata

A large amount of current and historic maps are archived as paper, Mylar, and other flat media. *Tablet digitizing* has been a primary data-conversion function due to its simplicity and relatively low capital cost. Most GISs include interfaces for tablet digitizing; some packages emphasize this functionality as a primary product attribute. Quality control requirements applied to digitized data include those for topological integrity (nodes and links), projection registration, and feature attribution. Source data copies are "scrubbed" to identify feature classes to be digitized. Digitizing options range from in-house digitizing for small projects to professional contracting for large jobs.

Scanning technologies have advanced such that existing map stock can be captured and processed using image-processing procedures. Although OCR (optical character recognition) and advanced processing can identify map features, it is usually required that manual review and identification of feature attributes be conducted. "Heads-up" digitizing for vector conversion is replacing tablet digitizing for many GIS practitioners where scanned maps or other image data are displayed in the GIS in the correct projection and coordinate system. Relevant features can then be digitized "on-screen" using a mouse or a pen, with the digitized features automatically assigned to the required image projection and coordinate system. The quality and precision of currently available display monitors greatly facilitate the use of "heads-up" digitizing, allowing operators to easily enter annotation and attribute information during the digitizing process.

Format conversions from other digital data sources can constitute a major portion of a GIS database development. One major source is conversion from CAD (computer-aided design) files. Most engineering organizations conduct their mapping and plan documentation using digital measurement and computerized technologies. For conversion to GIS, it is important that the CAD drawings be logically structured so that the features can be separated. In addition to digital map data, there are other data sources, including database files, raster maps and imagery, text reports, forms or service cards, and mechanical drawings.

COGO (coordinate geometry) is a technique for entering boundary information to a GIS by keyboard entry of distances, bearings, and curve calculations from field surveys and property titles. When distances and bearing are entered based on coordinate grids such as those found in state plane coordinate systems (SPCS), GIS uses this information to create a graphic representation of the lines (Figure 3.1).

Surveying has traditionally been the primary mapping tool for engineering projects. It is not uncommon for land surveys to be conducted entirely in digital formats. Electronic distance measurements can be taken directly to the computer, plots made, and uploaded to a GIS. The availability of automated measurement equipment has increased the survey data resolution, but knowledge of the principles of surveying is required to guide the survey.

GPS (global positioning system) has revolutionized field-data collection. GPS is based on a constellation of satellites, each of which broadcasts a unique signal. By reading the radio signals broadcast from as few as three of these satellites simultaneously, a receiver on Earth can pinpoint its location on the ground through a process called trilateration. The satellites and ground-based receivers transmit similarly coded radio signals so that the time delay between emission and receipt

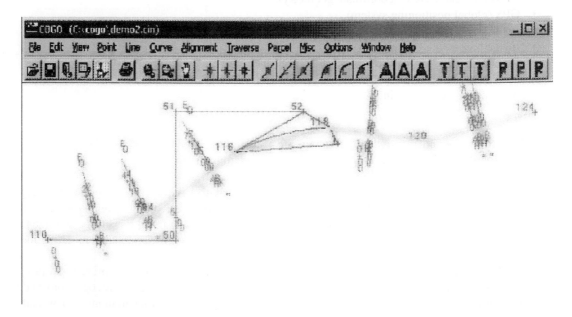

FIGURE 3.1   COGO program example display of the Colorado Department of Transportation (CDOT 2002).

gives the distance between the satellite and the receiver. If signals from three or more satellites are available, then trigonometry is used to calculate the location and elevation of the receiver. GPS data can be ported directly to the GIS software to provide location information and travel paths of field-data collections. Typical handheld GPS accuracy is about 10 to 15 m. Through correction procedures known collectively as "differential GPS," the accuracy level can be improved to a few centimeters for the more sophisticated instruments.

Photogrammetric data development is a primary means for developing high-resolution spatial databases for urban areas. Aerial overflights with high-resolution cameras provide the base data for photogrammetric processing. Using stereographic techniques, the analyst defines a photogrammetric model and develops point elevations, contours, and theme-feature identifications and tabulations. In the photogrammetric model, the locations of survey control points can be identified, and the mathematical relationship between control points and other features visible in the photos can be established.

Satellite remote sensing and image processing have become increasingly important, as platforms such as NASA's Landsat 7, Digital Globe's Quickbird and WorldView-1, GeoEye's IKONOS and OrbView, and the French SPOT satellite systems are providing enormous amounts of high-resolution data. Figure 3.2 shows typical high-resolution imagery. These orbital platforms collect and transmit data from various regions of the electromagnetic spectrum that, when analyzed using modern imaging-processing software, provide valuable information for a wide range of applications in water resources engineering, particularly when combined with airborne and ground-truth measurements. Satellite photogrammetry has greatly facilitated the mapping of selected areas with spatial resolutions of less than 1 m, and at a significantly lower cost than aerial photogrammetry. Further advantages of satellite imaging include the ability to obtain repetitive images in various seasons and under any weather conditions. Image-processing procedures are described further below.

FIGURE 3.2 Satellite imagery of Fairbanks, AK (with cloud). (*Source:* GeoEye (http://www.geoeye.com/.)

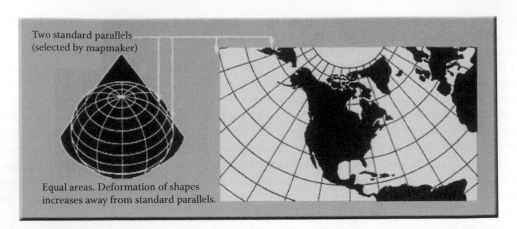

FIGURE 3.3 Coordinate reference systems arise from geometric mapping of the earth using projections onto three shapes that can be unrolled onto a flat map. (From USGS 1999.)

Object identification and editing are required to establish and maintain the features of a GIS database. All GISs require the capability to transform source data into the data structure of the system and to edit those files once they have been created.

Geometric transformations are applied to establish ground coordinates for a map. This requires that a registration correction be assigned to all data so that the overlay will correspond to the coordinate control and other map layers. The registration correction may be computed to a required acceptance level through least-squares analysis. Data sets having different origins, various units of measurement, or diverse orientations may be resolved to each other using appropriate linear mathematical transformations (called affine transformations) such as translations and scaling to create a common origin and rotation. Often a map layer is registered using a nonlinear rubber-sheeting procedure. Edge matching is required when the features crossing two or more maps do not match at the edge. Resolving mismatches between features located on two or more data layers requires a conflation procedure to reconcile the differences.

Projections involve the multiple ways that the oblate-spheroidal shape of the Earth is mapped to a planar surface (Figure 3.3). Locations on Earth are mapped by a system of latitude and longitude, which is then projected to a map sheet. There are many projections used to portray the Earth as a planar surface, and it is a requirement of a GIS to convert between projections so that data layers can be overlay compared. Coordinate referencing systems continue to improve as base monuments having known coordinates are established within high-accuracy reference networks (HARN).

A map projection requires the specification of an Earth datum, a projection method, and a set of projection parameters. Although the Earth is commonly thought of as a sphere, it is actually an oblate spheroid or ellipsoid. All latitude and longitude (geographic) coordinates are defined on an Earth datum, which includes a reference ellipse rotated around a defined axis of rotation. For the United States, two horizontal Earth datums are in common use: the North American Datum of 1927 (NAD 27) and the North American Datum of 1983 (NAD 83). The conversion from NAD 27 to NAD 83 moves a particular point in latitude and longitude coordinates between 10 and 100 m, depending on location.

There are three main methods of map projection: cylindrical, conical, and azimuthal (Snyder and Voxland 1989). The best-known cylindrical projection is the Transverse Mercator projection, which forms the basis of the Universal Transverse Mercator (UTM) coordinate system. The UTM is widely used in the United States for projections of states that have primarily a north-south orientation, such as California. The best-known conical projection is the Lambert Conformal Conic projection, used in the United States for projections of land masses with primarily east-west extents. Another conical projection is the Albers Equal Area projection, which preserves true Earth surface area on the flat map projection. An azimuthal map projection is one in which a flat-map surface touches the Earth's surface at one point; this approach is used for meteorological mapping and for views of Earth from space.

Several standardized projection systems exist for legal purposes. In the United States, the most important of these is the state plane coordinate system (SPCS), which defines for each state a set of one or more projection zones, and the map projection and parameters for each zone (Snyder and Voxland 1991). Because the state plane system was developed during the 1930s, its projections are based on the NAD 27 datum and have coordinates in units of feet. More modern standardized projection systems use the NAD 83 datum, and coordinates are in the units of meters.

Regardless of the projection, all maps should contain certain elements per cartographic practice. The main elements of a map are the frame, title, key (legend), scale, labels, and north arrow. These are often called annotations. In addition to these, graphic image, grids, graticules, and outline and filled boxes are also used as annotations. Annotations are text and graphic elements that help viewers interpret the information appearing on a map.

Attribute entry and editing involves the capture and management of the nonspatial data associated with objects such as point, line, and polygon features. Attributes are the characteristics of the objects. GIS functions for attribute data capture include forms for data entry that limit and check keyboard entries for appropriate format and range (e.g., no text for number fields). Import of attribute data from other digital files can be accomplished given standard formats. Scanning of attribute data and OCR (optical character recognition) processing can be used to convert paper records to digital formats. Number, type, date, engineering drawing, or picture can index these documents in a relational database. Regardless of the source, the GIS must provide functions to tag the attribute data to the graphic object; this usually occurs using an index or common key code for both the graphic and the attribute data. Geocoding is the process of tagging parcel identifiers to address and other tabular data on a property (Figure 3.4). GIS applications can be built that allow users to interactively point to a graphic object (e.g., parcel) and retrieve a database record for display (e.g., water-use records).

Metadata include information about the content, format, quality, accuracy, availability, and other characteristics of a GIS database. They are data about data. Metadata can help answer questions about GIS databases so users can decide which data sources may be useful for their needs. The Federal Geographic Data Committee (FGDC) has developed a standard for storing metadata called the Content Standard for GeoSpatial Metadata. This standard outlines major categories and specific data items for the metadata (http://www.fgdc.gov/metadata).

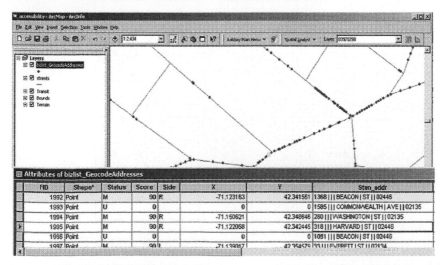

FIGURE 3.4   Geocoding of addresses in a table allowing their location on the map. (From Ferreira 2007.)

3.3 GIS DATA MODELS

3.3.1 OVERVIEW

A data model is a set of rules to identify and symbolize features of the real world (called entities) into digitally and logically represented spatial objects consisting of the attributes and the geometry. The attributes are characterized by thematic or semantic structures, while the geometry is represented by geometric-topological structures.

There are two basic categories of data involved: spatial and attribute. *Spatial data* include the locations of features, such as the latitude/longitude of dams, gauging stations, etc. Spatial data are often represented as objects such as points, lines, and polygons, which are used to represent the differing types of features. For example, the location of a well is a point, a stream path is a line (or vector), and a basin boundary is a polygon. Spatial data may also be represented as fields or images, such as might be derived from satellite imagery. *Attribute data* include numerical and character-type data that characterize the resource. Geographic data are characterized by a series of attribute and behavioral values that define their spatial (location), graphical, textual, and numeric dimensions (Worboys 1995). These include identifiers, names, and physical capacities for features of the water resources system, such as dams and reservoirs, pipelines, drainage basins, pumps, and turbines. Time-series data on river flows, reservoir releases, pumping rates, and other time variables are also managed in the attribute database.

3.3.2 RASTERS AND VECTORS

The two major types of geometric data models are raster and vector, as shown in Figure 3.5. These categories are sometimes related to how people think about things (objects) versus how they see

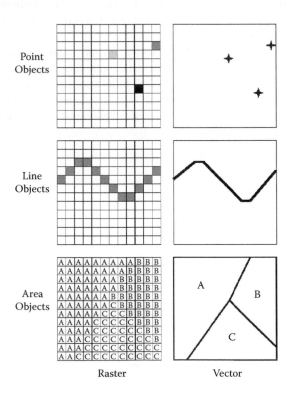

Raster   Vector

FIGURE 3.5 Raster and vector data models; the raster data model represents point, lines, and areas as collections of cells with values.

the world (field). The first general data model is the raster or field model, which is often defined on an x,y grid with each cell, or pixel (i.e., picture element), specifying the value of the data. The uniform grid is also referred to as a raster data structure. A field model is often used to represent variables that vary continuously over a region, such as temperature, rainfall, and elevation. In a raster, a point is given by a point identifier, its coordinates (i,j), and the attribute value. A line is given by a line identifier, a series of coordinates forming the line, and the attributes. A raster area segment is given by an area identifier, a group of coordinates forming the area, and the attributes.

A vector data structure maps regions into polygons, lines into polylines, and points into points (Figure 3.6). The vector model is used to represent spatial entities such as river and pipe networks and facilities (e.g., fire hydrants). These data often derive from land surveying and CAD (computer-aided design) drawings, as well as conversion of imagery data through processing. A key aspect of the vector data model is that the topology of relationships between features be established. Topology refers to the relationships or connectivity between spatial objects. The geometry of a point is given by two-dimensional (2-D) coordinates (x,y), while line, string, and area are given by a series of point coordinates.

- *Node*: an intersection of more than two lines or strings, or the start and end point of an arc with a node number
- *Arc*: a line or a string with an arc number, start and end node numbers, or left- and right-neighbored polygons
- *Polygon*: an area with a polygon number, or a series of arcs that form the area with directionality (e.g., positive in clockwise order)

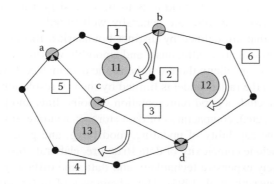

Chain Geometry

Chain	Start	Coordinates	End
1	(Xa, Ya)	(Xi, Yi) ... (Xj, Yj)	(Xb, Yb)
.	...	... ...	...
6	(Xb, Yb)	(Xi, Yi)...(Xj, Yj)	(Xd, Yd)

Polygon Topology

Polygon	Chain
11	1, 2, 5
12	6, −3, −2
13	4, −5, 3

Node Topology

Node	Chains
a	1, −5, −4
b	−1, 2, 6
c	−2, 3, 5
d	−3, 4, −6

Chain Topology

Chain	From	To	Left Polygon	Left Polygon
1	a	b	0	11
.	.	.	.	.
6	b	d	0	12

FIGURE 3.6   Vector topology.

TABLE 3.1
Vector versus Raster Data Structures

Data Structure	Advantages	Disadvantages
Vector	Compact storage	Complex data model
	Topology is explicit and powerful	Overlay operations complex and CPU intensive
	Represents entities	Requires attributes for entities
	Integrates with DBMS	Display and plotting can be time consuming
	Easy coordination of transformations	Display accuracy may be misleading
	Accurate graphics at all scales	
Raster	Simple data structure	Large data volumes
	Easy manipulation of attribute values	Accuracy limited by cell resolution
	Many analysis functions	Limited attribute representations
	Easy math modeling	Graphics coarse with zoom
	Many data forms available	Coordinate transforms difficult
	Economical technology	

A vector database must be topological before it can be reliably used for spatial queries or to support modeling. The vector data model is often preferred for sophisticated modeling projects because the objects can be manipulated in a logical manner. For example, a river network can be searched up or down.

Vector and raster structures both have advantages and disadvantages. Each approach tends to work best in situations where the spatial information is to be treated in a manner that closely resembles the data structure. Vector structures are generally well suited to represent networks, connected objects, and features that are defined by distinct boundaries. Raster structures work best when the attributes they represent are continuously and smoothly varying in space. The finer the grid used, the more geographic specificity there will be in the data matrix. However, the location precision for a map feature in the raster model is limited by the cell resolution. Generally, the advantages of vector structures include a good representation of point, line, and polygon features (streams, lakes, drainage divides, etc.); compactness of data storage; accurate graphics; relational representation of objects; and the capability of updating, modifying, and generalizing graphics and attributes. Disadvantages include complexity of data structure, elaborate processing for overlays and simulation, comparatively expensive technology and data, and difficulty in representing spatially varying attributes (Meijerink et al. 1994). The advantages of raster structures include the simplicity of the data structure, easy overlay and spatial analysis, availability of data, and comparatively cheap technology. Disadvantages include inefficient use of computer storage; inaccuracies in point, line, and area definitions; difficulty in establishing networks and topology; and unattractive visualization in low-resolution rasters (Meijerink et al. 1994). Table 3.1 summarizes the advantages and disadvantages of each data structure; Burrough and McDonnell (1998) provide a more comprehensive review.

The decision to use an object or field-data model is based on the requirements of the application, on tradition, and on the original source data. However, advances in GIS database technologies make the distinctions between the vector and raster data structures less compelling. Data-storage capacities continue to increase, which makes the issue of raster storage volume less critical, although the huge volumes of image data from various sensors still require difficult decisions on storage priorities. For many applications, the increasing resolution of imaging systems provides adequate accuracy for locating land features. Additionally, the long-term growth in computer capacities continues to mitigate processing issues.

Representations of the spatially varying characteristics of a watershed are often contained in a series of raster data sets, which may include such information as elevation, soil properties, and land use. However, the spatial resolution of raster elevation data may make it difficult to accurately locate smaller streams and other bodies of water using only the elevation data. The finest spatial resolution of widely available elevation data is 10 m; at such a scale, many smaller streams simply may not be identified, and drainage boundaries may be inaccurately determined. Vector stream data can help to eliminate some of these difficulties. Sources of raster and vector data for land surface and stream paths are described below.

3.4 DIGITAL DATA SOURCES FOR WATER RESOURCES

Data of various types are increasingly being made available for download from the U.S. Geological Survey, other federal agencies, and various corporations. Following are descriptions of some of these.

3.4.1 DIGITAL ELEVATION MODELS

Three-dimensional (3-D) surfaces are a special category of surfaces where the data are best suited for representation in 3-D form over an area. The most common example is land surface terrain represented by DEMs (digital elevation models), represented as elevation grids in the field model and as TINs (triangulated irregular networks) or contours in the vector model. A raster of elevation is alternatively called a digital terrain model (DTM). Other field data are often derived from satellite and other imagery; these data sources are increasingly available as the number and variety of imagery sensors grow. Digital elevation models are generally produced by photogrammetric techniques from stereo photo pairs, stereo satellite images, or interpolation of digitized elevation data. The most widely produced DEM structures are the square-grid DEMs and contour digital line graphs (DLGs).

DEMs as grid structures comprise a square grid with the elevation of each grid square (Figure 3.7a). Each grid element is called a pixel (for picture element). Location is established by the row and column locations within the grid, given information on the grid boundary coordinates. National Elevation Datasets are available online from the USGS EROS Data Center (http://ned. usgs.gov).

For TIN structures, a continuous surface is generated from interconnected triangles with known elevation values at the vertices of the triangles (Figure 3.7b). For each triangle, the location (x,y) and elevation (z) of the vertices are stored along with topological information identifying adjacent triangles. Triangles vary in size, with smaller triangles clustered in areas of

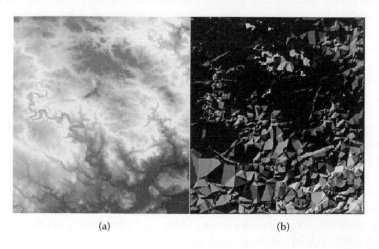

(a)   (b)

FIGURE 3.7 (See color insert following page 136.)   DEM represented as: a) grid, and b) TIN. (From Close 2003.)

rapidly changing topography and larger triangles in areas of relatively smooth topography. For TINs, the elevation data may be initially developed using photogrammetric procedures for spot elevations and break lines, such as streams and ridgelines. Contour-based structures consist of digitized contour lines defined by a collection of *x,y* coordinate pairs for contours of specified elevations. Contours are commonly obtained by computerized digitization of map contours or by interpolation from gridded TINs. Reverse interpolation is also used to generate grids or TINs from contours.

There are advantages and disadvantages associated with each of the DEM structures (DeBerry 1999; Moore et al. 1991). Square-grid DEMs are the most widely used because of their simplicity, processing ease, and computational efficiency. Disadvantages include the grid-size dependency of certain computed landscape parameters and the inability to adjust the grid size to changes in the complexity of landscape characteristics. TINs are preferred for complex topography, such as urban drainage features (e.g., street crown lines). However, the computations using TINs are more complex than for square-grid DEMs. Contour-based structures are considered by some to provide better visualizations of surface features than grid DEMs. However, contours are one-dimensional (1-D) features, and the representation of the 2-D landscape continuum with DLGs generally requires considerably more data than does representation by grid DEMs (Moore et al. 1991). The stated advantages and disadvantages of DEM structures are relative, in the sense that what is considered an advantage for data storage may be a disadvantage for data processing. Most GISs provide functions to convert between the different surface representations, depending on the application requirements.

The USGS has offered three DEM products for the conterminous United States for several years through the EROS Data Center (http://eros.usgs.gov):

1. Large-scale 7.5-min DEMs (approximately 1:24,000 scale) in both 30-m and 10-m resolutions
2. Intermediate-scale 15-min DEMs (approximately 1:100,000 scale) in a 60-m resolution
3. Small-scale 1° DEMs (approximately 1:250,000 scale) in a 90-m resolution

Since these data are offered in blocks, problems often arise in edge matching of adjacent blocks of DEM data, giving rise to slivers and seams when the blocks are merged together. To overcome these problems, the USGS has developed the National Elevation Database (NED) as a seamless mosaic of the best-available elevation data (http://ned.usgs.gov). Although 7.5-min elevation data for the conterminous United States are the primary data source, efficient processing and filtering methods have been applied to provide a consistent datum and projection, edge-matching, filling slivers of missing data at quadrangle seams, and conversion of all elevation values to decimal meters. As of this writing, resolution data at the level of 1 arc-second (approximately 30-m resolution) are available nationwide, with approximately 70% coverage for one-third arc-second (10-m resolution).

For highly detailed terrain mapping, such as that required to define floodplain details, the use of light detection and ranging (LIDAR) data has increased. These data are collected with aircraft-mounted lasers capable of recording elevation measurements at a rate of 2000 to 5000 pulses per second and have a vertical precision of 15 cm (6 in.). The LIDAR instruments only collect elevation data. To make these data spatially relevant, the positions of the data points must be known. A high-precision global positioning system (GPS) antenna is mounted on the upper aircraft fuselage. As the LIDAR sensor collects data points, the location of the data is simultaneously recorded by the GPS sensor. After the flight, the data are downloaded and processed using specially designed computer software. The end product is accurate, geographically registered longitude, latitude, and elevation (x,y,z) positions for every data point. These x,y,z data points allow the generation of a digital elevation model (DEM) of the ground surface.

A major cost for LIDAR, for example in floodplain mapping, is vegetation removal, which involves the postprocessing of the data to generate bare-Earth digital information. To get these bare-Earth elevation data, automated and manual postprocessing are used to eliminate points that impinged on elevation features. This creates data voids, so that the data set becomes more irregular with regard to point spacing on the ground. Automated postprocessing includes computerized procedures that detect elevation changes that appear to be unnatural. For example, rooftops are identified with relative ease because there are abrupt elevation changes between the yard and the rooftop. However, vegetation provides more difficult challenges for automated procedures. Manual postprocessing is more accurate (and more costly) and normally includes the overlay of data points on digital imagery, enabling the analyst to see where the laser points hit the ground.

3.4.2 DIGITAL LINE GRAPHS

Digitized stream and channel network data can be obtained from the U.S. Geological Survey (USGS) in the form of digital line graphs (DLGs). DLGs are available in several categories, including political boundaries, roadways, and hydrography (Figure 3.8). Hydrography data provide information on flowing water (streams and channels), standing water (lakes), and wetlands. The information is provided in the form of digital vectors that were developed from maps and related sources. The hydrographic data can be obtained for large, intermediate, and small scales. Large-scale DLGs are derived from the USGS 1:20,000-, 1:24,000-, and 1:25,000-scale 7.5-min topographic quadrangle maps; intermediate-scale DLGs are derived from the USGS 1:100,000-scale 30 × 60-min quadrangle maps; and small-scale DLGs are derived from the USGS 1:200,000-scale sectional maps of the National Atlas of the United States of America. The hydrography data contain full topological linkages in node, line, and area elements. Thus, channel connectivity information in upstream and downstream directions is available.

Surface drainage and channel network configuration are important landscape attributes for hydrologic modeling of runoff processes. Both attributes can be determined from field surveys, stereophotos, and detailed topographic contour maps. However, these approaches are resource and time consuming, particularly for large watersheds. A more expedient approach consists of purchasing previously digitized channel data or deriving the information from readily available digital elevation data.

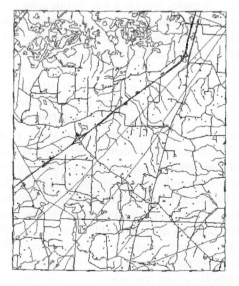

FIGURE 3.8 Large-scale (7.5-min) DLG boundary, hydrography and transportation layers. DLG hydrography data portray stream paths, ditches and canals, and lakes. (*Source:* http://edc.usgs.gov/products/map/dlg.html.)

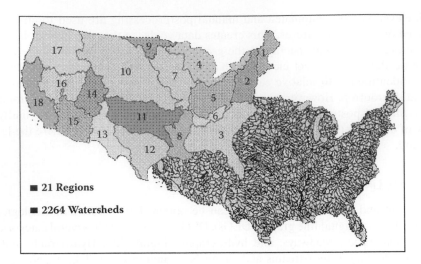

FIGURE 3.9   NHD watersheds are organized in a hierarchical manner for the various levels of scale. (*Source:* NHD 2000.)

3.4.3   NATIONAL HYDROGRAPHY DATASET

The National Hydrography Dataset (NHD) is the single-source geospatial database of U.S. surface waters. It is a complete, seamless, nationwide coverage provided at 1:100,000 scale in 2002, and a high-resolution NHD at 1:24,000-scale coverage released in 2007 (Figure 3.9). The NHD contains information about surface-water features such as lakes, ponds, streams, rivers, springs, and wells. Within the NHD, surface-water features are combined to form "reaches," which provide the framework for linking water-related data to the NHD surface-water drainage network. These linkages enable the analysis and display of these water-related data in upstream and downstream order. The NHD is based upon the content of USGS digital line graph (DLG) hydrography data integrated with reach-related information from the EPA Reach File version 3 (RF3). From RF3, the NHD acquires hydrographic sequencing, upstream and downstream navigation for modeling applications, and reach codes. The reach codes provide a way to integrate data from organizations at all levels by linking the data to this nationally consistent hydrographic network. The feature names are from the Geographic Names Information System (GNIS). Most of the information presented below is taken from the NHD Web site at http://nhd.usgs.gov, where interested users can obtain other documentation, tools, training materials, and technical support.

The NHD is designed to combine spatial accuracy with detailed features, attributes, and values to provide information on flow paths, permanent reach IDs, and hydrologic ordering for use in modeling (Perdue 2000). While initially based on 1:100,000 intermediate-scale data, nationwide coverage of large-scale 1:24,000 data sets was successfully completed in 2007. The NHD now provides a high-resolution, nationwide system for stream addressing, upstream/downstream flow direction modeling, and maintenance of hydrographic infrastructure. Although the NHD is a "seamless" data set, the data are currently accessed by sub-basin (i.e., eight-digit HUC [hydrologic unit code]-level watersheds; Figure 3.10). The NHD data are distributed as ESRI shapefiles, personal geodatabases, or ArcInfo® coverages, which supplant the less convenient SDTS (spatial data transfer standard). The NHD data sets are compatible with the Arc Hydro data model and tools operating as extensions in the ArcMap® interface to ESRI's ArcGIS® (described below). A convenient map viewer is available (http://nhdgeo.usgs.gov/viewer.htm) for locating and downloading available hydrography data sets at various resolutions.

In the upper left of Figure 3.10, "the irregular shapes on and around the shaded image of Kentucky are the hydrologic cataloging units that are in the state of Kentucky. On the right is a

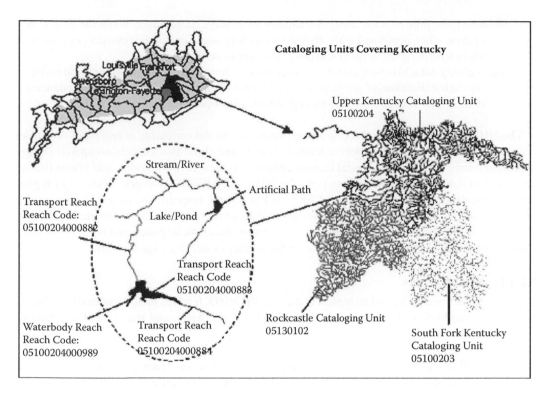

FIGURE 3.10 Example NHD for Kentucky. In the upper left, the irregular shapes on and around the shaded image of Kentucky are the hydrologic cataloging units that are in the State of Kentucky. On the right is a closer high-level view of three cataloging units, showing their hydrographic features and labeled with the names and numbers of the units. In the oval, a small area is enlarged to show examples of individual features with their reach codes. Note that the reach codes incorporate the eight-digit number of the cataloging unit in which they lie. (*Source:* http://erg.usgs.gov/isb/pubs/factsheets/fs10699.html).

closer high-level view of three cataloging units, showing their hydrographic features and labeled with the names and numbers of the units. In the oval, a small area is enlarged to show examples of individual features with their reach codes. Note that the reach codes incorporate the eight-digit number of the cataloging unit in which they lie" (USGS 1999b).

Characteristics of the NHD include the following:

- It is a feature-based data set that interconnects and uniquely identifies the stream segments or "reaches" that make up the nation's surface-water drainage system.
- Unique reach codes (originally developed by the USEPA) are provided for networked features and isolated water bodies.
- The reach code structure is designed to accommodate higher resolution data.
- Common identifiers uniquely identify every occurrence of a feature.
- Data are in decimal degrees on the North American Datum of 1983.
- Names with GNIS identification numbers are included for lakes, other water bodies, and many stream courses.
- It provides flow direction and centerline representations through surface-water bodies.
- The data support many applications, such as:
  Making maps. Positional and descriptive data in the NHD provide the starting point for making many different kinds of maps.
  Geocoding observations. Much like street addresses provide a way to link data to a road network, the NHD's "reach code" provides the means to link data to water features.

Modeling the flow of water along the nation's waterways. Information about the direction of flow, when combined with other data, can help users model the transport of materials in hydrographic networks, among other applications.

Maintaining data. Many organizations would like to share the costs of improving and updating their collections of geographic data. Unique identifiers and other methods encoded in the NHD help to solve technical problems of cooperative data maintenance.

The NHD is implemented in ArcInfo as a georelational model consisting of two types of entities: those that contain a spatial component (called *themes*) and those that contain no spatial component (called *tables*). The NHD spatial themes consist of routes, regions, nodes, and points that are composed of selected spatial elements from the NHD or NHDPT coverages. Routes and regions uniquely identify and manage groups of lines and polygons, respectively, as single entities. The route, region, and node NHD spatial themes are subsets of the spatial elements in the NHD coverage. For example, the DRAIN route contains streams/rivers, artificial paths, and other NHD linear features from the NHD coverage that make up the surface-water drainage network.

3.4.3.1   NHD Features

A feature is a defined entity and its representation. In the NHD, features include naturally occurring and constructed bodies of water, paths through which water flows, and related entities. Features are classified by type, may be described by additional characteristics, and are delineated using standard methods. Features are classified by type. These feature types, such as stream/river, canal/ditch, and lake/pond, provide the basic description of the feature. Each type has a name and a definition. For example, the three most frequently encountered feature types and corresponding definitions are

- STREAM/RIVER: a body of flowing water
- LAKE/POND: a standing body of water with a predominantly natural shoreline surrounded by land
- CANAL/DITCH: an artificial open waterway constructed to transport water, to irrigate or drain land, to connect two or more bodies of water, or to serve as a waterway for watercraft

3.4.3.2   NHD Reaches

A reach is a continuous, unbroken stretch or expanse of surface water. In the NHD, this idea has been expanded to define a reach as a significant segment of surface water that has similar hydrologic characteristics, such as a stretch of stream/river between two confluences, or a lake/pond. Reaches also are defined for unconnected (isolated) features, such as an isolated lake/pond. Once a reach is defined for a segment of water and assigned a reach code, the reach will rarely be changed, if at all. Three types of reaches are in use: transport, coastline, and water body.

A transport reach represents the pathway for the movement of water through a drainage network. These reaches are also used to encode the direction in which water flows along the reach when the direction is known. They provide a basis on which locations of observations can be geocoded and linked to the drainage network. Lines delineate transport reaches. Only lines that delineate features of the types canal/ditch, pipeline, stream/river, artificial path, and connector delineate transport reaches. For transport reaches for which the direction of flow is known, the lines are oriented in the direction of the flow of water.

A reach code is a numeric code that uniquely labels each reach. This 14-digit code has two parts: the first eight digits are the hydrologic unit code (HUC) for the sub-basin in which the reach exists; the last six digits are assigned in sequential order, and arbitrarily among the reaches. Each reach code occurs only once throughout the nation. Reach codes can serve to geocode an observation to a reach or to a position along a reach.

Flow relations among transport and coastline reaches encode drainage network connectivity among reaches independently from their delineations. Relations among transport reaches define a connected hydrographic network and encode the direction of water flow among reaches. This connectivity enables hydrologic sequencing of reaches (what is upstream and downstream of a given point in the hydrographic network) and navigating the network in an upstream or downstream direction. Figure 3.11 illustrates the network-structure identifier scheme.

The stream level is a numeric code that identifies each main path of water flow through a drainage network. Stream level is assigned by identifying the terminus of a drainage network. The lowest value for stream level is assigned to a transport reach at the end of a flow and to upstream transport reaches that trace the main path of flow back to the head. The stream-level value is incremented by 1 and is assigned to all transport reaches that terminate at this path (i.e., all tributaries to the path) and to all transport reaches that trace the main path of the flow along each tributary back to its head. The stream-level value is incremented again and is assigned to transport reaches that trace the main path of the tributaries to their heads. This process is continued until all transport reaches for which flow is encoded have been assigned a stream level. For example (see Figure 3.12), the Mississippi River terminates at the Gulf of Mexico.

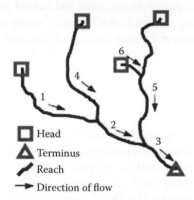

| Head |
| Terminus |
| Reach |
| Direction of flow |

Flow Relations		
Common Identifier for First Reach	Common Identifier for Second Reach	Direction Description
1	2	In
4	2	In
6	5	In
5	3	In
2	3	In
0	1	Network Start
0	4	Network Start
0	6	Network Start
0	5	Network Start
3	0	Network End

FIGURE 3.11 Flow relations illustrating in, out, network start, and network end directions. A common identifier value of 0 represents a null entry. (*Source:* NHD 2000.)

FIGURE 3.12 Stream-level assignments along the Mississippi River. (*Source:* NHD 2000.)

NHDPlus (http://www.horizon-systems.com/nhdplus) is an attempt to develop an integrated set of geospatial data sets that combine the best features of the intermediate-scale NHD data set (1:100,000), the National Elevation Dataset (NED), the National Land Cover Dataset (NLCD), and the Watershed Boundaries Dataset (WBD). NHDPlus is distinguished by improved NHD layers, inclusion of elevation-derived catchments, mean annual flow-volume and -velocity estimates for each NHD flow line in the conterminous United States, improved network naming, and numerous value-added attributes (VAAs). The VAAs include stream order, retrieval of all flow lines and catchments upstream of a given flow line using queries, stream-level paths sorted by hydrologic order for stream-profile mapping, and cumulative catchment attributes such as drainage areas and land-cover distributions.

The NHD network functionality is also providing the basis for StreamStats (http://water.usgs.gov/osw/streamstats), which was developed cooperatively by the USGS and the Environmental Systems Research Institute, Inc. (ESRI). StreamStats is a GIS-based Web application for generating important stream-flow statistics and basin characteristics at USGS data-collection stations as well as ungauged sites. Stream-flow statistics such as the 100-yr flood, mean annual flow, and 7-d 10-yr low flow (7Q10) provide important information for dam, bridge, and culvert design; water supply planning and management; appropriation and permitting of new water uses; wastewater and industrial discharge permitting; hydropower facility design and regulation; and habitat protection for endangered species. StreamStats is developed using the ArcHydro toolset, thereby allowing stream networks to be navigated in order to locate stream gauging stations, dams, and point discharges. The eventual goal is to allow other Web or GIS applications to access StreamStats functionality remotely by use of Web services.

3.4.4 SOILS DATA

Soils data are available from the soils-mapping agencies, typically those dealing with the agriculture sector. In the United States, soil survey data are available in digital formats from the Natural Resources Conservation Service (NRCS), including the State Soil Geographic (STATSGO) and Soil Survey Geographic (SSURGO) databases. Differences in the level of detail of the two soils data sets are illustrated in Figure 3.13. The mapping scale for the STATSGO map is 1:250,000 and was created by generalizing more-detailed soil survey maps. The SSURGO digitizing duplicates the original soil survey maps at mapping scales ranging from 1:12,000 to 1:63,360. SSURGO is linked to the National Soil Information System (NASIS) attribute database. The attribute database gives the proportionate extent of the component soils and their properties for each map unit. Examples of information that can be queried from the database are water capacity, soil reaction, electrical conductivity, and flooding; building site development and engineering uses; cropland, woodland, rangeland, pastureland, and wildlife; and recreational development. Also, both data sets are available in the USGS Digital Line Graph (DLG-3) optional distribution format, and both are

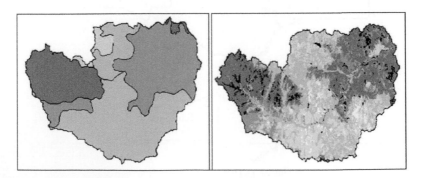

FIGURE 3.13   Level of spatial detail of soils data in the (a) STATSCO and (b) 30-m county level (SSURGO) data sets. (From Reed and Maidment 1998.)

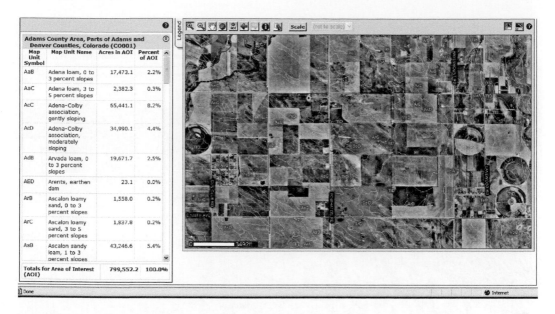

FIGURE 3.14 Soils data are available for Web download. (*Source:* http://soils.usda.gov/.)

also available for download from the NRCS Web site (http://soils.usda.gov/). A concern with the SSURGO data is that the edges of adjacent surveys may not match exactly due to differences in interpretation by the field-mapping scientist.

Soils data (Figure 3.14) can be downloaded from the NRCS Web Soil Survey (http://soils.usda. gov/). This Web site provides soil data and information produced by the National Cooperative Soil Survey. The NRCS has soil maps and data available online for more than 95% of the nation's counties and anticipates having 100% in the near future. The soils databases involve both graphic and attribute data. Graphic data are DLGs of the soil-type boundary polygons. Attribute data include the Official Soil Series Descriptions (OSD) that define the soil series, taxonomic classification, detailed soil-profile description, range in characteristics, drainage and permeability, and use and vegetation, to name just a few.

3.4.5 LAND-USE DATA

Digital land-use data can be obtained from city and county governments that maintain a digital parcel database. In most of these types of databases, a land-use classification is attached to the parcel information. These land-use codes will vary from agency to agency, but all should have the different classes of residential, commercial, industrial, agricultural, and undeveloped land represented. When available, these local parcel databases are the best source of land-use information, since they are usually maintained on a constant basis by the local assessor's office, are positionally accurate, and are recorded as large-scale data at the detailed parcel level. Planning or zoning databases may also be maintained by some local agencies that provide future scenarios of possible land uses that may affect the hydrologic characteristics of an area.

The U.S. Geological Survey has a National Land Cover Database (NLCD) product that is derived from thematic overlays registered to 1:250,000-scale base maps and a limited number of 1:100,000-scale base maps (Homer et al. 2007). Figure 3.15 illustrates a typical product. These data are available for the conterminous United States and Hawaii, with land-use codes that are derived from a modified Anderson Level II scheme (Anderson et al. 1976). These data are in the UTM projection and are available in a vector polygon format or in composite grid-cell format (4-ha [10-acre] cell size). The Multi-Resolution Land Characteristics (MRLC) Consortium is a

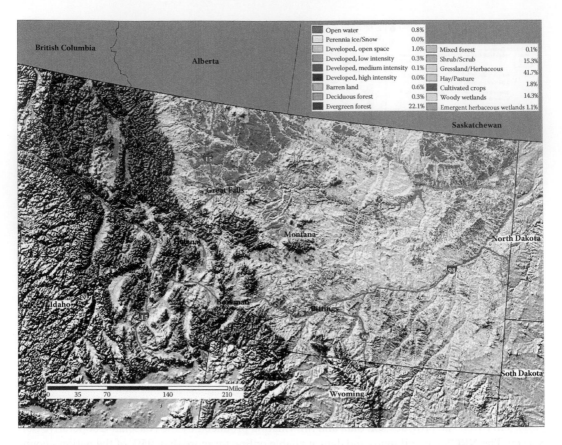

Open water	0.8%
Perennia ice/Snow	0.0%
Developed, open space	1.0%
Developed, low intensity	0.3%
Developed, medium intensity	0.1%
Developed, high intensity	0.0%
Barren land	0.6%
Deciduous forest	0.3%
Evergreen forest	22.1%

Mixed forest	0.1%
Shrub/Scrub	15.3%
Gressland/Herbaceous	41.7%
Hay/Pasture	1.8%
Cultivated crops	14.3%
Woody wetlands	
Emergent herbaceous wetlands	1.1%

FIGURE 3.15 (See color insert following page 136.)   Land-use–land-cover map for Montana is an example of one of the products developed from LandSat 7 imagery for all states in the United States (*Source:* http://landcover.usgs.gov/.)

group of federal agencies that first joined together in 1993 to purchase Landsat 5 imagery for the conterminous United States and to develop a land-cover data set called the National Land Cover Dataset (NLCD 1992). In 1999, a second-generation MRLC consortium was formed to purchase three dates of Landsat 7 imagery for the entire United States (MRLC 2001) and to coordinate the production of a comprehensive land-cover database for the nation called the National Land Cover Database (NLCD 2001). The MRLC data set provides topographic, census, agricultural, soil, and wetland data as well as other land-cover maps at resolutions of 30 m (see http://www.mrlc.gov/).

Land-use and land-cover areas are classified into nine major classes: urban or built-up land, agricultural land, rangeland, forestland, water, wetland, barren land, tundra, and perennial snow or ice. Each major class is composed of several minor classes (e.g., forestlands are further classified as deciduous, evergreen, or mixed). This classification system (Anderson et al. 1976) was reviewed by a committee of representatives from the USGS, NASA, the Soil Conservation Service (SCS), the Association of American Geographers, and the International Geographical Union. The classification system (Table 3.2) was designed to be used with data obtained from remote sensors on aircraft and satellites.

Satellite remote sensing can provide valuable sources of data for delineating drainage basins and stream networks as well as inventories of surface-water bodies and storages. Satellites using the visible and near-infrared regions of the spectrum can provide detailed information of the land characteristics, and SPOT, with its stereo capability, can even provide topographic information (Gugan and Dowman 1988). Side-looking airborne radar (SLAR) and satellite synthetic-aperture radar (SARs) can produce very detailed maps of basin characteristics, even in traditionally cloudy

---

TABLE 3.2
Land-Use and Land-Cover Categories Used by the U.S. Geological Survey

1. Urban or built-up land	**5. Water**
11 Residential	51 Streams and canals
12 Commercial	52 Lakes
13 Industrial	53 Reservoirs
14 Transportation, communications, and utilities	54 Bays and estuaries
15 Industrial and commercial complexes	**6. Wetland**
16 Mixed urban or built-up land	61 Forested wetland
17 Other urban or built-up land	62 Nonforested wetland
2. Agriculture land	**7. Barren land**
21 Cropland and pasture	71 Dry salt flats
22 Orchards, groves, vineyards, nurseries, and	72 Beaches
ornamental horticulture areas	73 Sandy areas other than beaches
23 Confined feeding operations	74 Bare exposed rock
24 Other agriculture land	75 Strip mines, quarries, and gravel pits
3. Rangeland	76 Transitional areas
31 Herbaceous rangeland	77 Mixed barren land
32 Shrub and brush rangeland	**8. Tundra**
33 Mixed rangeland	81 Shrub and brush tundra
4. Forestland	82 Herbaceous tundra
41 Deciduous forestland	83 Bare ground tundra
42 Evergreen forestland	84 Wet tundra
43 Mixed forestland	85 Mixed tundra
	9. Perennial snow or ice
	91 Perennial snowfields
	92 Glaciers

Source: Anderson et al. (1976); http://www.npwrc.usgs.gov/resource/habitat/research/data.htm.

---

areas and areas with heavy vegetation growth. Interferometric SAR can also provide quantitative measures of topography.

Interpretation of land-use information from satellite imagery or aerial photographs is another means of obtaining land-use and land-cover data. The techniques of supervised and unsupervised classification of satellite imagery yield clusters of different spectral classes that are assigned into different land-use types. The same technique can be used with aerial photos that are digitally scanned.

Research by Ragan and Jackson (1980) and Bondelid et al. (1982) has shown that determining the degree of urban land use or various categories of agricultural or forest can be determined accurately through remote sensing and used as variables in urban runoff models or the SCS Runoff Curve Number. Studies have shown (Jackson and Rawls 1981) that for planning studies, the Landsat approach is cost effective. The authors estimated that the cost benefits were on the order of 2.5 to 1 and can be as high as 6 to 1 in favor of the Landsat approach. These benefits increase for larger basins or for multiple basins in the same general hydrological area. Mettel et al. (1994) demonstrated that the recomputation of PMFs (probable maximum floods) for the Au Sable River, using the HEC-1 hydrologic model and updated and detailed land-use data from Landsat, resulted in 90% cost cuts in upgrading dams and spillways in the basin.

Vegetation indices such as the NDVI (normalized difference vegetation index) and LAI (leaf area index) have been shown to be related to evaporation coefficients such as a crop coefficient (defined as the ratio of actual evaporation to reference crop evaporation) and a transpiration coefficient (defined as the ratio of unstressed evaporation and reference crop evaporation) (Tucker 1979).

3.5 GEODATABASES

3.5.1 OVERVIEW

Database management systems (DBMS) are computer programs for storing and managing large amounts of data. Required functions of a DBMS include: (a) consistency with little or no redundancy; (b) maintenance of data quality, including updating; (c) self-descriptive with metadata; (d) a database language for query retrievals and report generation; (e) security, including access control; and (f) shareability among users. Most DBMS are designed to handle attribute data. A special characteristic of a geodatabase is the join between spatial and attribute data for water resources system features.

There are four basic ways of organizing data that also reflect the logical models used to model real-world features. These are (1) hierarchical, (2) network, (3) relational, and (4) object-oriented (Figure 3.16).

3.5.1.1 Hierarchical Database Structure

A hierarchical database structure is used when the data have a parent-child or one-to-many relation, such as various levels of administration, soil series within a soil family, or pixels within a polygon. The advantages of a hierarchical model are its simplicity, high access speed, and ease of updating. However, the disadvantage is that linkages are only possible vertically but not horizontally or diagonally; that means that there is no relation between different trees at the same level unless they share the same parent. An extension of the hierarchical model used for image data is the *quadtree*, which is used to access a small part of a large raster image or map area. A quadtree approach first divides a total map area into 4, 16, 32, ... segments and records whether the quadrant

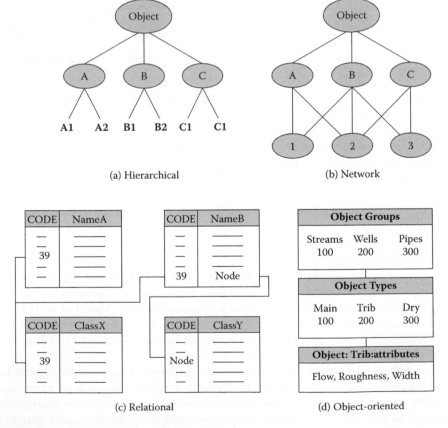

FIGURE 3.16 Database structures.

is homogeneous (i.e., cells all have the same attribute value); if so, then that quadrant is stored, and no further subdivision is required. Nonhomogeneous quadrants are further divided until all nonuniform quadrants have been reduced. The smallest unit would be a single pixel. Quadtrees are one method for reducing the storage required for image data sets.

3.5.1.2 Network Database Structure

The network database structure takes advantage of linkages between features that can be specified beforehand. Networks are composed of nodes and connecting links (or arcs), and are a somewhat specialized data model for representing transportation and river networks. The *valency* of a network is the number of links at the node; 4-valent nodes are most common in street networks, and 3-valent nodes are common in hydrology (i.e., two tributaries join to form the main stream). Network structures have an inherent advantage in that the various mathematical procedures have been developed for efficient solution of problems of route finding and resource allocation.

In networks, each data element can be explicitly encoded with a pointer that directs it to all of the other pieces of data to which it relates. Networks apply in special cases, such as river and transportation networks, where the features interact with the linked features. This avoids the ponderous hierarchical structure and provides for very fast access. However, networks can get quite complicated and are reserved for special situations.

3.5.1.3 Relational Database Structure

Relational database structures have become the norm over the past several decades through development of the commercial DBMS products such as Oracle, Informix, and others. The relational database links the complex spatial relationships between features using tables consisting of records having a set of attributes. Each table (called a relation) consists of records, which are sets of fields, each containing an attribute. In the relational structure, each entity is defined in terms of its data records and the logical relations that can be established between the attributes and their values. A most important aspect of the relational database is to build a set of key attributes that serve as a unique identifier. This allows dependence between attributes to be established as well as avoiding loss of information when records are inserted or deleted.

3.5.1.4 Object-Oriented Database Model

An object-oriented (O-O) database model uses functions to model spatial and nonspatial relationships of geographic objects and their attributes. An object is an encapsulated unit that is characterized by attributes, a set of orientations, and rules. In O-O databases, data are defined in terms of a series of unique objects, which are organized into groups of similar phenomena (known as object classes). Relationships between different objects and different classes are established through explicit links. The characteristics of an object may be described in the database in terms of its attributes (its state) as well as a set of procedures that describe its behavior. These data are encapsulated within an object, which is defined by a unique identifier in the database.

There have been conceptual advances in the integration of vector and raster data with commercial relational database management systems (RDBMS). In combination with spatial data, the interest has turned to spatial database management systems (SDBMS). A hybrid relational DBMS approach allowed separation of graphic and attribute data handling. Examples of the hybrid structures include the Arc-Node-RDBMS, Compact raster-RDBMS, Quadtree-RDBMS, and Object-RDBMS (Shekhar and Chawla 2003). The object-oriented paradigm allows storage of the rules of operation for, say, a reservoir to be embedded with the object. The attraction of a commercial RDBMS is to realize advantages of this software to handle multiple transactions and maintain persistence for the data. Also, beyond the spatial and attribute data, RDBMS have been developed to handle temporal, visual, and other multimedia forms of data. These reliability capabilities become increasingly important when an organization migrates to an enterprise level, where data are centralized and served to users in different departments.

3.5.2 GEODATABASE DATA MODELS

The geodatabase is the spatial data storage format for ESRI ArcGIS software and presents several differences from the shapefile and coverage in its design and functionality. It is a relational database and provides the ability to store spatial and attribute data and the relationships that exist among them. Rules, relationships, and topological associations provide the framework that enables the geodatabase to model real-world entities and events. A geodatabase stores point, line, and polygon spatial geometry called *feature classes* (similar to a shapefile or coverage). Feature classes can be grouped into feature data sets. Tables without spatial representation can be stored in the geodatabase as well, and are common in the design of a relational database management system.

A "data model" is conceptual and is used to guide the storage of geospatial and time-series data within a standardized framework. There are many ArcGIS data models; for Arc Hydro (see Section 3.5.3), the focus is on hydrologic geospatial and time-series data. Other ArcGIS data models that may be of interest for water resources are Agriculture, Basemap, Biodiversity, Environmental Regulated Facilities, Geology, Groundwater, Land Parcels, Local Government, Pipeline, Raster, Transportation, and Water Utilities (water supply, sewer, storm water; see http://support.esri.com/index.cfm?fa=downloads.dataModels.matrix).

State-of-the-art GIS software systems utilize an object-oriented approach for linking the attribute database to a topologically consistent spatial database. For example, the geodatabase framework for data storage and management incorporated in ArcGIS (ESRI, Inc.) provides an ideal structure for the representation and modeling of river basin network objects in a geographic information system. The ESRI geodatabase supports a wide range of data types required in a spatial decision-support system for river basin management (Figure 3.17). These data types include:

- Attribute tables of network objects, such as reservoir area-capacity elevation tables and operational guide curves
- Geographic features representing reservoirs, demands, river channels, canals, pipelines, wells, etc.
- Satellite and aerial imagery for accurate spatial representations of river basin features
- Surface modeling data, such as calculation of terrain and channel slopes, channel distances, travel time calculations, etc.
- Survey measurements for precise georeferencing of all important river basin network objects

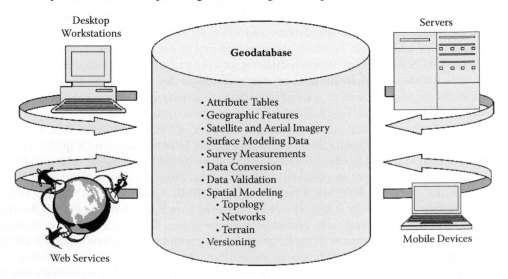

FIGURE 3.17 Geodatabase framework supports multiple feature types, databases, and toolboxes with access by various user devices.

CASE (computer-aided software engineering) tools have become common for use in geodatabase design. They allow the user to create diagrams and designs using the Unified Modeling Language (UML). The diagrams created can represent and eventually create actual geodatabase components while visually showing relationships and information concerning the design. UML model diagrams are extremely useful in GIS transportation modeling because the process usually involves the migration of a large amount of data from differing sources with differing needs relating to technical associations and properties. The creators of the ArcGIS data models have provided a UML model that can be changed and manipulated to fit the user's customized needs.

Geodatabase design and functionality also allow the user to model data to contain behavior through attribute domains, subtypes, and relationships. This functionality plays a large role in the design and methods found in the ESRI data models, so a solid understanding of these concepts is beneficial in understanding the model design. The following are concepts that can be used with geodatabase functionality:

- *Domains*: These are valid attribute values for a chosen field in a feature class or nonspatial attribute table. Domains are defined at the top level of the geodatabase design and therefore can be assigned to multiple feature classes, tables, and fields within a geodatabase or single feature class. Domains are extremely valuable for preventing errors in data entry, for keeping valid values standardized, and for validating previously entered attribute values. Once the domain is established in the geodatabase properties, it can be assigned to fields located anywhere within the geodatabase.
- *Subtypes*: Records that represent features within a geographic database provide the user with the ability to accurately describe each individual feature by assigning attributes that differentiate a given feature from other features. Many times, features within a feature class can be categorized into values that represent features that will almost always have the same attribute values for certain fields. Subtypes provide the ability to group these similar features into categories that can have different domains, default values, and relationship rules.
- *Relationships*: These define associations between objects within the geodatabase. A relationship can be defined between two feature-class tables, between a feature-class table and a nonspatial table, or between two nonspatial tables. Relationships are valuable because, once stored, information from one table can be accessed through another table or feature class. Relationship rules are beneficial in modeling features as they actually behave. A relationship class is the definition of a relationship stored between two tables. The class contains properties such as the relationship name, cardinality, origin, and destination tables. Relationship classes can be defined as simple or composite, with their dependence on one another as the deciding factor. Simple relationships are created for tables that exist independently of one another. Composite relationships are set for destination objects that cannot exist without origin objects.
- *Geometric networks*: A geometric network is made up of line and point features described as edges and junctions. Networks give the data more "intelligence" because the data have rules about their connectivity. For instance, if an edge represents a street and a junction represents an intersection of the street, the network would know to automatically move connected streets if the intersection were moved. Both junctions and edges can be simple or complex. Simple edges are split if a junction is added midway. Complex edges are always connected to two junction features at their endpoints, but have the ability to be connected to junction features at any point along the length of the line without splitting the feature. Another important feature of geometric networks as they relate to water resources is a concept called *weights*. Weights can be assigned to a network to associate the weight information with the cost of flow along network features. Common weights associated with pipe networks might include the length of the pipe and roughness.

- *Linear referencing and route features*: New releases of ArcGIS are incorporating the ability to use linear referencing systems and routing capabilities within the geodatabase format. GIS users commonly use x and y coordinates to reference features to their location. With a linear referencing system, the user has the added ability to reference attributes or events to linear data by a measure on the data, such as length from the beginning of a segment. For example, staff in the field collecting information on sewer conditions might GPS the location of the object to get its coordinates or use pipe imagery or other relative data to digitize its approximate location. With linear referencing, the data collector would know the exact location of the point by measuring its distance from a junction (manhole) or the starting point of a pipe. When inputting the GIS data, the user would enter the FeatureID and measure information to reference the object to the route.
- *Dynamic segmentation*: There are many times when attributes of linear features do not match segments that contain records that hold the attributes. An example is pipe condition. A GIS database of sewer pipes might contain several segments of a pipe, but the pipe-condition attribute information (poor, fair, or good) might actually span from the middle of one segment to the middle of another segment. This phenomenon is sometimes dealt with by using the average pipe condition for a segment. Though the information is usable by using an average quality, accuracy of the information is lost by calculating an average as opposed to showing the attributes of the condition at its exact location. Dynamic segmentation provides a solution to this issue by using references in event tables to symbolize features on a map. The features follow measured values along pipe segments, allowing flexibility in starting and ending point locations.

3.5.3 Arc Hydro Data Model

The Arc Hydro data model is a conceptual and software development for the ArcGIS software package of the ESRI (www.esri.com). It was developed by the Center for Research in Water Resources (CRWR) at the University of Texas and has been documented in the book *Arc Hydro: GIS for Water Resources* (Maidment 2002) and in a number of other CRWR publications (e.g., Obenour and Maidment 2004). Although implemented in a commercial GIS environment, the data model and toolset are in the public domain and available free of charge (ESRI; http://www.esri.com/software/arcgis/geodatabase/).

Standardization is important because it allows for sharing of data and applications between multiple water agencies, and provides a framework for integration with hydrologic computational programs of various types. Various software tools have been developed to facilitate population of the database with information on watersheds, stream networks, channels, structures, gauge stations, and land surface properties. The geographic features can be linked to time-series data (e.g., stream flow records) in the database, thus providing the foundation for simulation modeling for urban watersheds and river basins as well as groundwater systems. Arc Hydro applications with integration of computational models are described in other chapters of this book, including Chapter 5 (GIS for Surface Water Hydrology), Chapter 6 (GIS for Groundwater Hydrology), Chapter 9 (GIS for Floodplain Management), Chapter 10 (GIS for Water Quality), and Chapter 12 (GIS for River Basin Planning and Management).

Arc Hydro has three main components (Obenour and Maidment 2004):

- a standardized format for storing geographic and time-series hydrologic data
- logical data relationships among geographic features (or "objects")
- a set of tools for creating, manipulating, and viewing hydrologic data

The standard Arc Hydro data model was created using the Unified Modeling Language (UML) to create a series of tables representing various hydrologic themes. UML is a generalized language

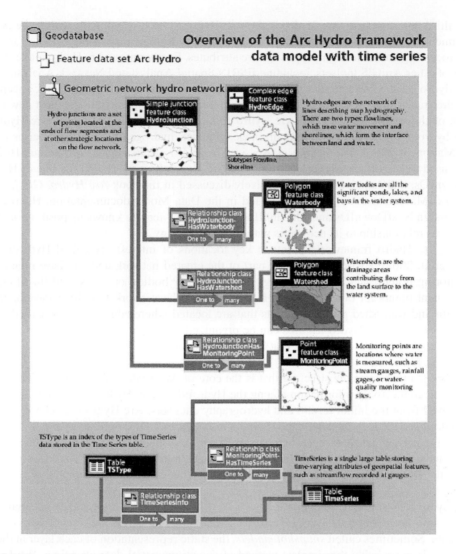

FIGURE 3.18   Arc Hydro framework with time series (Maidment 2002. With permission.)

and graphics toolset for developing object-oriented software. The UML uses mostly graphical notations to express the design of software projects. Figure 3.18 shows the Arc Hydro framework with time series.

Tables in the Arc Hydro data model represent spatial entities, such as watersheds and monitoring points; these are called *feature classes*. Tables that represent nonspatial data, such as time series, are called *object classes*. Inside these tables (or "classes") are standardized field names such as JunctionID, HydroCode, and FlowDir. Fields with an "ID" or "Code" suffix are used for identifying individual hydrologic objects. Fields with the ID suffix are integer fields, and fields with the Code suffix are text fields. In addition to identification fields, there are also fields that store hydrologic characteristics, such as FlowDir (which stores the direction of flow in a stream line) and AreaSqKm (which stores the area of a watershed).

Arc Hydro's data relationships were also created using UML. These relationships are used to associate hydrology-related entities. For instance, a watershed may be related to an outlet point (to which the watershed drains), a stream gauge may be related to a point along a river, or a set of precipitation records may be related to a rain-gauge station. With these relationships established, the user may easily query related hydrologic information.

The third component of Arc Hydro is the Arc Hydro toolset, which runs within the ArcMap environment. This toolset can be used for a number of hydrologic processing routines, including raster processing and the assignment of feature attributes. Furthermore, Arc Hydro makes use of a number of other ArcGIS toolsets, including ESRI's Spatial Analyst and Network Analyst.

Arc Hydro is a flexible data structure that contains both spatial and temporal data. Conceptually, Arc Hydro traces the flow of a drop of water where it falls on a catchment to the most downstream location of the stream system. The Arc Hydro framework is a basic version of Arc Hydro that divides hydrologic data into network, drainage, channel, hydrography, and time-series components (Maidment 2002). The feature classes in Arc Hydro are managed with unique HydroIDs that are used to keep track of every feature. The Arc Hydro framework and the role of HydroID, NextDownIDs, and JunctionIDs are extensively discussed in the book *Arc Hydro: GIS for Water Resources* (Maidment 2002) and documented in the Data Model documentation. HydroIDs are used to assign NextDownIDs and JunctionIDs, so that every feature knows its position in the network and its relationship to its neighbors without GIS topology.

In the Arc Hydro framework, points at key locations of interest are called HydroJunctions (Olivera et al. 2002). HydroJunctions are part of the created network and can have relationships to monitoring points such as stream-gauge stations, water bodies, or areas, and therefore serve as attachment points for other features. HydroEdges are the network flow lines developed from the terrain and watershed products. Points that are located where edges meet are called Hydro_NetworkJunctions. Arc Hydro features can be organized as simple objects with attributes but no spatial data (time series), as features that have spatial coordinates and attributes, and as network features.

Network features form the structure that is the core of Arc Hydro's functionality. The network contains the connectivity information among the HydroEdges and HydroJunctions, both of which were derived from the initial terrain and hydrography data sets. Arc Hydro uses elevation data to develop drainage areas in the landscape (Maidment 2002). In a GIS, processing raster data can be automated to a great extent except where terrain is relatively flat. The regularity of the raster data allows standardized methods to be allowed for drainage area derivation. In ArcGIS, this method is called the D8, or eight-direction pour-point method. This method is the simplest flow direction method.

Arc Hydro represents spatiotemporal objects by discretizing temporally related spatial objects into the same layer. Several layers representing time steps make up the temporal accession of spatial data. Sometimes called *snapshot models*, the static representation of each layer at different time intervals is probably the simplest method of depicting spatial data over time. While this is a rudimentary approach to spatiotemporal modeling, the approach has the advantage of grouping time-series data and associated spatial objects in the same database.

Data development for Arc Hydro includes hydrography data on the stream paths and watershed boundaries. Two data sets provide the primary inputs: the National Hydrography Dataset (NHD) and the National Elevation Dataset (NED). The NHD has become the U.S. standard in hydrography data sets. The NHD provides hydrological ordering, navigation, and unique feature identifiers for features at the 1:100,000 scale, considered a medium-resolution data set with an accuracy of 50.9 m for 90% of the data. Another key feature is centerline delineation through water bodies, a development that allows simplified implementation of the Arc Hydro data model. The NED is a raster product developed by the USGS incorporating over 50,000 digital elevation models (DEM) into a seamless, nationwide elevation coverage. The NED can be downloaded and clipped to individual watersheds using the hydrologic unit (HU) polygons; this process is efficiently conducted using a GIS (e.g., ArcView 3.2a®) Avenue script entitled Grid.ClipToPoly. Also, the NHD route.ch data set can be intersected with the 1:250,000-scale watershed units.

Terrain preprocessing is conducted using the NED and NHD data sets for each watershed. Figure 3.19 illustrates various products generated. The Arc Hydro data model, as implemented within the ArcGIS, performs the following terrain-preprocessing procedures:

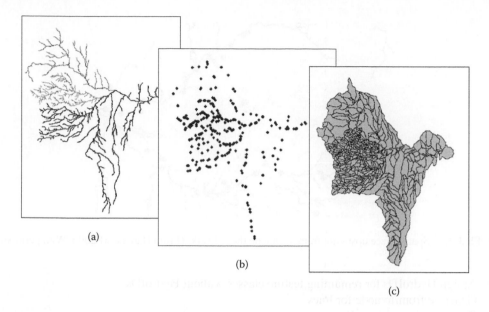

(a)

(b)

(c)

FIGURE 3.19 Results of terrain processing: a) DrainageLines merged into a regional feature class, b) regional DrainagePoint feature class, c) Catchment feature classes merged into a regional layer. (From Hattendorf, 2004. With permission.)

- DEM reconditioning: imposes linear features onto the elevation data
- Fill sinks: ensure that water has drainage to an outlet
- Flow direction: upstream to downstream
- Flow accumulation: accumulated number of cells upstream from a given cell
- Stream definition: stream path begins at a set area upstream (e.g., 5000 cells)
- Stream segmentation: segments between junction nodes
- Catchment grid delineation: assigns grid cells to a catchment
- Catchment polygon processing: defines the catchment boundary, assigns HydroIDs
- Drainage line processing: establishes NextDownID topology of network
- Adjoint catchment processing
- Drainage point processing: verified to lie on drainage lines
- Slope: magnitude of slope
- Aspect: direction of slope

Network generation for each individual watershed is executed after the terrain preprocessing. The Arc Hydro Network Tools Data Management menu allows several choices for data input:

- HydroEdge: a complex edge feature representing a line in the hydro network
- HydroJunction: a hydro network junction
- SchemaLink: a line in a schematic network connecting hydro features
- SchemaNode: a point in a schematic network connecting hydro features

Watershed links are defined as the connection between a Watershed SchemaNode and its outlet Junction SchemaNode. Junction links tie together Junction SchemaNodes. At this point, flow direction has been stored in the data set. Given establishment of the network, geodatabase retrievals can be conducted. For example, the network upstream from a selected node can be retrieved (Figure 3.20). Feature and attribute data on the selected network can then be reported.

Attribute tools in Arc Hydro provide functionality not requiring a network but which assign attributes to features of the network. These include:

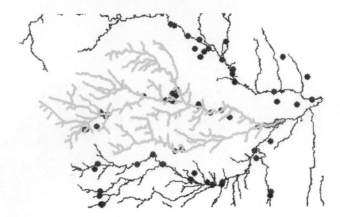

FIGURE 3.20   Spatial trace upstream from a point on the network. (From Hattendorf 2004. With permission.)

- Assign HydroIDs for remaining feature classes without HydroIDs
- Generate from/to node for lines
- Find next downstream line
- Calculate length downstream for edges
- Calculate length downstream for junctions
- Find next downstream junction
- Store area outlets: connects areas to junctions
- Consolidate attributes: calculates the total area of catchments related to a HydroJunction
- Accumulate attributes
- Display time series
- Get parameters

Time-series creation in the geodatabase is the next step. The table feature class called TimeSeries is created by importation, and a relationship is created between the TSType table and the TimeSeries table. Time-series data were prepared in advance with certain columns present in the data set. The TSType table was first edited in MS Access to include such variables as Streamflow or Electrical Conductivity (EC). The variable units, interval regularity, data type, and origin were recorded. The TimeSeries table is linked to TSType by TSTypeID. The TimeSeries table includes a FeatureID column as a text column. The FeatureID is set equal to the HydroID of the feature the TimeSeries is attached to. In ArcCatalog®, the Load Data command is used to bring the TimeSeries data into the geodatabases. The spatial MonitoringPoint data are then loaded. Each MonitoringPoint is connected to a HydroJunction by setting the JunctionID in MonitoringPoint equal to the HydroID of the HydroJunction. The MonitoringPointHasTimeSeries relationship is then set up in ArcCatalog by creating a new relationship. Spatial data and time-series tables can then be imported into ArcMap for further manipulation and use. To query for a selected time series at a certain location, a join based on the MonitoringPointHasTimeSeries relationship was created.

Queries are performed in the table SelectByAttributes query. Because the attribute tables are joined, a limited SQL (structured query language) can be used. To extract all data for a certain date, the query is formatted in this manner:

"TimeSeries.TSDateTime = date '7/31/2002' AND "MonitoringPoint.OBJECTID" Is Not Null".

To develop a series of time-slice data sets, the results of each query can be exported to a separate table. A joined MonitoringPoint and TimeSeries query table has a one-to-many relationship, i.e., one location has many time-series records.

3.5.4 CUAHSI HYDROLOGIC INFORMATION SYSTEM

A hydrologic information system (HIS) is a combination of hydrologic data, tools, and simulation models that support hydrologic science, education, and practice. The Consortium of Universities for the Advancement of Hydrologic Science, Inc. (CUAHSI; http://www.cuahsi.org) has prototyped the HIS as a geographically distributed network of hydrologic data sources and functions that are integrated using Web services so that they function as a connected whole (Maidment 2005). The HIS addresses the problem that environmental data are collected by numerous federal, state, and local agencies as well as by academic scientists. Although the Internet has improved access to these disparate data sources, gathering the data required for most hydrologic studies requires visiting multiple sites, each with its own access protocols and data-exporting formats. Similarly, the Internet provides a means for individual scientists to publish their data (as increasingly required by funding agencies), yet many scientists do not have the expertise to provide sophisticated interactive data-retrieval sites.

To address these problems, the CUAHSI HIS seeks to provide the following services:

- *Data discovery*: A map-based viewer will display the data collected by various entities in one location, including both government-collected and university-collected data.
- *Data delivery*: Through the use of programmatic calls, users will be able to retrieve data directly into databases, spreadsheets, and analysis packages using a single syntax, regardless of data source.
- *Data publication*: Academic scientists will be able to easily publish the data they have collected, so that they appear within the common data viewer and respond to the same data-retrieval calls as government sources.
- *Data curation*: An HIS data center (HISDAC) will provide a repository for archival data. These data can be viewed and delivered using the same mechanisms as described above, so that their delivery is seamless to the user.

A *digital hydrologic observatory* is a comprehensive information depiction of a river basin that describes its natural environment, its hydrologic measurements, simulation models of its processes and phenomena, and conceptual frameworks for thinking about its hydrologic functioning. A digital hydrologic observatory is produced by the application of the Hydrologic Information Model to a hydrologic region defined by a river basin or aquifer boundaries.

Cyberinfrastructure is the combination of computer tools, telecommunications, database structures, and distributed computer networks that collectively support advancements in science and engineering through integrated information access and processing. The cyberinfrastructure for a digital hydrologic observatory is shown in Figure 3.21. Users of the digital hydrologic observatory enter through a digital hydrologic observatory portal, which is an Internet-based computer interface that provides a local user with access to information resources scattered across a distributed domain of many remote computers, data sources, formats, and software tools. Underneath the portal will reside the digital hydrologic observatory information repository, which will be a combination of a relational database containing point observation data and GIS data, and a structured file system storing remote sensing, netCDF, and hydrologic simulation model files.

The HIS toolkit is a collection of software tools and resources developed by CUAHSI partners to address various aspects of hydrologic analysis. Currently, the tools include:

- HydroObjects: a .net DLL with COM classes that support hydrology applications; in particular, the HydroObjects library assists programs without direct access to Web services in downloading data from CUAHSI WaterOneFlow services
- Arc Hydro Groundwater Toolbar (ArcGIS Toolbar): an ArcScene toolbar with tools for creating 3-D features such as GeoVolumes, GeoSections, and BoreLines, which are part of the Arc Hydro groundwater data model

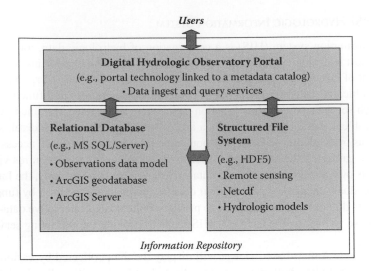

FIGURE 3.21 Cyberinfrastructure for a digital hydrologic observatory (Maidment, 2005. With permission.)

- GeoLearn: a stand-alone application for remote-sensing data
- River Channel Morphology Model (ArcGIS toolbar): an ArcGIS toolbar for creating an analytical description of river channels in three dimensions
- Time Series Analyst: a stand-alone Web application developed at Utah State University; designed to provide users with plotting and export functionality for data at any U.S. Geological Survey (USGS) stream-flow or water-quality monitoring station in the United States
- Watershed and Stream Network Delineation (ArcGIS toolbar): explains how to use the major functionality available in the Arc Hydro tools for raster analysis; the Arc Hydro tools are used to derive several data sets that collectively describe the drainage patterns of a catchment
- HydroSeek: allows data discovery over multiple and heterogeneous data sources; it relies on several Web technologies such as AJAX and OWL; CUAHSI Controlled Vocabulary and several other service ontologies comprise an important part of HydroSeek

REFERENCES

Anderson, J. R., E. E. Hardy, J. T. Roach, and R. E. Witmer. 1976. A land use and land cover classification system for use with remote sensor data. Geological Survey Professional Paper 964. Washington, D.C.: USGS.

Bondelid, T. R., R. H. McCuen, and T. J. Jackson. 1982. Sensitivity of SCS models to curve number variation. *Am. Water Resour. Assoc. Water Resour. Bull.* 18 (1): 111–116.

Burrough, P. A., and R. A. McDonnell. 1998. *Principals of geographical information systems.* Oxford: Oxford University Press.

CDOT (Colorado Department of Transportation). 2002. COGO user guide and command reference. Denver: Colorado Department of Transportation. http://www.dot.state.co.us/ECSU/cogo/COGOManual.pdf.

Close, C. 2003. Converting digital elevation models into grids and TINs for use in ArcView and ArcGIS. http://www.lib.uwaterloo.ca/locations/umd/digital/DEM_Documentation.doc.

DeBerry, P., ed. 1999. GIS modules and distributed models of the watershed: A report from the ASCE Task Committee. Reston, Va.: ASCE.

Ferreira, J. 2007. Lecture 9: Georeferencing, digitizing & advanced raster operations. 11.520: A workshop on geographic information systems. Department of Urban Studies and Planning, Massachusetts Institute of Technology, Cambridge. http://web.mit.edu/11.520/www/lectures/lecture9.html.

Gugan, D. J., and I. J. Dowman. 1988. Accuracy and completeness of topographic mapping from SPOT imagery. *Photogrammetric Rec.* 12 (72): 787–796.

Hattendorf, M. J. 2004. Arc Hydro for the lower South Platte River basin: A regional application of Arc Hydro in support of water quality studies. Master of Engineering report, Dept. Civil Engineering, Univ. Colorado at Denver.

Homer, C., J. Dewitz, J. Fry, M. Coan, N. Hossain, C. Larson, N. Herold, A. McKerrow, J. N. VanDriel, and J. Wickham. 2007. Completion of the 2001 National Land Cover Database for the conterminous United States. *Photogrammetric Eng. Remote Sensing* 73 (4): 337–341.

Jackson, T. J., and W. J. Rawls. 1981. SCS urban curve numbers from a LandSat data base. *J. Am. Water Resour. Assoc.* 17 (5): 857–862.

Jantzen, T. L., and D. R. Maidment. 2007. Implementation of a state hydrologic information system. CRWR online report 07-1. http://www.crwr.utexas.edu/online.shtml.

Maidment, D. R., ed. 2002. *Arc Hydro: GIS for water resources.* Redlands, Calif.: ESRI Press.

Maidment, D. R., ed. 2005. Hydrologic information system: Status report. Report to the National Science Foundation (NSF) under Grant Nos. 03-26064, 04-12975, and 04-47287. Arlington, Va.: NSF. http://www.cuahsi.org/docs/HISStatusSept15.pdf.

Meijerink, A. M. J., H. A. M. Brouwer, C. M. Mannaerts, and C. R. Valenzuela. 1994. Introduction to the use of geographic information systems for practical hydrology. UNESCO International Hydrological Programme, Publication No. 23. Venice: UNESCO.

Mettel, C., D. McGraw, and S. Strater. 1994. Money saving model. *Civ. Eng.* 64 (1): 54–56.

Moore, I. D., R. B. Grayson, and A. R. Ladson. 1991. Digital terrain modeling: A review of hydrological, geomorphological, and biological applications. *Hydrological Processes* 5 (1): 3–30.

NHD (National Hydrography Dataset). 2000. The National Hydrography Dataset: Concepts and contents. Washington, D.C.: U.S. Geological Survey. http://nhd.usgs.gov.

Obenour, D. R., and D. R. Maidment. 2004. Arc Hydro Developments for the lower Colorado River basin. CRWR online report 04-6. http://www.crwr.utexas.edu/online.shtml.

Perdue, D. 2000. National Hydrography Dataset: Concepts and contents. Washington, D.C.: U.S. Geological Survey. http://nhd.usgs.gov/techref.html.

Ragan, R. M., and T. J. Jackson. 1980. Runoff synthesis using Landsat and the SCS model. *J. Hydraulics Div. ASCE* 106: 3–14.

Reed, S. M. 1998. Use of digital soil maps in a rainfall-runoff model. CRWR online report 98-8. http://www.crwr.utexas.edu/online.shtml.

Shekhar, S., and S. Chawla. 2003. *Spatial databases: A tour.* Englewood Cliffs, N.J.: Prentice-Hall, Pearson Education.

Snyder, J. P., and P. M. Voxland. 1989. An album of map projections. Professional paper 1453. Washington, D.C.: U.S. Geological Survey.

Tucker, C. J. 1979. Red and photographic infrared linear combinations for monitoring vegetation. *Remote Sensing Environ.* 8 (2): 127–150.

USGS. 1999a. Map projections. U.S. Geological Survey fact sheet 087-99. http://erg.usgs. gov/isb/pubs/fact sheets/fs0879 9.html.

USGS. 1999b. The National Hydrography Dataset. Fact sheet 106-99. USGS; http://erg.us gs.gov/isb/p ubs/factshee ts/fs10699.html.

Worboys, M. F. 1995. *GIS: A computing perspective.* Bristol, Pa.: Taylor & Francis.

4 GIS Analysis Functions and Operations

4.1 OVERVIEW OF GIS ANALYSIS FUNCTIONS

A primary difference between a GIS and other types of information systems is the spatial analysis functions of a GIS. Particularly in water resources engineering, there is an emphasis on map-oriented analyses. Beginning with map representations of the land, engineers seek to plan and design facilities that accommodate terrain relief. For example, the terrain governs the layout of a water supply pipeline so that gravity flow conditions dominate, thereby minimizing pumping. Prior to GIS, computations were based on data extracted from maps by hand. With GIS data in digital formats, these computations can now be conducted automatically.

In a sense, GIS spatial and nonspatial databases provide an abstract model of the world. Indeed, GIS visualization capabilities can be used to portray depictions of the world that are understandable to most people. The concept of a "model" can be taken as this visualization. The GIS can be used to portray the Earth in a physically realistic manner so we can "look" at the terrain, the steepness and direction of slope, and vegetative cover. In so doing, we gain insight on the character of the land in a manner much like we would if we were to walk or fly over that land. An extension of the model concept is that the GIS can be used to predict how the land might look in the future given changes that might be applied. This "forecasting" capability is of great interest and value in engineering planning and design. For example, using mathematical models integrated with the GIS databases, we can simulate the depth of flood runoff, the drawdown of water levels due to pumping wells, or the predicted population distribution of proposed development.

GISs provide a rich suite of intrinsic functions that perform analyses using attributes of spatial data. In many respects, these intrinsic functions provide unprecedented capabilities (i.e., no historical manual equivalent) that are difficult and time consuming if performed manually. Terrain processing for watershed delineation is but one example of a GIS function that can be conducted better and much more easily than can be done manually. As will be shown in subsequent chapters, the range and sophistication of spatial analyses applied to water resources problems is extensive. Moreover, integration of conventional water resources analysis procedures into the GIS sphere has extended the realm of GIS to include advanced surface modeling, simulation, and optimization functions heretofore not often recognized by GIS practitioners. Also, the water resources field now includes a wide range of decision-support systems for planning and operations that involve a dominant spatial dimension.

The art and science of using a GIS entails combining the available analysis functions with the appropriate data to generate the desired information. GIS practice therefore requires some schema of design to ensure that the effort is focused on answering the appropriate questions. Here, the GIS database and the associated modeling and visualization tools provide enhancements to the traditional engineering design process. The GIS provides a powerful means to manage data, conduct analyses, and communicate planning and design outcomes to the various "publics" concerned with these outcomes. This communication dimension of GIS is particularly important in environmental and water resources engineering, because much of this work concerns public resources having significant impacts over extensive areas on a large number of interest groups.

TABLE 4.1
GIS Analysis Functions

Spatial data capture and maintenance
    Tablet or "heads-up" digitizing
    Scanning
    Format conversion
    COGO
    Photogrammetric data development
    Remote sensing and image processing
    Surveying
    GPS
    Object identification and editing
    Geometric transformations
    Projection conversions
    Attribute entry and editing
    Metadata

Geometrics and measurements
    Intersections
    Distances, weighted distances
    Perimeters, areas

Spatial and aspatial queries; classifications
    Selection
    Windowing
    Show, find, browse, report
    Sort, recode, restrict, compute
    Relations and relational joins (SQL)
    Classification and generalization
    Principal-components analysis

Neighborhood operations
    Search
    Proximity
    Buffer
    Filters
    Slope, aspect, convexity
    Finite differences, finite elements

Spatial arrangement and connectivity functions
    Topologic "clean"
    Adjacency
    Contiguity
    Networks
    Optimal path

Surface operations
    Regular and irregular tesselations
    Volume; cut and fill
    Spread and seek
    Intervisibility
    Illumination; perspective view
    Watershed and terrain modeling

Overlays and map algebra
    Overlays (raster and vector)
    Map algebra
    Cartographic modeling
    Suitability assessment

Spatial statistics
    Descriptive statistics
    Patterns and units
    Thiessen polygons
    Trend surface
    Interpolation
    Multivariate analysis

Image processing
    Radiometric and geometric correction
    Image enhancement
    Image classification

Display, interfaces, integration
    Map design
    Symbology
    Color representation
    Interface
    Reports
    Visualization and animation
    Model integration
    Internet and network access

Management models
    Simulation
    Multiple criteria
    Optimization (LP, NLP, DP)
    Multiple objectives
    Decision-support systems

Table 4.1 lists the functions included in each of the 11 general categories of analysis:

1. Spatial data capture and maintenance
2. Geometrics and measurements
3. Spatial and aspatial queries; classifications
4. Neighborhood operations
5. Spatial arrangement and connectivity functions
6. Surface operations

7. Overlays and map algebra
8. Spatial statistics
9. Image processing
10. Display, interfaces, integration
11. Management models

4.2 SPATIAL DATA CAPTURE AND MAINTENANCE

GIS functions for spatial data capture were reviewed in Chapter 3 (GIS Data and Databases). These include functions for digitizing, scanning, surveying, satellite image processing, and projections. Although sometimes not thought of as analysis functions, these data-capture and -development functions incorporate a wide variety of spatial analysis techniques that are fundamental to accurate and efficient GIS operations. As an example, interactive spatial data editing is aided by "point and click" operations. The seemingly simple act of pointing the cursor at a graphic object on the screen involves quite sophisticated geometric computations of proximity, point-in-polygon, and object identification.

Testing whether a point is inside a polygon is a basic operation in computer graphics and GIS. One definition of whether a point is inside a region is the Jordan curve theorem (Haines 1994). Essentially, it says that a point is inside a polygon if, for any ray from this point, there is an odd number of crossings of the ray with the polygon's edges For the example illustrated in Figure 4.1, there are three crossings, so the point is inside the polygon.

4.3 GEOMETRICS AND MEASUREMENTS

Given that the spatial data are coordinate referenced to a common projection, it is possible for the GIS to conduct various geometric computations and measurements. Calculating intersections of lines and finding features that cross a line or polygon are common geometric functions. Also, measurements such as lengths, perimeters, and areas are routinely computed for line and polygon features. A measurement can represent distance, time, address, or any other event at a given point along a linear feature. It can also be defined as a value that defines a discrete location along a linear feature. It is possible, however, that different measurements can be obtained, depending on whether raster or vector data are used. In a raster GIS, there is more than one way to compute the distance between two points. The commonly used Euclidian distance is computed using Pythagorean geometry for a right-angled triangle. Alternatively, a "Manhattan" or block distance can be calculated along the raster cell sides. A vector GIS computes distances by the Euclidian method or by adding the line segment lengths along a line. In either case, the distances are approximate due to the nature of the data and their representation. For small distances on the order of the cell size or vector line segment, distance measurements can have significant errors. "Weighted" distances may be computed as some function of map theme values. For example, a weighted-distance map can be

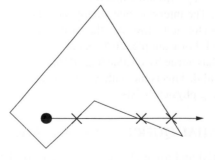

FIGURE 4.1 Point-in-polygon problem.

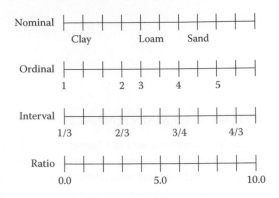

FIGURE 4.2 Measurement scales for assignment of values to mapped attributes.

computed as some function of terrain slope where steep slopes are assigned a higher difficulty than flat terrain. Multiple themes may be used as well (e.g., slope and soil type).

Perimeter measurements in a raster GIS are computed as the number of cell sides that make up the boundary of a feature multiplied by the known resolution of the raster grid. Raster GIS area is computed as the number of cells of a feature multiplied by the known area of the grid cell. A vector GIS computes perimeters as the sum of the straight-line segments of the boundary. Vector polygon areas are computed as the sum of areas of simple geometric shapes formed by subdividing the feature of interest.

Distance and area computations will be fundamentally influenced by the projection used. As noted in Chapter 2, there are a variety of projections used for transforming the Earth's geographic coordinates onto a flat surface. Three primary projection categories are conformal, equivalent, and compromise. *Conformal* projections preserve the property of local shape, so that the outline of a small area such as a county or lake is correct. With the conformal map projections, such as the Lambert Conformal Conic and the Mercator projections, lines of latitude and longitude meet at right angles, thereby preserving directions around any given point. *Equivalent* or equal-area projections preserve the property of area. The primary example is the Albers Equal Area projection, in which all parts of the Earth's surface are shown with their correct area, as on a sphere or ellipsoid. *Compromise* projections preserve distances, but only along one or a few lines between places on the map. Examples include the simple conic and the azimuthal equidistant projections. The smaller the map scale, the more significant the map projection considerations become. Errors are significant at the 1:24,000 scale, and at smaller scales of 1:1,000,000 the errors are large.

Map data are often encoded with information through the assignment of scaled values. It is important to understand when this is being done and to acknowledge the influence such scaling may have on analysis. Data may be represented by four general measurement scales: nominal, ordinal, interval, and ratio (Figure 4.2). Nominal is simply an identifier by which we can identify equivalence (e.g., soil type). Ordinal is a partitioning and ranking of the data, but the ranking number only establishes equality and order. The interval scale attempts to place the values on a meaningful scale so that computations can be performed. However, the base value of the scale may not be known, so computations may not be valid. For example, 20°C is not twice as hot as 10°C, because 0°C is not absolute zero. The ratio scale has a true base value (e.g., 0.0), and mathematical operations involving the ratio of two numbers are valid. This is the scale we use when measuring soil infiltration capacity or computing flood runoff using physically based hydrologic principles.

4.4 SPATIAL AND ASPATIAL QUERIES; CLASSIFICATION

Conducting queries on a spatial and aspatial database to retrieve data is an essential function for a GIS. Queries on the GIS database are often the primary justification for development of a GIS and are useful at all stages of GIS analysis for error checking and reviewing results. Spatial queries are interactive in that the user selects a feature by pointing to a map feature using the mouse cursor, and the GIS returns tabular data about that feature. The "pointing" operation is a complex geometric computation of nearness to features such as points, lines, or polygons. Polygon identification involves a fundamental computation of "point-in-polygon" and is dependent on the topological integrity of the spatial database. Windowing entails the selection of all features within a rectangle or irregular polygon that the user defines. The interactive identification function is often accompanied by visual feedback by highlighting of the selected feature.

Aspatial queries concern the feature attributes, such as the type and size of a water pipe or the date of last maintenance work. Results of an aspatial query may be displayed graphically. For example, a query may ask to find all of the water pipes on which maintenance was conducted during the past five years, and the results displayed on the map. Various search-and-retrieval commands are used, including "show attributes," "show records," "find," "browse," "generate a report," "sort," "recode," "restrict," and "compute" (Clarke 2002).

Queries can be combined to identify entities in a database that meet two or more spatial and aspatial criteria. Boolean operators are used to combine the queries, such as the AND, OR, NOT, and XOR operators. Boolean operations apply where there is a two-way choice and are a special case of set theory. Selections may be based on integer, real, and interval values, and can be set to find those features that meet a restricted range of values.

Relations are the foundation of the relational database model. Here, the database can consist of several flat files or tables, and each can contain different attributes (or fields) associated with a record. In the relational database, keys link files with different attributes, and records with a common key relate to one feature in the GIS. A relational join involves combining two or more tables to form a single table useful for data analysis and reporting. A spatial join summarizes the attributes of features based on polygons in another theme that fall inside, and then joins the summary data to these polygons so that the polygons can be symbolized and queried based on these data. The query language SQL (Structured Query Language) is the extant standard for aspatial queries on relational databases, and this approach has been extended to the spatial domain as well.

Using these set-oriented commands, the user can accomplish classification and generalization operations that can be displayed as thematic maps highlighting the attribute(s) of interest. Classification procedures are a primary means for resolving spatial data into patterns that we can understand. In some GISs, the procedure is implemented as a reclass command. Simple classifications are often made to display continuously varying properties such as elevation, population density, and so on. Generalization involves a combination of the number of classes to reduce the detail to more general classes. Burrough (1986) discusses the various classification methods for a single variable, including exogenous, arbitrary, ideographic, and serial class intervals. Exogenous classification uses threshold levels considered relevant or standard. The arbitrary approach is similar to exogenous, but allows nonstandard classification. Idiographic classifications are based on "natural breaks" in the data. Serial class intervals can be based on statistical analysis of the data and may be equal frequency intervals, equal arithmetic, or some other mathematical normalization of the data (Figure 4.3). Burrough (1986) points out that there is a tendency in inventory and mapping to classify the data soon after the original observations are made. This can result in a loss of data that could be avoided with appropriate use of a GIS. He advocates storage of the original data and cautions against archiving of data in classified form, as the classification can be subject to bias by different analysts.

Principal-components analysis (PCA) is a mathematical technique for examining the relations between a number of map themes, coverages, or layers. The original data are transformed into a set of new properties called principal components, which are linear combinations of the original variables and derived from the correlation matrix. Typically applied in satellite image processing to determine the information content across sensor bands, principal-components analysis can also be applied to any other multivariate data sets for an area. Burrough (1986) describes extensions of PCA, such as cluster analysis.

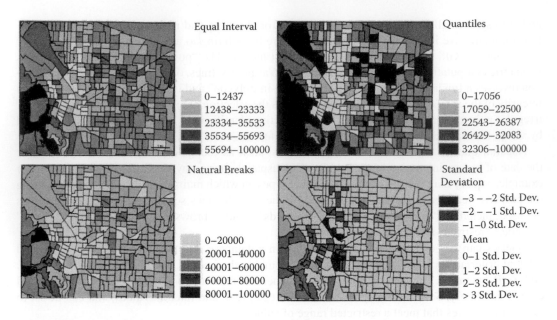

Equal Interval

0–12437
12438–23333
23334–35533
35534–55693
55694–100000

Quantiles

0–17056
17059–22500
22543–26387
26429–32083
32306–100000

Natural Breaks

0–20000
20001–40000
40001–60000
60001–80000
80001–100000

Standard Deviation

−3 – −2 Std. Dev.
−2 – −1 Std. Dev.
−1–0 Std. Dev.
Mean
0–1 Std. Dev.
1–2 Std. Dev.
2–3 Std. Dev.
> 3 Std. Dev.

FIGURE 4.3 Comparison of classification methods for thematic map data.

4.5 NEIGHBORHOOD OPERATIONS

A neighborhood operation applied to a point analyzes conditions for an area surrounding a given location. Sometimes called a "search" function, all points or cells that fall within that neighborhood are used in the calculations. A new map is formed with values computed by some function of a target location and the neighborhood around the target. The types of summary statistics that can be calculated for a neighborhood are counts and tallies, minimum, maximum, mean, median, sum, range, standard deviation, majority, and variety. The shape of the neighborhood can be a rectangle, circle, or other search window as specified, such as a zone.

Proximity assigns areas of proximity to features found in the active theme, creating a grid theme as output. Each cell location in the grid theme is allocated to the closest feature as determined by Euclidean distance. Thiessen polygons are formed from a point coverage as the regions closest to each point.

Buffer operations find features within a specific distance of other features. If a point is buffered, a circular zone is obtained, whereas buffering lines and polygons creates new areas (Figure 4.4). Buffering is conceptually simple but can be applied in a variety of ways when combined with other layers to identify complex relationships between themes. For example, buffering around wells and water supply sources can be used to identify areas that should be protected from waste disposal and nonpoint source pollution.

Filters are commonly applied to elevation data to determine slope, aspect, and other terrain characterizations. The terrain data may be represented by a DTM (digital terrain model) or TIN (triangulated irregular network), depending on the data model. Filters are applied to grid-cell data to determine *slope* (i.e., the rate of change of elevation). Slope is computed as the maximum gradient of the cells surrounding a given cell, usually using a 3 × 3 and D8 (eight direction) filter that is successively moved over the map (Figure 4.5).

Aspect is the horizontal direction of the maximum slope (i.e., the facing direction) and may be expressed as the real number of the direction, or it may be expressed as one of the nine main compass directions (i.e., N, NE, E, SE, S, SW, W, NW, and flat). The second differential of *convexity*

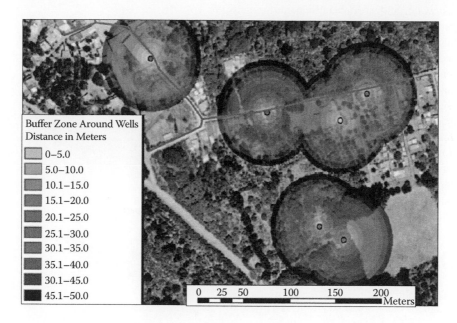

Buffer Zone Around Wells
Distance in Meters

- 0–5.0
- 5.0–10.0
- 10.1–15.0
- 15.1–20.0
- 20.1–25.0
- 25.1–30.0
- 30.1–35.0
- 35.1–40.0
- 30.1–45.0
- 45.1–50.0

0 25 50   100   150   200
                                Meters

FIGURE 4.4   Example of buffer operations for defining zones for well head protection from contamination.

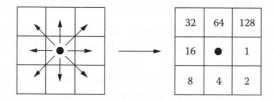

32	64	128
16	●	1
8	4	2

FIGURE 4.5   D8 direction scheme for terrain processing.

(i.e., the rate of change of slope) is useful in geomorphology. Filters may also be applied to smooth out variations in a raster image; this is a function routinely applied in satellite image processing. Computations of slope and aspect for TIN surfaces are based on three-dimensional (3-D) trigonometry for the TIN facet, where the x, y, and z coordinates are known for each apex of the facet. Figure 4.6 illustrates slope determination from an elevation raster or digital elevation model (DEM).

A more complicated type of filter function is that used for finite-difference and finite-element numerical procedures. These are mathematical procedures applied to spatial-phenomena data, where the mathematical expressions quantify fundamental physical principles, such as conservation of mass and energy. Water resources applications include surface and groundwater hydraulics, for example. The expressions are adapted and simplified in each case according to the special features of the problem being addressed (i.e., boundary conditions). In the method of finite differences, the solution region is envisaged as an ensemble of grid points; the governing differential equations are replaced by difference equations with respect to the function values at the grid points; and the problem is then transformed into an algebraic structure. Finite-element methods are applied where the solution region is characterized as a collection of subregions (finite elements), which also leads to a set of algebraic relations between adjacent subregions.

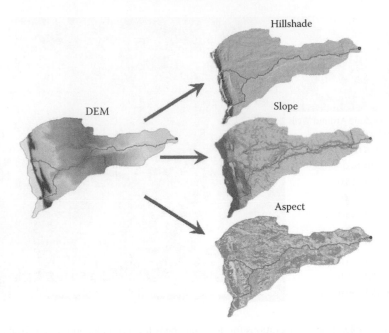

FIGURE 4.6 (See color insert following page 136.) Slope, aspect, and hillshade can be derived from an elevation raster layer.

4.6 SPATIAL ARRANGEMENT AND CONNECTIVITY FUNCTIONS

Topology, or the logical connectedness of a vector data set, must be established so that unique features, connectivity, and spatial arrangement can be established. This is a data-capture and -maintenance function and is commonly required when CAD files are imported into a database or map digitizing is conducted as the means for data capture. Depending on tolerances, vectors may overshoot or undershoot intersections, or polygons may not close. A vector GIS typically provides routines to "clean" an imported data set to connect over- and undershoots and to close unjoined polygons. Also, the problem of "island" polygons requires that polygons completely enclosed by another polygon be identified. Unresolved and redundant vectors and polygons may be identified so that ambiguities can be resolved manually, perhaps through heads-up digitizing.

Creation of topologically correct polygons requires information on how the connected points and lines define a boundary. Here, a vector coverage of polygons is required to not overlap (planar reinforcement), and the adjacency of map features is established. Contiguity functions determine characteristics of areas that are connected. The relationship between features is an important factor in vector databases. For example, census data may be referenced to the street network so that traversing the network in a specified direction permits identification of census block enumeration districts on the left and right of the street.

A *network* is a data structure composed of nodes (junctions), connecting arcs (edges), and information about the relationships between them. Data on arc capacities, travel times, and other attributes are tagged to the node and arc features. Common applications include routing and allocation of materials or flows (Figure 4.7). Network models find wide application in water resources management for piping and water supply systems. Mass-balance and energy equations are programmed into network modeling functions, and solution procedures take advantage of the topologic structure implicit in the data structure. More than just simulation, networks can be solved to identify the optimal path that minimizes travel time or other objective functions. Network analyses can be conducted using raster data sets, but the complexities of system representations are somewhat limited with the grid-cell approach.

FIGURE 4.7 Optimal routes for distribution of reclaimed water, Southern California Comprehensive Water Reclamation and Reuse Study. (*Source:* Lynch et al. 1998. With permission.)

4.7 SURFACE OPERATIONS

Surface operations include a variety of GIS functions that apply to both grid-cell and vector representations of surfaces. Most commonly, the surfaces are representations of terrain using DEMs or TINs, but the procedures are often applied to surfaces created for other variables. Grid-cell or raster DEMs are sometimes called a *regular tessellation* and the TINs an *irregular tessellation* representation of terrain. Sometimes, surfaces may be represented using 3-D mathematical models such as multiquadratic polynomials, cubic splines, or Fourier series. These surfaces are called *parametric* representations because the surface information resides in the parameters of the model and can be generated on command. Parametric models have the advantage that data-storage requirements are low and certain computations easier, but particular details of the surface may be lost due to the imprecision of the mathematical model.

Automated extraction of watersheds or surface drainage, channel networks, drainage divides, and other hydrographic features from DEMs is a standard surface-processing routine in modern GIS. The eight-direction, or D8, method (Figure 4.5) is the most common approach to identifying the direction of flow from a grid cell. Using an iterative approach similar to the spread and seek functions, the D8 defines the drainage network from raster DEMs based on an overland flow analogue. The method identifies the steepest downslope flow path between each cell of a raster DEM and its eight neighbors, and defines this path as the only flow path leaving the raster cell. The method also accumulates the catchment area downslope along the flow paths connecting adjacent cells. The drainage network is identified by selecting a threshold catchment area at the bottom of which a source channel originates and classifying all cells with a greater catchment area as part of the drainage network. This drainage network identification approach is simple and directly generates connected networks. Figure 4.8 illustrates an example. Watershed functions are discussed in more detail in Chapter 5 (GIS for Surface-Water Hydrology).

Cut and fill computations are often required for land and highway development to estimate the volume of material needed to be excavated or imported. Accuracy of the computation is required for estimating the cost of the construction. Surface models formed for the "before" and "after" conditions are differenced to obtain the volume of the cut or fill.

A spread, or cost distance function, is applied to a surface to determine travel time or cost. The surface is normally a grid-cell structure where the cells have been encoded with values of travel

FIGURE 4.8   DEM processing routines are applied to extract stream networks and watersheds.

time, cost, or friction, perhaps derived from the slope and other variables relating to the phenomenon of interest. The spread function involves designation of one or more starting points from which an iterative procedure of moving out from a point is used to calculate the cost of moving to an adjacent cell. The procedure then steps out to the adjacent cells, and a cumulative total of the minimum cost for getting to that cell is recorded for each cell. The spread function is robust in that barriers can be inserted into the grid coverage, and the spread function will work its way around that barrier. Once completed for grid-cell coverage, the minimum cost path from any point to the source point can be determined by moving from the destination point along the minimum cost path. This is called the *seek* or *stream* function and is used for watershed delineation (see discussion below). This procedure is actually a variation of Dykstra's algorithm, which is a particular implementation of a generalized optimization procedure called dynamic programming (Hillier and Lieberman 2001). The path distance function extends the cost distance function to include horizontal and vertical factors in determining minimum cost paths, as well as considering actual distances on sloped surfaces. For example, possible horizontal factors might include the influence of wind speed and direction on the travel times of spreading wildfires from the points of ignition (Davis and Miller 2004). Example vertical factors could include slope directions in the routing of overland flow in a watershed.

Terrain models are also the primary input for *intervisibility*, or line-of-sight, analyses. Line-of-sight maps are derived from DEMs (as grid DEMs or TINs) using ray-tracing techniques emanating from a viewpoint. Locations that can be seen or not seen from the viewpoint are identified, and a viewshed map is created. Multiple viewpoints can be accommodated and the results combined. Intervisibility analyses are used for a variety of applications, including roadway layout (to preserve scenic viewsheds), radar antenna beam blockage, and microwave-tower siting studies. Often, ancillary data on vegetation and building heights are added to the DEM to account for that influence on intervisibility.

Illumination functions simulate the effect of a light (or the sun) shining on a 3-D surface. The surface portrayal generated takes account of the light source position, the terrain surface, and its viewing orientation. The shaded relief image generated is popular with users, as it illustrates how the Earth might look if one were flying over the land in an airplane. Often, the shaded relief image is draped with another coverage such as land use or land cover, rendering a natural view of the land as it is or might be (Figure 4.9). The perspective view is obtained when the viewing position is oblique from the vertical. Specialized routines for simulating rapid movement over the land provide a means for flyovers, much like the popular computer game simulations.

FIGURE 4.9   Thirty perspectives of Mount St. Helens.

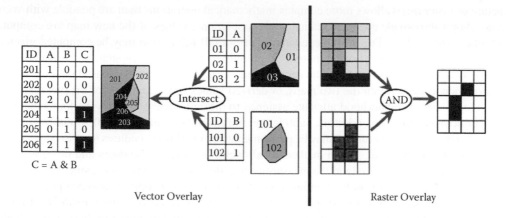

Vector Overlay | Raster Overlay

FIGURE 4.10   Example vector and raster overlay operations of the union of two input data layers.

4.8   OVERLAYS AND MAP ALGEBRA

Overlay operations are a primary analysis function category that sets GIS apart from any other data-management and -analysis software. Overlay was the conceptual procedure used by McHarg (1969) in his original application of minimum-impact highway route finding using Mylar maps. Overlay operations may be logical or mathematical and are applicable to both raster and vector spatial data coverages. *Logical overlays* involve application of Boolean logic to the values of a map to derive another map containing those areas that meet a specified set of conditions. For raster coverages, this is accomplished using progressive applications of reclass operations to convert the map to a binary format where the subset of conditions is true, and then combining the map sets to derive the desired result. Robinove (1977) summarized the foundations for the logical form of overlay.

Vector overlays involve determination of the intersection of an input polygon or line-vector map with an overlay polygon feature map. Intersections, point-in-polygon, and clipping operations are key to this procedure. Vector overlays are computationally intensive and challenging for many GIS developers, involving complex data structures and operations. As illustrated in the intersection, or logical AND operation, between two polygon map layers in Figure 4.10, vector overlay generally results in the generation of new features and attributes in the spatial database.

Raster overlay operations (Figure 4.10) are commonly referred to as *map algebra* or "mapematics," a term coined by Berry (1995). Map layers or coverages usually represent a single theme, and

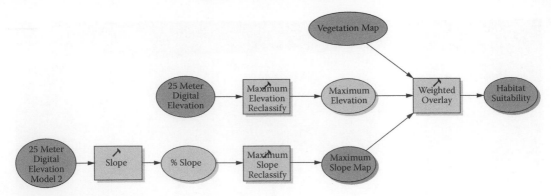

FIGURE 4.11   Suitability model for prairie dog habitat involves staged applications of GIS analysis functions per scientific criteria. Example shown created using ArcGIS® Model Builder.

algebraic operations may be applied to a single coverage or to multiple coverages to generate a new coverage where every value is some function of the original map value. The simple grid-based data structure of raster maps allows more complex mathematical operations than are possible with vector overlay. Applied to single coverage, called *scalar* operations, values of the new map are computed much like a spreadsheet. For example, estimates of rainfall infiltration may be computed given an initial map of soil hydraulic conductivity. For multiple map algebra computations, two maps may be multiplied by each other to form the derived map. Computations may be based on a one-to-one cell correspondence between the two maps, or derived values of the features for the two maps may be used (e.g., some neighborhood attribute assigned to the cell values). Besides the four fundamental algebraic operations (i.e., addition, subtraction, multiplication, and division), a number of other algebraic and statistical operations may be performed using overlay procedures (e.g., average, power, rank, minimum, maximum). When mathematical operations such as products and ratios are computed, it is important to understand the measurement scales for the mapped data, since the computations may be invalid. For example, nominal and ordinal data do not support ratio computations.

Extension of the overlay and map algebra concepts leads to *cartographic modeling* (Tomlin 1990)—the use of basic GIS manipulation functions in a sequential manner to solve complex spatial problems. For example, a sequence of GIS functions may be applied to identify the "best" routes for a highway by taking into account terrain, soils, land cover, and other factors. For these complex model exercises, it is helpful to illustrate the sequencing of operations using a flow chart (Figure 4.11).

Suitability modeling involves the determination of areas "suitable" for a given purpose. This is a general class of cartographic models where technical and policy-related judgments are incorporated into the procedures to "weight" the overlays in a manner that incorporates the judgments. The theoretical foundations of suitability modeling are established in multicriteria decision analysis (MCDA) literature (e.g., Malczewski 1999). MCDA problems involve a set of alternatives that are evaluated on the basis of often conflicting and incommensurate criteria while seeking the maximization of some objective. Where several objectives are involved, as is typical, then a multiple-objective problem emerges, where the goal is to clarify trade-offs between the objectives. Suitability modeling can be considered as a precursor to management modeling, as described in Section 4.12.

4.9   SPATIAL STATISTICS

Spatial statistics is a rich area of GIS analysis functions, one that has developed historically due to the common need to interpolate from point values to unsampled locations or to a regular lattice over an area of interest. More recently, the need to extract information from satellite imagery has led to the development of additional spatial statistical routines.

Descriptive statistics are applied to data without regard to location in order to explore and summarize the data. They can be applied to collections of point data as well as images or field data. Computations

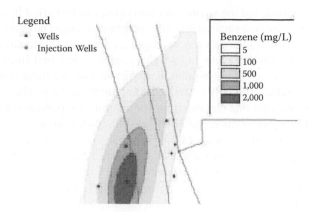

Legend
* Wells
* Injection Wells

Benzene (mg/L)
☐ 5
☐ 100
☐ 500
▨ 1,000
▨ 2,000

FIGURE 4.12 Application of natural neighbor spatial interpolation method for benzene concentration mapping of contaminant plume. (Fleetwood 2008. With permission.)

of the average, range, and class frequencies are conducted as basic reductions of the data. Histograms are graphic representations of frequency distributions, where aggregating the frequencies in an ascending or descending manner tabulates the cumulative distribution. The mean, variance, and standard deviation provide estimates of central tendency, and skewness characterizes asymmetry relative to the symmetric normal distribution. Box and whisker plots provide a schematic graphic useful for portraying important features of the data, including the center, interquartile range, tails, and outliers.

Procedures for interpolation between data points derive fundamentally from the observation that points near each other will be more similar than points far apart. Given a pattern of data points, it has been common for the analyst to assign boundaries to delineate units of homogeneous conditions. The resulting "stepped" model is common for choropleth maps and maps of land use, geology, soils, and vegetation. A concern with these maps is whether they are objective renditions of the data, since different analysts can define different boundaries.

A common reproducible interpolation method is the Thiessen polygon, or "nearest neighbor" method, also known as Voronoi tesselations. This method identifies boundaries between data points based strictly on distance and assigns values to areas based on the data values of the nearest data point. The Thiessen method is entirely dependent on the arrangement of the data points, sometimes leading to odd shapes, since it makes use of only one data point in assigning a value to the area.

Various interpolation procedures are applied to model continuous changes over an area (Figure 4.12). Least-squares polynomial regression, spline functions, Fourier series, or moving averages including kriging may define the interpolated surface. Global methods such as trend surfaces and Fourier series analysis use all of the observations, but may lose details of local anomalies. Local fitting techniques, such as splines and moving averages, can retain local anomalies, since values are estimated from neighboring points only. Trend surfaces describe gradual regional variations by polynomial regression, typically in two dimensions. Polynomials of quadratic or higher order can be used. Although trend surface analysis produces smooth functions, the physical meaning may be lost, and the original data points may not be reproduced. Often, trend surface analysis is used to identify regional trends, after which local interpolators may be used to enhance definition of details in certain locations.

Fourier series describe two-dimensional (2-D) variations by a linear combination of sine and cosine waves. Although quite flexible in application, Fourier series seem restricted in application for complex surfaces. Local interpolators such as splines and moving averages can retain local variations in the original data set. Splines are piecewise functions exactly fitted to a small number of data points exactly while providing continuous joins between one part of the curve and another. Splines retain local features and are computationally efficient, but they provide no estimation of errors. Moving averages are the most commonly used interpolation method, particularly when a weighted moving average is used. With the weighted approach, the weights are computed as some function of distance to the nearest data points, and the window for averaging can be varied from large to small to emphasize scale effects. The inverse squared distance-weighting method has been used often and is provided in many GISs. Optimal interpolation, or *kriging*, addresses shortcomings of the other averaging techniques in that it provides averaging weights that provide a best linear unbiased estimate of the value of a variable at the unknown point. Kriging accounts for three components of the data: a constant trend component, a spatially correlated component, and a residual error term. It is knowledge of the residual error term that makes kriging popular in comparison with other interpolation methods for smoothly varying phenomena, such as groundwater surfaces.

Multivariate analyses apply to single images or images taken in combination. Regression analysis applied to two images typically generates a scatter diagram, trend line, a tabular summary of the regression equation, the correlation coefficient, and results of tests of significance. Regression is often applied for change-analysis applications using two images of the same area taken at different times. Cross-tabulation provides another comparison technique, but in this case for qualitative data. Cross-tabulation generates a table that lists the frequency with which each combination of the categories on the two images occurs. Also, various statistics, such as a chi-square statistic, may be obtained to measure the degree of association.

4.10 IMAGE PROCESSING

Image-processing functions have been developed to extract information from satellite imagery, although many of the procedures can be applied to other grid data sets as well. Jensen (1996) describes image-processing functions and techniques in some detail. Radiometric and geometric corrections are preprocessing operations applied to satellite imagery to correct for errors in the sensor and to obtain an appropriate projection for the imagery. Radiometric errors arise from atmospheric attenuation and topographic effects, among others. Geometric errors arise from satellite-scan skew, Earth rotation, and panoramic distortion effects in the satellite equipment.

Image-enhancement functions are applied primarily to improve the appearance of an image for human visual analysis. Point operations modify the brightness values of each pixel independently of its neighbors. Local operations modify the value of each pixel in the context of its neighbors. Contrast enhancement, or stretching, expands the original brightness values to make use of the total range of the display device and may be linear or nonlinear. Histogram equalization seeks to distribute an equal number of pixel values to each interval. Spatial convolution filtering may be applied to enhance low- and high-frequency detail and edges in an image. Various convolution masks are 3×3 or larger filters designed to enhance edges and can be applied to highlight the directions of edges. Fourier analysis is a mathematical technique for separating an image into its various spatial frequency components. Principal-components analysis is used to reduce the dimensionality of remotely sensed multiband imagery (e.g., seven Thematic Mapper bands) into just two or three transformed principal-components images. Vegetation indices have been applied to satellite imagery to reduce multiple bands to a single number that predicts or assesses vegetation canopy conditions such as biomass, productivity, leaf area, photosynthetically active radiation consumed, and percent of vegetative ground cover. Texture transformations seek to characterize the variability of tonal features using local statistical operators to create a new image of these values. The texture image is then used as an additional band in image-classification operations.

Image classification is accomplished using multispectral classification methods that transform raw reflectivity data into information on land-cover classes. There are a variety of classification algorithms, including (a) hard classification using supervised or unsupervised approaches, (b) classification using fuzzy logic, or (c) hybrid approaches using ancillary (collateral) information.

Supervised classification involves a priori identification and location of land-cover types, such as urban, agriculture, or wetland, through a combination of field work, aerial photography, and other mapping. Specific sites, called training sites, having known spectral characteristics are located in the image and are used to train the classification algorithm for application to the remainder of the

Marshes ■ Water ■ Farm land ■ Aquatic bed ■ Wet meadow

FIGURE 4.13 (See color insert following page 136.) Image-processing techniques are used to classify land characteristics. Example shown identifies wetland areas in mixed agricultural landscape. (Hsu and Johnson 2007).

image. This is a hard-classification scheme, since each pixel is assigned to only one class. In *unsupervised classification*, the identities of land-cover types are not known a priori, and training-site data are not collected or are unavailable. Here, cluster analysis is performed using functions available in a GIS or image-processing software that groups pixels with similar spectral characteristics according to certain statistical membership criteria.

Fuzzy set classification takes into account the inherent imprecision of the imagery reflectance values by assigning a membership index indicating the "degree of truth" that each pixel is a member of each class. Ancillary data on elevation or other attributes may be used with any of the three approaches to enhance the classification process. Object-classification approaches take account of pixel-collection shapes and adjacent pixel characteristics (Figure 4.13). Assessment of errors or accuracy of the classification is accomplished using an error matrix, where errors of commission and omission are tabulated for the known pixels. Overall accuracy is computed by dividing the total number of correct classifications by the total number of pixels in the error matrix.

Other (*hybrid*) discrete multivariate analytical techniques may be applied to the error matrix. There are major efforts to develop land-use–land-cover (LULC) maps worldwide using image-processing techniques. In the United States, the USGS Land Cover Institute has developed maps of each state for land cover, percent forest canopy, and percent impervious surface, all derived from the National Land Cover Database (USGS 1999); this program is discussed in more detail in Chapter 3 (GIS Data and Databases).

4.11 DISPLAY, INTERFACES, INTEGRATION

GIS color-graphic display capabilities are perhaps the most important reason for the popularity of modern spatial data management systems. It is the visual portrayal of spatial data and analysis results that helps communicate the character of our natural and built environment. Combined with user-friendly interfaces and interaction functions, GIS provides a flexible means for those of varying skill levels to access and view spatial data and learn about management plans and engineering design proposals.

Map design is an important factor for the accurate and consistent portrayal of spatial information. Some basic cartographic principles apply whether the map is displayed on the color-graphic screen (soft copy) or printed on a color plotter. The basic cartographic elements of a map include a title, legend, scale, and north arrow. The legend translates the symbols into words. The scale expresses the ratio between ground coordinates and the map space, and is referred to as the representative fraction (RF) or natural scale. It is also useful to include a graphic or bar scale as well as a ratio, since the

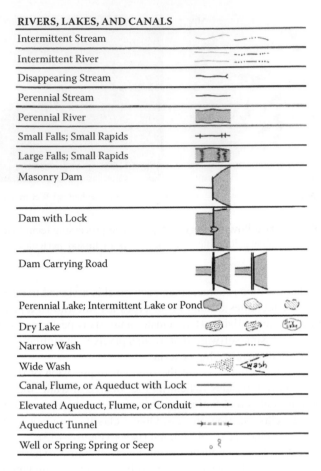

RIVERS, LAKES, AND CANALS

Intermittent Stream	
Intermittent River	
Disappearing Stream	
Perennial Stream	
Perennial River	
Small Falls; Small Rapids	
Large Falls; Small Rapids	
Masonry Dam	
Dam with Lock	
Dam Carrying Road	
Perennial Lake; Intermittent Lake or Pond	
Dry Lake	
Narrow Wash	
Wide Wash	
Canal, Flume, or Aqueduct with Lock	
Elevated Aqueduct, Flume, or Conduit	
Aqueduct Tunnel	
Well or Spring; Spring or Seep	

FIGURE 4.14 USGS symbology for hydrographic features. (*Source:* http://erg.usgs.gov/isb/pubs/booklets/symbols/.)

map display may be arbitrarily reduced or magnified by the user. The map should have a border or "neat line" defining the visual frame for the map, and a coordinate reference graticule or grid should also be shown. Text inserts for place names, credits, and other annotations are required to explain what the map is about, and these should be consistent in font type and size. An inset places the map coverage into its regional context.

A typical GIS provides basic symbology functions to represent features in an abstract manner. Various graphic standards have been adopted to represent features that are distinctive to certain applications. For example, symbols for the USGS topographic quadrangle maps are extant standards (Figure 4.14). Similarly, industry-specific symbols for valves and pumps are standardized for the waterworks industry.

Color representation is also an important factor in map design and symbology, as colors have inherent appeal and meaning to the viewer. For example, color sequencing to show values should be even, flowing from dark to light, with dark being high and light low. Color changes are appropriate to distinguish between opposites on the same map (e.g., above/below a statistical mean). Color balance is essential, and all colors of the rainbow should rarely be used. Cartographic conventions should be adhered to, such as north at the top of the map. Also, contours are frequently brown; water features are cyan (blue); roads are red; and vegetation and forest areas are green. Areas colored red generally suggest some type of critical alert to the viewer, whereas green is generally considered to represent "safety."

User interfaces provide the means for user access and interaction. Most GISs provide the standard modern interfaces using command icons, windows, menu lists, and mouse selection functions. Interaction modes are provided for novice users who would make selections from menus and icons. Expert users may want to bypass the user interface to speed up and/or automate command selections. For these users, a programming language environment is provided, perhaps as macro utilities that call and execute the commands, or as a higher level programming language, such as MS Visual Basic for Applications (VBA) or Python, through which command sequences can be enabled. Most GISs provide functionality for customizing the interface. For implementation in a large organization, the usability of the interface needs to be established through evaluation feedback and training.

Report generation is another key feature of a GIS. The GIS typically provides a capability to interactively select map features, retrieve tabular data associated with those features, and print those tables for reporting. *Visualization* is a relatively new dimension of computing engendered by faster computers and higher resolution display devices. Scientific visualization attempts to merge the processing power of the human brain with the display and processing power of the computer. The result is an enhanced capability for identification of patterns and relationships visible in the data but not detectable using ordinary descriptive and statistical methods. GIS software is advancing to 3-D display and animation capabilities that support visualizations of complex environmental phenomena over time.

Model integration involves the use of the GIS in an integrated model-development environment. Intrinsic GIS analysis functions can be combined in modeling sequences so that a user can conduct modeling exercises in a user-friendly way by accessing compatible data sets and executing the model in a rapid and iterative manner. Exogenous models can be merged with the GIS modeling package as a seamless integrated modeling environment for specialized applications. This data and model integration capability of a GIS finds wide application in environmental and water resources engineering and is a primary focus of this book.

Internet and network access expand the horizons of GIS beyond just the single workstation. It has become routine, for example, to access and download digital map data over the Internet. Sharing files and GIS databases over a network is also commonplace. The network may be internal to an organization (intranet) or wider through connections to distant offices, colleagues, and clients. Also, an organization or individual can create a Web page and serve map products and data. The possibilities for GIS data and product sharing continue to expand through use of the Internet.

4.12 MANAGEMENT MODELS

4.12.1 BACKGROUND

Models are used extensively by engineers as aids in the description, analysis, and design phases of problem solving; to facilitate communication of information to others; and as a means for storing information for future reference and use. Often, the modeling medium is the symbolism of mathematics derived from flow charts and schematic descriptions of the system of interest. *Management modeling* refers to the use of mathematical models to solve management problems such as those encountered in engineering planning, management, and design. In a management model, the computational engine may include the analysis capabilities of a GIS that is embedded in a procedure seeking to maximize some objective or criterion function through specification of design or decision variables—those input data for which there is a choice. Objective functions are mathematical statements that measure performance. For example, it may be desired to maximize firm or reliable water supply amounts through the development of reservoirs at various locations. Among the many (possibly infinite) options available for reservoir siting, sizing, and operational rules, that unique option (or subset of options) that maximizes water supply is sought. Also, there are constraints that limit how much water supply can be developed based on hydrologic conditions or site limitations. The problem is typically made more complicated by the existence of multiple objectives or criteria, such as attempting to maximize water supply while minimizing adverse environmental impacts. These objectives may compete such that enhancing one objective requires detracting from another, requiring analysis of trade-offs between the conflicting objectives.

Two general categories of analysis techniques used for water resources design studies are simulation and optimization. These techniques may be founded on known physical laws, such as Newton's second law and the equations of conservation of mass and energy as applied in hydrology and hydraulics. The models may also incorporate economic dimensions to support accountings of benefits and costs. River basin models typically use time series of historical data on flows and climate, and forecast demands to generate new time series of system outputs.

4.12.2 SIMULATION

Simulation is the process of conducting experiments with a model of the system that is being studied or designed. Simulation or *descriptive* models help answer *what if* questions regarding performance of alternative operational strategies. They can accurately represent system operations and are useful for Monte Carlo analysis in examining the long-term reliability of proposed operating strategies. In GIS, a model is constructed to assess the possible outcomes of alternative actions. The model may involve a variety of GIS analysis functions sequenced into an analysis process or simulation model. Simulation models may be lumped or distributed, depending on the spatial detail, and deterministic or stochastic, depending on whether elements of probability are incorporated. Also, the model may be cast in a multicriteria manner requiring "weights" or preference factors to be applied as part of the simulation process. In an analysis, the system model is generally fixed, and the objective is to determine the system response when a set of input variables is allowed to take on different values. Some inputs are a function of natural conditions and are uncontrollable (e.g., climate), while other inputs are controllable (e.g., operating policies); these are called decision variables. The simulation process is then basically an iterative, or trial-and-error, procedure involving user judgments of performance per the objectives to guide changes in the decision variables. Output from a given simulation run is examined, modification of the controllable input data is made, another run is made, and so on (Figure 4.15). A major advantage of the simulation approach is that, once the simulation model has been formulated, it can be used to test alternative designs under a wide spectrum of environmental conditions. Also, simulation modeling is descriptive in nature and therefore is perhaps easier and more understandable to users. Shortcomings of simulation are that, due to its trial-and-error character, it can take a long time, and there is no guarantee that a "best" solution or design has been identified.

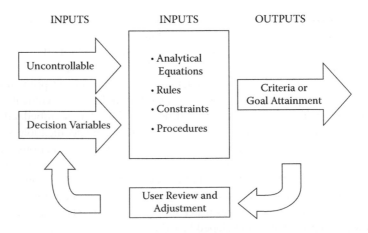

FIGURE 4.15   Schematic diagram of a simulation model with user controlled iterative search loop.

Many simulation models are customized for a particular system, but there is also substantial usage of public-domain, general-purpose models such as HEC 5 (Hydrologic Engineering Center), which has been updated as HEC-ResSim to include a Windows-based graphical user interface (Klipsch and Hurst 2007). Spreadsheets and generalized dynamic simulation models such as STELLA (High Performance Systems, Inc.) are also popular (Stein et al. 2001). Other similar system dynamics simulation models include POWERSIM (Powersim, Inc.), applied by Varvel and Lansey (2002), and VENSIM (Ventana Systems, Inc.), applied by Caballero et al. (2001). Unfortunately, descriptive models are ill suited to *prescribing* the best or optimum strategies when flexibility exists in coordinated system operations.

4.12.3 Optimization

Optimization or *prescriptive* models offer an expanded capability to systematically select optimal solutions, or families of solutions, under agreed-upon objectives and constraints. Optimization procedures differ from simulation in that they attempt to identify the best set of decision variables in a single run, thereby avoiding the trial-and-error search mode. Optimization is a process for determining the "best" solution according to the specified objective and the constraints of the system being studied. In contrast to simulation, optimization procedures seek to define the input decision variables yielding the best performance as related to the objective.

Various mathematical procedures have been developed through operations research that profess to yield optimal solutions to a stated problem. Mathematical optimization procedures include linear, nonlinear, and dynamic programming and their variants, as well as various heuristic procedures such as genetic algorithms. For certain data structures and model formulations, such as networks, solution procedures have advanced to provide efficient and robust solutions. Labadie (2004) provided a detailed review of reservoir system optimization models, focusing on stochastic optimization of multireservoir systems. Model formulation includes specification of an objective function to be maximized (or minimized) and various constraints that must be acknowledged due to physical and policy factors. As discussed by Labadie (2004), optimization models can be extended to the multiobjective case, which requires subjective analysis of preferences between the various criteria.

As illustrated in Figure 4.16, optimization models can be considered as simulation models embedded within an optimal or heuristic search algorithm for systematic selection of the optimal or near-optimal solution. Optimization procedures are often used in a "screening" mode, since it can be difficult to incorporate adequate system detail in the mathematical optimization model. For example, linear programming models require that system interactions and dynamics must be

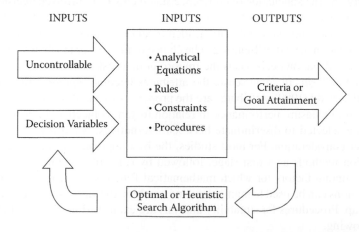

FIGURE 4.16 Schematic diagram of an optimization model with a mathematically based optimal or heuristic search algorithm.

represented as linear relationships, which may preclude realistic analyses of water resource systems that are inherently nonlinear, such as those including hydropower. Since simulation models have an advantage in that higher levels of detail can be achieved, many complex systems are analyzed by the coordinated use of simulation and optimization models, with each being used for that part of the study for which it is best suited.

4.12.4 MULTIPLE-CRITERIA EVALUATION

Multiple-criteria evaluation (MCE) involves a set of alternatives that have multiple attributes and objectives. Usually the attributes are measured in different ways, and comparisons between the attributes must be made in units that are incommensurate. For example, the criteria may be measured in units as different as dollars and the number of acres of wetland converted to residential land use. Such units are not readily compared or easily combined into a single index of an alternative's overall value. Also, the objectives are usually conflicting, in that gaining performance for one objective may require giving up performance for another objective. There is a rich literature on MCE and its variants—multiattribute decision making and multiple objective decision analysis (e.g., Malczewski 1999).

MCE procedures seek to provide a means for analyzing alternatives, with the intent of searching out their comparative advantages and disadvantages while setting down the findings of such analyses in a logical framework. The outcome is a ranking of the alternatives. Inherent in the evaluation process is that there is more than one set of values or preferences relevant in choosing among proposals. The values drive the selection of "weights" applied to define the relative importance of certain attributes and criteria in the evaluation. Particularly in public works, there is a need to identify how individuals and groups value alternative actions, and to include procedures for this in the planning and design process.

Suitability analysis is a general category of MCE applied to land resource inventories, including water resources. A suitability map shows the spatial pattern of requirements, preferences, or predictors of some activity. Although use of the word *suitability* is often restricted to land development, usually in association with the seminal work of McHarg (1969), *Design with Nature*, the analytical concepts involved are much more general. The output of a land-suitability analysis is a set of maps, one for each land use, showing which level of suitability characterizes each unit of land. There are two necessary components of any method: (a) a procedure for identifying parcels of land that are homogeneous and (b) a procedure for rating these parcels with respect to the suitability of each land use. Development of suitability maps requires a mix of theoretical considerations relating to the conduct of "rating" suitability and the solicitation of preference statements from involved decision makers.

Hopkins (1977) reviewed methods for generating land-suitability maps and identified four major categories: gestalt, mathematical combination, identification of regions, and logical combination. These methods are summarized in Table 4.2. The choice of *factors*, or attributes, by which to evaluate alternative land-allocation decisions is the important first task of suitability analysis. These are also known as the *criteria*. In some sense, the attributes become representative of objectives, as described by Malczewski (1999). Attributes are the properties of a real world that can be measured, and these are used to measure performance in relation to some objective. In general, it is required that the factors be selected to discriminate between alternatives. A *constraint* serves to limit the alternatives under consideration. For most studies, the best approach is to use the linear and nonlinear combination methods as a first stage, followed by rules of combination. First, incorporate the relationships among factors for which mathematical functions, either linear or nonlinear, are known. GIS functions can be used to manipulate and combine the various attribute maps to generate the suitability map. Procedures for suitability analysis are examined in the case studies included in the chapters following.

An important area of application related to MCE is modeling of land-use changes. The spatial allocation of new development is a key factor in anticipation of required infrastructure for water

TABLE 4.2
Methods for Generating Land-Suitability Maps

Gestalt method

A gestalt is a whole that cannot be derived through consideration of its parts

Homogeneous regions are determined directly by an analyst's judgment

Dependent on a domain expert, whose judgment is not explicit and is difficult to confirm

Mathematical combination

Requires rating of each factor for each land use

Ordinal combination assigns a relative rating to each factor

    Rated factor maps are overlay added to obtain composite suitability

    Ordinal combination addition is mathematically invalid but often used

Linear and nonlinear combination methods reclassify attributes on interval scale (e.g., 1 to 10)

    Requires that each scaled factor map be "weighted" per its importance

    Ratings for each factor are multiplied by the weight for the factor

    Suitability map is computed as the sum of the multiplied ratings

Importance weights are generated by ranking, rating, pairwise comparison, or trade-off analysis

Identification of regions

Explicitly identifies homogeneous regions for which suitability ratings can be assigned

Factor combination involves complete logical intersection of boundaries from each factor map

    Works for studies having only a few factors; many factors too cumbersome

Cluster analysis identifies regions by successively pairing the most similar sites or groups of sites, based on an index of similarity across the set of factors

Logical combinations

Rules of combination

    Assign suitability to sets of combinations of types rather than to single combinations

    Expressed in terms of verbal logic rather than in numbers and arithmetic

    Not necessary to evaluate each combination separately as in the factor combination method

    Not necessary to find a precise mathematical statement of the relationships among factors as in the nonlinear combination method

Hierarchical combination

    Combinations of strongly interdependent factors are rated for suitability as combinations

    Higher order combinations from these subsets of factors are rated, with each lower order combination now treated as an integrated whole

    Sequence is repeated until a rating is obtained that includes all relevant factors

Source: Hopkins (1977).

---

supply and sewers as well as environmental impacts of various kinds. There are various approaches to modeling land-use change, including (a) multicriteria evaluation (MCE) models, (b) Markov chain models, (c) logistic regression-based models, and (d) cellular automata models. All of the methods are dependent on collating spatial data for differing times so that the character of changes can be identified and codified. Output may be the likelihood of each cell converting to a given class (at an unspecified time) or the predicted land use at one or more (specified) dates. The level of detail of the data is also relevant. For example, regional-scale data can support land-use-change forecasts that are relatively coarse. Parcel-level data are required to forecast land conversions at a high level of detail. In addition, some models attempt to represent market factors of land demand and pricing as well as the influence of land management strategies. A study by the U.S. EPA (2000) summarized the various land-use-change models. There is considerable ongoing research on CA models, and applications are being developed for major urban areas in the United States (Figure 4.17) and worldwide.

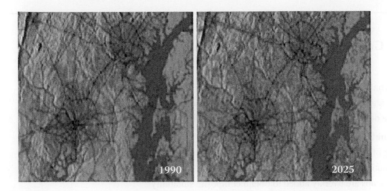

FIGURE 4.17 (See color insert following page 136.)   Land-use-change projections for 2025, made with a land-use-change model, show likely areas of new urban growth in yellow (high probability) and in greens (light green is moderate probability; dark green is low probability). (*Source:* USGS 1999.)

4.12.5 DECISION-SUPPORT SYSTEMS

A decision-support system (DSS) is a "system that supports technological and managerial decision making by assisting in the organization of knowledge about ill-structured, semi-structured or unstructured issues" (Sage 1991). The term *unstructured* refers to the nature of the decision problem; initially, the alternatives, their impacts, or the procedures for evaluation and selection may not be fully understood. The primary components of a DSS include subsystems for database management, model management, and dialog generation. In the context of a GIS, each of these components exists, although the interface for dialog generation may need to be customized for a given organization or decision-support situation. Within the GIS field, there are many examples of the use of GIS software to provide decision support, and some argue that a GIS is a DSS. However, GISs are general-purpose systems, are not focused on a particular decision, and lack the support that customized models can provide.

Decision-support systems encompass a wide variety of approaches that can aid in decision making through the use of computers, database management systems, and visual interactive modeling techniques. Water resources management DSS data subsystems have been strongly influenced by the low-cost availability of GIS data sets and the proliferation of low-cost microprocessor data-collection platforms (DCPs). Interactive displays are used to aid data validation and review tasks, as well as to provide a medium for communicating results of analyses to non-computer-oriented managers. The user-friendly criterion is most important in allowing system managers to bypass intermediary computer operators—a situation that is inconvenient and lessens user confidence. DSSs differ from traditional record-keeping and transaction-processing uses of computers primarily because they require a symbiosis between the users and the system in order to be effective. That is, a DSS has a decision focus in contrast to the information and data focus associated with earlier data-processing and management-information systems.

The integration of GIS with decision-support systems is referred to as a spatial decision-support system (SDSS) (Walsh 1992). Spatial DSSs make second-order uses of spatial data and provide additional processing or integration with the nonspatial models required to fully support the decision maker. Software advances are making GIS increasingly appropriate as a generator for an SDSS. It is now easier to interact with models using GIS as a sophisticated interface for spatial information. Even limited-functionality GIS software provides the ability to zoom and to display or highlight different features. GIS provides database support designed to allow for effective storage of spatial data. Furthermore, GIS software provides a link between the interface and database to allow the user to easily query spatial data. However, for the full range of potential uses of a spatial data in decision making, a GIS is not a complete DSS until problem-specific models are integrated that address the organization's decision needs in a flexible and interactive manner.

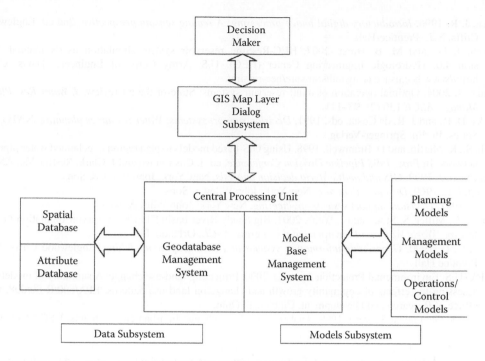

FIGURE 4.18 Generic components of a spatial decision-support system.

As depicted in Figure 4.18, there are direct linkages between the spatial database incorporated in a GIS and the analysis modules of the SDSS, allowing the user to access both spatial and attribute data in a flexible manner without extensive training. GIS provides general capabilities for capturing and manipulating spatial data, and when incorporated into a structure for interactive use with a user-friendly interface, the SDSS can support users in semistructured or poorly structured problem solving (Loucks and da Costa 1991).

REFERENCES

Berry, J. K. 1995. *Beyond mapping: Concepts, algorithms and issues in GIS.* Fort Collins, Colo.: GIS World Books.

Burrough, P. A. 1986. *Principles of geographic information systems for land resource assessment.* Monographs on soil and resources survey No. 12. New York: Oxford Science Publications.

Caballero, Y., P. Chevallier, A. Boone, and J. Noilhan. 2001. River flow modeling for a tropical high-altitude mountain: A case study of the Rio Zongo Valley in the Royal Cordillera, Bolivia. Paper presented at the 6th scientific assembly of the International Association of Hydrological Sciences. Maastricht, Netherlands.

Cai, Z. 2003. Use of GIS in analyzing environmental cancer risks as a function of geographic scale. Research report. Department of Mathematics, University of Arizona, Tucson.

Clarke, K. C. 2002. *Getting started with geographic information systems.* 4th ed. Englewood Cliffs, N.J.: Prentice Hall.

Davis, B., and C. Miller. 2004. Modeling wildfire probability using a GIS. In *Proc. American Society for Photogrammetry and Remote Sensing.* Annual Meeting of the ASPRS. Denver.

Fleetwood, J. 2008. Development of a Geodatabase and Geoprocessing Tool for Environmental Data Management. Masters Report. Dept. Civil Engineering, Univ. of Colorado Denver. April.

Haines, E. 1994. Point in polygon strategies. In *Graphics gems.* Vol. IV, ed. P. Heckbert, 24–46. New York: Academic Press.

Hillier, F. S., and G. J. Lieberman. 2001. *Introduction to operations research.* New York: McGraw-Hill.

Hopkins, L. D. 1977. Methods for generating land suitability maps: A comparative evaluation. *J. Am. Inst. Planners* 43 (4): 386–400.

Hsu, C., and L. Johnson. 2007. Multi-criteria wetlands mapping using an integrated pixel-based and object-based classification approach. In *Proc. AWRA Annual Conference.* Albuquerque, N.M.

Jensen, J. R. 1996. *Introductory digital image processing: A remote sensing perspective.* 2nd ed. Englewood Cliffs, N.J.: Prentice-Hall.

Klipsch, J. D., and M. B. Hurst. 2007. HEC-ResSim, reservoir system simulation user's manual, version 4.0. Hydrologic Engineering Center (HEC), U.S. Army Corps of Engineers, Davis, Calif. http://www.hec.usace.army.mil/software/hec-ressim/.

Labadie, J. 2004. Optimal operation of multi-reservoir systems: State-of-the-art review. *J. Water Res. Plann. Manage. ASCE* 130 (2): 93–111.

Loucks, D. P., and J. R. da Costa, eds. 1991. *Decision support systems: Water resources planning.* NATO ASI Series. Berlin: Springer-Verlag.

Lynch, S., K. Martin, and D. Bramwell. 1998. Using GIS-based models to plan regional reclaimed water pipeline network. In *Proc. 1998 Pipeline Division Conference*, ed. J. Castronovo and J. Clark. Reston, Va.: ASCE.

Malczewski, J. 1999. *GIS and multicriteria decision analysis.* New York: John Wiley & Sons.

McHarg, I. L. 1969. *Design with nature.* New York: John Wiley & Sons.

Sage, A. 1991. *Decision support systems engineering.* New York: John Wiley & Sons.

Stein, S., C. Miller, S. Stout, and J. Webb. 2001. Big Sandy River basin STELLA reservoir regulation model. In *Proc. World Water and Environmental Congress, ASCE.* Orlando, Fla.

Tomlin, C. D. 1990. *Geographic information systems and cartographic modelling.* Englewood Cliffs, N.J.: Prentice Hall.

USEPA (U.S. Environmental Protection Agency). 2000. Projecting land-use change: A summary of models for assessing the effects of community growth and change on land-use patterns. EPA/600/R-00/098. EPA Office of Research and Development. Cincinnati, Ohio.

USGS (U.S. Geological Survey). 1999. Analyzing land use change in urban environments. USGS fact sheet, 188-99. http://landcover.usgs.gov/urban/info/factsht.pdf.

Varvel, K., and K. Lansey. 2002. Simulating surface water flow on the Upper Rio Grande using PowerSim 2001. SAHRA–NSF Science and Technology Center for Sustainability of Semi-Arid Hydrology and Riparian Areas, second annual meeting. University of Arizona, Tucson.

Walsh, M. R. 1992. Toward spatial decision support systems in water resources. *J. Water Resour. Plann. Manage.* 109 (2): 158–169.

Weiss, D. 2008. Geography 391: Quantitative methods in geography. Lecture 3 Notes. Department of Geography, University of North Carolina, Chapel Hill.

5 GIS for Surface-Water Hydrology

5.1 INTRODUCTION

Hydrologic models attempt to represent various components of the precipitation-runoff process in a watershed (Figure 5.1). The processes illustrated begin with precipitation, which can fall on the watershed's vegetation, land surface, and water bodies. Precipitation as snow would require additional accounting for snow-melt processes (not shown). In a natural hydrologic system, much of the water that falls as precipitation returns to the atmosphere through evaporation and transpiration from vegetation. At the land surface, the water may pond, and depending upon the soil type, ground cover, antecedent moisture, and other land surface properties, a portion may infiltrate into the soil. Impervious land surfaces restrict infiltration and increase runoff. Excess precipitation that does not pond or infiltrate moves by overland flow to a stream channel. Infiltrated water is stored temporarily in the upper layers of soil. From there, it may rise back to the surface by capillary and vegetative action, move horizontally as interflow, or percolate vertically to the groundwater aquifer. Interflow eventually moves into the stream channel. Water in the aquifer moves slowly, but eventually some returns to the channels as base flow. The stream-channel network collects and routes the overland flow, interflow, and base flow to the watershed outlet.

The watershed-runoff processes summarized are all inherently spatial in character, so there is a strong motivation to use GIS tools to organize the data and formulate hydrologic models. Surface-water hydrology is perhaps the area for which GIS has been most applied in the water resources and environmental fields. The advent of digital data products and software for processing spatial information has prompted a change in the way we look at hydrologic systems, and it has made it possible to more precisely describe watershed characteristics and runoff response to precipitation inputs.

There is a movement away from the so-called lumped-parameter models to the more spatially distinct or "distributed" modeling approaches that represent fundamental physical processes. The general availability of digital elevation models (DEMs), triangulated irregular networks (TINs), digital line graphs (DLGs), digital soils and land-use data, radar-rainfall and satellite imagery, and real-time gauge reporting—and the GIS software to process these—has contributed to an increased awareness of the spatial distribution of hydrologic processes. In a sense, there is a "supply-push" situation, in that availability of high-resolution data is pushing us toward consideration of increasing spatial detail in our modeling efforts. However, in spite of increased data and computing power, there remains the challenge of calibrating the models so that they accurately reflect recorded runoff responses.

This chapter addresses various aspects of surface-water hydrologic modeling using GIS. Hydrologic-database development procedures are reviewed, including conversion of available data archives and remote sensing. Digital terrain-modeling techniques are described in some detail, as these often provide the primary means for basin delineation and drainage network extraction. Spatial interpolation methods are often used to address the common situation of determining hydrologic-model input data characteristics based on point-data collections (e.g., rainfall from gauges). Common GIS applications are presented for simulating water-budget accounting and watershed flood runoff. Examples of GIS and hydrologic-modeling integration are also discussed. There is some overlap between this chapter and Chapter 8 (GIS for Wastewater and Stormwater Systems)

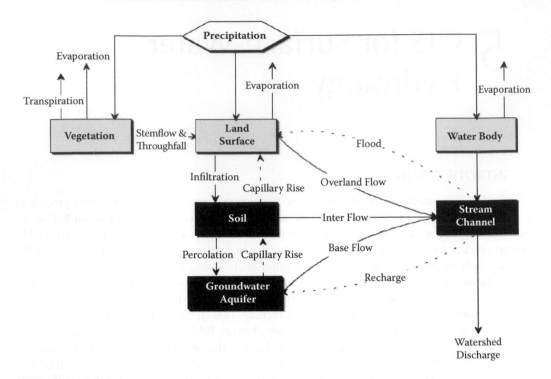

FIGURE 5.1   Generalized diagram of the runoff process at local scale. (From HEC 2000.)

and Chapter 10 (GIS for Water Quality). Both of these chapters extend the surface-water hydrology concepts to address urban runoff and water-quality aspects.

5.2   SURFACE-WATER HYDROLOGIC DATA

5.2.1   Overview

A variety of data are required for surface-water hydrologic studies and modeling. Table 5.1 lists categories and types of data. Hydrologic and watershed data originate from a multitude of government agencies, such as the U.S. Geological Survey, which have collected and processed data into standardized formats. Often, the data are available for download from Internet Web sites. The data have various basic formats as raster, vectors, and associated alphanumeric attributes. Other data may require field collection and processing. Satellite imagery is an example of this type, as image processing is required to identify updated land-use and land-cover (LU/LC) characteristics. Field surveys may also be required to identify details on stream-channel characteristics, such as channel shape and roughness. In a number of instances, GIS functions are applied to develop the data into formats usable for surface-water hydrologic modeling; digital terrain-processing tools are an example of this. GIS database operations aid in collating the various data sets into a coherent database supportive of hydrologic analyses and modeling. Raster and vector data for water resources applications are described in Chapter 3.

This section presents a review of the types of data used for surface-water hydrology studies and typical sources and methods for establishing these data in a GIS.

5.2.2   Digital Elevation Model Data

Most surface-water hydrologic applications begin with raster data of the terrain due to the wide availability of DEMs (Figure 5.2) and intrinsic GIS software functions to conduct digital terrain

---

TABLE 5.1
Surface-Water Hydrologic Data

Terrain	**Streamflow records**
Digital elevation models	Long term
Slope and aspect	Storm runoff events (calibration)
Watersheds and subcatchments	Statistics (e.g., frequency, peak flows)
Drainage networks	**Land use**
Hydrography	Land uses (types and percent impervious)
Stream paths and network topology	Land cover
Stream names and IDs	Population
Reaches	Zoning and master plans
Channel data (e.g., shape and roughness)	Forecast land use
Junctions	**Water control works**
Lakes and wetlands	Dams and reservoirs
Ditches and canals	Weirs
Soils	Detention basins
Permeability	Diversions
Layer depths	
Soil textural percentages	**Satellite and aerial imagery**
Soil water content	Land use/land cover (LU/LC)
	Vegetation
Precipitation and climate	Surface temperature
Rain-gauge data	Soil moisture
Gauge locations and contexts	Snow cover & water equivalent
Statistics (e.g., intensity-duration-frequency)	**Administrative boundaries**
Radar-rainfall imagery	Political boundaries
Snow	Water management districts
Temperature	
Evaporation and transpiration	
Potential ET	
Consumptive use	

---

processing (discussed in Section 5.3.2). The sources for DEM data, such as the National Elevation Dataset (NED; http://ned.usgs.gov/), are described in Chapter 3 (GIS Data and Databases). Topography plays a primary role in the distribution and flux of water and energy within the natural landscape. Examples of interest include surface runoff, infiltration, evaporation, and heat exchange that take place at the ground–atmosphere interface. Quantitative assessment of these processes depends on the topographic configuration of the landscape, which is one of several controlling factors. Also, natural processes that vary with location over an area can often be more readily modeled using a field approach. For example, a two-dimensional (2-D) surface rainfall-runoff model can be based on cellular-type operations, where the physics of overland flow are represented for each cell and its immediate neighbors.

With DEMs, the smallest mappable unit is dependent on the grid resolution, so a point can only be located somewhere within a cell, and a stream channel can be no smaller than 30-m wide, assuming that is the cell resolution. For example, a 7.5-min USGS DEM has grid cells with a nominal 30-m size, although 10-m data are available in most of the conterminous United States. These location precision limitations of raster data can be a disadvantage, particularly for small areas in urban drainage applications. However, for many watershed studies, raster data serve quite well in providing surface definition far superior to data previously available. As long as the area being mapped is large relative to the size of the grid cells, then any errors associated with watershed areas will

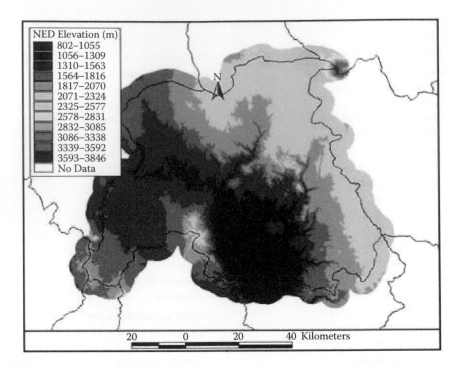

FIGURE 5.2 Digital elevation model data provide a preferred basis for development of a surface-water hydrologic model. (*Source:* http://www.nssl.noaa.gov/projects/basins/.)

be small. In addition, the simplicity of data management and processing in the raster data model often make raster data the first choice for hydrologic modeling. Processing of DEM data to derive watershed boundaries and stream paths is described in more detail below.

For highly detailed terrain mapping, such as is required to define floodplain details, there is increasingly wide use of light-detection and ranging (LIDAR) data. This data type is described in Chapter 3 (GIS Data and Databases), and applications of LIDAR for floodplain mapping are described in Chapter 9 (GIS for Floodplain Management).

5.2.3 HYDROGRAPHIC VECTOR DATA

Hydrographic vector data of surface-water systems are also common and may have been developed as features in the original mapmaking process or derived from DEM processing. A primary data set on stream vectors is the National Hydrography Dataset (NHD), which is described in Chapter 3. For example, the NHD contains the hydrography coverages that represent stream centerlines. An important aspect of vector data is topology, feature connectivity, area definition, and adjacency. With vector data, it is possible to maintain these important topologic relationships, whereas with raster data it is not. Also, vector data permit tagging of nonspatial attribute data to the map features. Maintenance of topologic relationships and tagging to feature attributes are important in surface-water hydrology. For example, these relations can be used to schedule computations through a linked stream network having attributes of channel roughness and shape. Drawbacks to the vector data include the increased complexity of data management and processing. Also, the precise vector data portrayals imply an accuracy of data that may not be warranted.

Vector digitizing is another method of obtaining vector information from existing maps. In this manner, one can choose the level of detail that is most appropriate to meet the specific needs of the project. If no adequate map exists, or if field verification of a map is desired, then a field survey could be performed to create a data set containing the desired stream and river network information in the watershed. File formats for the USGS and EPA data sets are standard formats that most GISs can interpret, and while getting digitized or surveyed data into a suitable format and proper geographic coordinate system is not difficult, it is an issue that should be dealt with before digitizing or surveying occurs.

5.2.4 Soils and Soil Moisture Data

A fundamental concern with hydrologic modeling of watersheds is the type of soils and the status of soil moisture and associated infiltration dynamics. Sources for soils data are primarily the NRCS STATSCO and SSURGO data sets, which are described in Chapter 3 (GIS Data and Databases).

Precipitation as rainfall and snowmelt has strong interactions with the soil layers, which determine the amount of excess precipitation and subsequent surface runoff. Accordingly, infiltration and soil moisture are important components of the hydrologic budget and models. The rate at which infiltration occurs is influenced by such factors as the physical properties of the soil, rainfall intensity, status of the soil moisture, type and extent of vegetal cover, the condition of the surface crust, and temperature. Also, the governing factors vary greatly in space due to the spatial heterogeneity of soil, land, and rainfall characteristics. These factors vary during and between precipitation events, thus creating a dynamic process that has been difficult to measure and model reliably. Further, measurement of soil moisture and infiltration has been problematic, particularly in real time. Modeling of soil moisture is described in Section 5.3.6.

Soil moisture can be measured by in-ground sensors at specific locations. The Illinois State Water Survey has made soil-moisture measurements in the top 2 m of the soil profile since 1981; these are currently made twice per month at 18 sites (Schaake et al. 2004). The Oklahoma Climatological Survey (http://climate.ok.gov/research/) has installed more than 100 soil-moisture sensor sites at depths of 5, 25, 60, and 75 cm, which collect readings every 30 min. The sensors measure soil moisture by heating and then taking the temperature of a wire, pausing for a few seconds, and then taking the temperature a second time. Since heat dissipates more slowly in water, wet soils will have a smaller change in temperature measured by the sensor than dry soils. This temperature difference allows hydrological variables such as soil water content, soil matrix potential, and fractional water index (FWI) to be calculated. Figure 5.3 shows a typical image of soil moisture at a depth of 5 cm during the early part of the summer. New inexpensive instrumentation has been developed that has

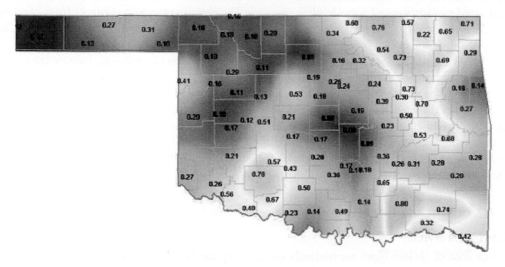

FIGURE 5.3 Soil moisture map for Oklahoma obtained by interpolating point measurement site values. The darker areas show drier soils, while the lighter areas indicate more moist soils. (*Source:* http://climate.ok.gov/research/.)

made accurate, real-time measurements of volumetric soil-moisture content possible. The Campbell Scientific 616 Water Content Reflectometers have been employed by NOAA researchers for the last six years in the Russian and American River Basins located in northern California. The volumetric soil water observations obtained provide a basis for tracking soil moisture levels in the watersheds. In combination with satellite microwave sensors and hydrologic models, there is potential for more spatially distinct soil-moisture estimates across a watershed.

Recent advances in remote-sensing technology have shown that soil moisture can be quantitatively estimated using microwave technology under a variety of topographic and vegetation cover conditions. A summary assessment of the state of the art of remote sensing of soil moisture was presented by the NRC (2007). Microwave emissions from the land surface have a strong sensitivity to several hydrologic variables, including soil and snow water content, as well as other land-surface characteristics such as vegetation water content, soil texture, and surface roughness (Jackson 1999). Remote sensing of the moisture content of surface soils has been most successful using passive microwave instruments. The capability currently exists to retrieve surface (0–2 cm) soil-moisture estimates from the space-based Advanced Microwave Scanning Radiometer-EOS (AMSR-E) instrument at frequencies of about 6 GHz (C band) (Jackson and Hsu 2001). However, previous research has shown that frequencies in the 1–3-GHz (L band) range are better suited for soil-moisture retrievals because the microwave emissions emanate from within a deeper soil layer (0–5 cm) (Jackson et al. 1995, 2002a, 2002b). The inherent spatial resolution of AMSR-E soil-moisture data is 60 km, yet higher resolution (e.g., 10 km) is better for hydrological applications. Also, land-based systems such as ground-penetrating radar offer promise for monitoring deeper soil water content, and airborne L-band passive microwave instruments have proven soil-moisture-mapping capabilities (Wang et al. 1989; Jackson et al. 1995, 2002a, 2002b). A study by Santanello et al. (2007) demonstrated the utility of near-surface soil-moisture estimates derived from passive (L-band) microwave remote sensing for soil moisture. The methodology was also applied using a new soil-moisture product from active (C-band) radar imagery with much lower spatial and temporal resolution.

5.2.5 LAND-USE AND LAND-COVER DATA

Land-use and land-cover information is used in hydrologic modeling to estimate surface roughness or friction values, since it affects the velocity of the overland flow of water. Land-use information, coupled with the hydrologic characteristics of soils on the land surface, can also provide measures of expected percolation and water-holding capacity. The amount of expected runoff from vegetated land-use types, such as forest, is not only affected by the surface and soil physical properties, but also by the uptake capacity of the vegetation present.

Sources of land-use data, described in Chapter 3 (GIS Data and Databases), include the USGS-EPA National Land Cover Dataset (NLCD; http://landcover.usgs.gov/), which has been developed with time stamps for 1992 and 2001 and is appropriate for most watershed modeling (Figure 5.4). The NLCD is distributed as 30-m-resolution raster images in an Albers Equal-Area map projection. The NLCD 2001 product series includes independent per-pixel estimates of imperviousness and tree canopy coverages. Updates to public domain land-use–land-cover data may be developed through independent image processing of satellite imagery. Local governmental agencies may also have land-use maps developed for their jurisdictions.

5.2.6 CLIMATE AND PRECIPITATION DATA

5.2.6.1 Overview

Climate data of various types are routinely collected by the National Weather Service, other agencies, and citizen volunteers at specific locations. The primary data archive for climate and other meteorological data is the National Climatic Data Center (NCDC; http://www.ncdc.noaa.gov). Land-based observations archived by the NCDC contain various meteorological elements that,

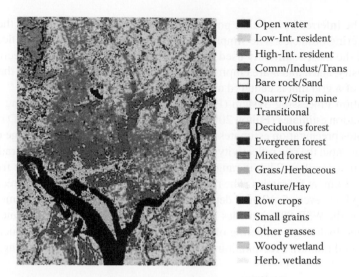

■	Open water
░	Low-Int. resident
■	High-Int. resident
■	Comm/Indust/Trans
☐	Bare rock/Sand
■	Quarry/Strip mine
■	Transitional
▤	Deciduous forest
■	Evergreen forest
▤	Mixed forest
▨	Grass/Herbaceous
	Pasture/Hay
■	Row crops
▤	Small grains
░	Other grasses
░	Woody wetland
	Herb. wetlands

FIGURE 5.4 (See color insert following page 136.)   National Land Cover Dataset is available nationwide and is used for watershed modeling studies. (*Source:* http://landcover.usgs.gov/.)

over time, describe the climate of a location or region. These elements include temperature, dew point, relative humidity, precipitation, snowfall, snow depth, wind speed, wind direction, cloudiness, visibility, atmospheric pressure, evaporation, soil temperatures, and various types of weather occurrences such as hail, fog, thunder, etc. This also includes data observed by weather radars. The weather observations vary in their time interval from subhourly to hourly to daily to monthly. Summaries are then produced from these data. Land-based data from around the world are archived at NCDC. Relevant available precipitation data from NCDC include: (a) cooperative station daily summary data, (b) hourly rainfall rates recorded at first-order NWS meteorological stations, and (c) quarterly (15 min) rainfall rates for selected stations around the United States. Also, various commercial vendors produce CD-ROMs containing NCDC precipitation data including daily, hourly, and quarterly resolution rainfall data.

5.2.6.2 Radar-Rainfall Estimation

Radars send out short pulses of radio waves that bounce off particles in the atmosphere, and the energy is reflected back to the radar dish. A computer processes the returned signals and, through algorithms, can make conclusions about what kinds of particles it "saw," including the directions they are moving in (Doppler effect) and the speed of their movement. Weather radar data are available from the National Weather Service (NWS) Weather Surveillance Radar Doppler units (WSR-88D) throughout the United States. Each of these units provides coverage of a 230-km-radius circular area. The WSR-88D radar transmits an S-band signal that is reflected when it encounters a raindrop or another obstacle in the atmosphere. The power of the reflected signal, which is commonly expressed in terms of reflectivity, is measured at the transmitter during 360° azimuthal scans centered at the radar unit. The WSR-88D radar transmits horizontal pulses, which give a measure of the horizontal dimension of the cloud (cloud water and cloud ice) and precipitation (snow, ice pellets, hail, and rain particles). Over a 5- to 10-min period, successive scans are made with 0.5° increments in elevation. The reflectivity observations from these scans are integrated over time and space to yield estimates of particle size and density in an atmospheric column over a particular location. To simplify data management, display, and analysis, the NWS digitizes and reports reflectivity for cells in a Hydrologic Rainfall Analysis Project (HRAP) grid. Cells of the grid are approximately 4 × 4 km.

The basic signal-processing algorithms for radar-rainfall estimation were described by Fulton et al. (1999) and detailed in OFCM (2006). Given the reflectivity, the rainfall rate for each of the

HRAP cells can be inferred because the power of the reflected signal is related to the sizes and densities of the reflecting obstacles. The simplest model to estimate rainfall from reflectivity is a $Z–R$ relationship, and the most commonly used of these is $Z = aR^b$, in which Z = reflectivity factor, R = the rainfall intensity, and a and b are empirical coefficients. Thus, as a product of the weather radar, rainfall for cells of a grid that is centered about radar units can be estimated.

The Precipitation Processing System (PPS) is the first of two serial processing algorithms that compute precipitation estimates (Fulton 2002). It runs at each of the 160 WSR-88D radars and automatically computes precipitation estimates on a polar grid out to a maximum range of 230 km from the radar, using as input the raw reflectivity factor measurements collected by the radar. The rainfall grid is a fixed 1° in azimuth by 2 km in range, and therefore grid cells range in size from about 1 km^2 at close ranges to 8 km^2 at far ranges. The rainfall accumulations, with internal precision of 0.1 mm of rainfall, are updated every 5–10 min, depending on which of several scanning modes the radar is operating in. Thus the WSR-88D provides high-resolution rainfall estimates in time and space suitable for distributed hydrologic modeling of small catchments. Precipitation products are estimated ground-accumulated rainfall. Precipitation products available include the following:

- One-hour precipitation (N1P/78)
  A display of estimated 1-h precipitation accumulation on a 1.1 × 1.1-mile grid. This product is used to assess rainfall intensities for flash flood warnings, urban flood statements, and special weather statements.
- Three-hour precipitation (N3P/79)
  Same as 1-h precipitation, except for a 3-h period.
- Storm-total precipitation (NTP/80)
  A map of estimated storm-total precipitation accumulation continuously updated since the last 1-h break. This product is used to locate flood potential over urban or rural areas, to estimate total basin runoff, and to provide rainfall data 24 h per day.
- Digital precipitation array (DPA/81)
  An array format of estimated 1-h precipitation accumulations on the 1/4 LFM (4.7625-km HRAP) grid. This is an eight-bit product with 255 possible precipitation values. This product is used to assess rainfall intensities for flash flood warnings, urban flood statements, and special weather statements.

The PPS currently generates four rainfall products updated every 5–10 min: the 1-h precipitation accumulation product (OHP), the 3-h precipitation accumulation product (THP), the storm-total precipitation accumulation product (STP), and the hourly digital precipitation array (DPA) product. The first three products are graphical products of rainfall depth in inches that have been quantized into 16 rainfall-data levels. The graphical OHP and STP image products and time loops of these products can be accessed in real time at http://weather.noaa.gov/radar/mosaic/DS.p19r0/ar.us.conus. shtml. An example of an STP product is shown in Figure 5.5.

The fourth precipitation product is the DPA. This is currently the only rainfall product produced by the PPS that is suitable for follow-on quantitative applications such as hydrologic modeling because it has full 256 data levels in units of logarithm of rainfall depth, unlike the quantized 16 data levels in the graphical products above. It is a running 1-h rainfall accumulation product, updated every 5–10 min, on a national polar stereographic grid called the NWS HRAP (Hydrologic Rainfall Analysis Project) grid that is nominally 4 km on a side in mid-latitudes. This DPA product and HRAP grid are described in Fulton (2002) and Reed (1998). Instructions for using the HRAP grid within commercial GIS packages are described at http://www.nws.noaa.gov/oh/hrl/distmodel/hrap.htm.

Given the quantity of rainfall data required for hydrologic modeling, weather radar has provided improved rainfall estimates over large areas, particularly the location and timing. However, there are many sources of radar-rainfall estimation error that must be acknowledged and which are difficult to quantify. Fulton (2002) conducted a broad review of the deficiencies of the Precipitation

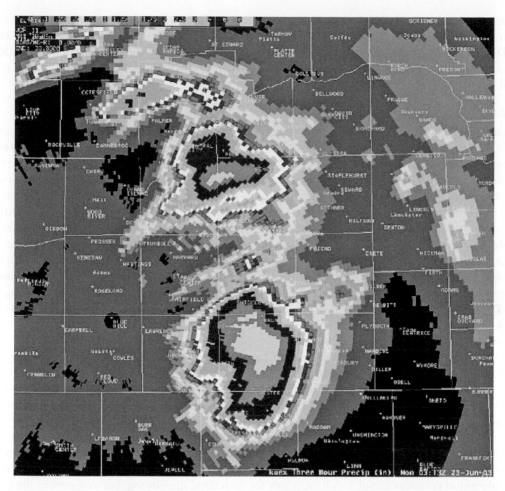

FIGURE 5.5 Radar-rainfall data are being collected nationwide by the WSR-88D system. Image displays example of 3-h rainfall accumulation product. (*Source:* OCFM 2006; http://www.ncdc.noaa.gov/oa/radar/radarresources.html.)

Processing System (PPS) algorithms. These include: (a) parameter optimization, (b) bright band and snow, (c) range degradation, (d) reflectivity calibration and clutter suppression, (e) anomalous propagation, (f) local gauge-radar biases, (g) attenuation, and (h) dual polarization. The Z–R parameters a and b must be selected to convert the radar reflectivity factor into rainfall rate estimates. However, there is no unique Z–R relation for a particular radar, season, geographic region, or storm type. For these reasons, the practical application of radar-rainfall estimates should include validation against available rain gauge observations. Verification research has shown that the radar-rainfall Z–R data uncertainty can be reduced significantly through real-time bias correction based on rain gauges. Fulton (2005) found 63% to 25% reductions are possible with bias correction (generally overestimates). Despite these measures, the uncertainties of radar-rainfall estimates remain a significant source of uncertainty in hydrologic model outputs.

Adding dual-polarization capability to the WSR-88D radars has been demonstrated to improve precipitation discrimination and estimation, extending the functionality of the current NEXRAD system (http://www.nssl.noaa.gov/research/radar/improvements.php). In contrast to the WSR-88D radars, polarimetric radars, also called dual-polarization radars, transmit radio wave pulses that have both horizontal and vertical orientations. The additional information from vertical pulses greatly improves the definition of storm properties. Dual-polarization radars can produce rainfall estimates with considerably less error than the single-polarization observations provided by the current WSR-88D network. Current plans are for the U.S. NWS to upgrade some or all of the WSR-88D radars to dual-polarization capability; this is expected to increase the accuracy of precipitation estimates, particularly in storms with high rainfall rates or when the radar beam encounters ice, hail, or melting snow.

5.2.6.3 Satellite Estimation of Rainfall

Satellite imagery can provide useful information on rainfall distribution over large areas and inaccessible regions. However, direct measurement of rainfall from satellites for operational purposes has not been generally feasible because the presence of clouds prevents direct observation of precipitation with visible, near-infrared, and thermal-infrared sensors. The visible and infrared images from the polar orbiting satellites, including the NOAA-N series and the Defense Meteorological Satellite Program, and geostationary satellites such as GOES, GMS, and Meteosat, provide information only about the cloud tops rather than cloud bases or interiors. These satellites provide frequent observations (even at night with thermal sensors), and the characteristics of potentially precipitating clouds and the rates of changes in cloud area and shape can be observed. From these observations, estimates of rainfall can be made that relate cloud characteristics to instantaneous rainfall rates and cumulative rainfall over time. Improved analysis of rainfall can be achieved by combining satellite and conventional gauge data.

There are a number of references to satellite precipitation estimation methods and products in the hydrological and meteorological literature. These include GOES IR (Vicente et al. 1998), TAMSAT (Grimes et al. 1999; Thorne et al. 2001), TOVS (Susskind et al. 1997), GPCP (Huffman et al. 1997; Huffman et al. 2001), and PERSIANN (Sorooshian et al. 2000). The GOES-IR technique (Vicente et al. 1998) uses the Geostationary Operational Environmental Satellite-8 and -9 in the infrared (IR) 10.7-mm band to compute real-time precipitation amounts based on a power-law regression algorithm. This regression is derived from a statistical analysis between surface-radar-derived instantaneous rainfall estimates and satellite-derived IR cloud-top temperatures collocated in time and space. The auto-estimator was used experimentally for almost 3 years to provide real-time instantaneous rainfall-rate estimates; average hourly estimates; and 3-, 6-, and 24-h accumulations over the conterminous 48 United States and nearby ocean areas. Accuracy assessments determined that the algorithm produces useful 1–6-h estimates for flash flood monitoring, but exaggerates the area of precipitation, causing overestimation of 24-h rainfall total associated with slow-moving, cold-topped mesoscale convective systems. The SAB analyses have also shown a tendency for underestimation of rainfall rates in warm-top stratiform cloud systems.

5.2.6.4 Snow

Snow is another hydrologic variable that has been successfully measured for large regions using aerial and satellite remote sensors. Ground-based snow surveys are also routinely collected at sites by the Natural Resources Conservation Service (NRCS; http://www.nrcs.usda.gov/programs/snowsurvey/). The National Operational Hydrologic Remote Sensing Center (NOHRSC; http://www.nohrsc.noaa.gov/) ingests daily ground-based, airborne, and satellite snow observations from all available electronic sources for the conterminous United States. These data are used along with estimates of snowpack characteristics generated by a physically based snow model to generate the operational, daily NOAA National Snow Analyses (NSA) for the conterminous United States.

The extent of snow cover can be determined with satellite visible and near-infrared (VIS/NIR) data and can be observed in remote regions that are generally inaccessible during the winter months. Although snow cover does not provide the most important information about the snow, that of its depth and water content, these can sometimes be inferred from time-series snow-cover maps known as snow-cover depletion curves, or from simple regression equations (Rango 1993). The microwave region is the area that offers the most promise for retrieving important snow data. Depending upon the wavelength, estimates of the depth, water content, and the amount/presence of liquid water in the snowpack are possible.

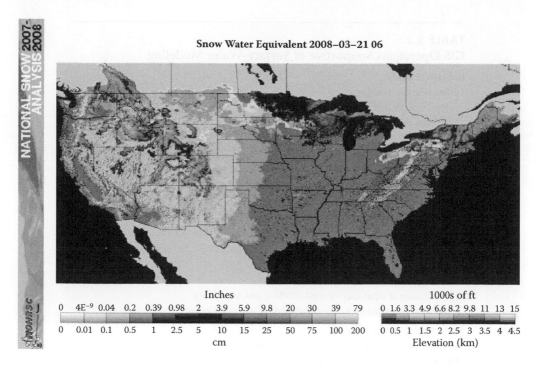

FIGURE 5.6 Snow-water equivalent map generated by the NWS National Operational Hydrologic Remote Sensing Center. (*Source:* http://www.nohrsc.noaa.gov/.)

The NOHRSC snow model is an energy-and-mass-balance, spatially uncoupled, vertically distributed, multilayer snow model run operationally at 1-km² spatial resolution and hourly temporal resolution for the nation. Details of the model are provided by Cline and Carroll (1998) and Bitner et al. (2002). Ground-based and remotely sensed snow observations are assimilated daily into the simulated snow-model state variables. NOHRSC NSA output products are distributed in a variety of interactive map, text discussion, alphanumeric, time-series, and gridded formats (Figure 5.6). NSA product formats include: (a) daily national and regional maps for nine snowpack characteristics; (b) seasonal, two-week, and 24-h movie-loop animations for nine snowpack characteristics; (c) text summaries; (d) a suite of interactive maps, text, and time-series products; and (e) selected gridded snow products for the continental United States. The NSA provides information about snow water equivalent, snow depth, surface and profile snowpack temperatures, snowmelt, surface and blowing snow sublimation, snow–surface energy exchanges, precipitation, and weather forcings in multiple formats.

5.3 GIS FOR SURFACE-WATER HYDROLOGY MODELING

5.3.1 OVERVIEW

GIS analysis and database functions provide extensive means for developing surface-water hydrologic model data sets and modeling operations. Table 5.2 lists some of these GIS operations. A primary area of application is processing of digital terrain data to derive landscape features pertinent to hydrology such as stream paths and drainage divides. GIS databases are created to help organize the multitude of spatial and nonspatial attribute data needed for surface-water hydrologic studies. Intrinsic GIS surface and network analysis functions provide fundamental capabilities for deriving surface-water modeling products. Examples of GIS analysis and database functions are described in this section.

TABLE 5.2
GIS Operations Supportive of Surface-Water Modeling

Data development
  Digital terrain modeling
  Slope and aspect; flow direction
  Pit filling and sinks
  Area and flow accumulation
  Stream paths and drainage networks
  Watersheds and subcatchments
  Contours generation

Data management
  Database of spatial data keyed to location and depth
  Capture of archived site inventory data and conversion to GIS formats
  Collation of watershed attribute data on areas, slopes, LU/LC, and stream flows
  Design of model network and flow-control works
  Automatic formulation of model input data
  Visualization of input data for error checking and consistency
  Statistical interpolations to assign field data to watershed extent

Surface-water-system modeling
  Establishing watershed model boundary conditions
  Systematic assignment of model parameters
  Soil moisture accounting using map algebra on grid cells
  Development of time-area unit hydrographs
  Routing overland flows using finite-difference or finite-element equations
  Routing channel flows through stream network
  Interactive model simulations
  Sensitivity analyses aided by GIS-based parameter changes

Model-output review
  Display of model outputs in color-coded map-oriented formats (with animations)
  Map and graphical comparisons of watershed simulation results with field calibration data
  Model reporting and archive

5.3.2 DIGITAL TERRAIN MODELING

Digital representations of landscape topography as digital elevation models (DEMs) or digital terrain models (DTMs) incorporate arrays of elevation values so that terrain features can be evaluated using specialized numerical algorithms and the GIS visualizations rendered. Landscape features such as slope, aspect, flow length, contributing areas, drainage divides, and channel network can be rapidly and reliably determined from DEMs even for large watersheds (Garbrecht and Martz 1999). A concise review of digital terrain-processing methods was presented in DeBarry (1999); a more expanded treatise is provided by Vieux (2004). DEMs are commonly stored in one of three data structures: grids, contours, or triangulated irregular networks (TINs). These DEM data structures are described in Chapter 3 (GIS Data and Databases).

Automated extraction of surface drainage, channel networks, drainage divides, drainage networks and associated topologic information, and other hydrography data from DEMs has advanced considerably over the past decade and is now routinely a part of most GIS software packages. The automated techniques are faster and provide more precise and reproducible measurements than traditional manual techniques applied to topographic maps. The major issues with DEM-derived hydrographic data are related to the resolution and quality of the DEM and to the derivation of

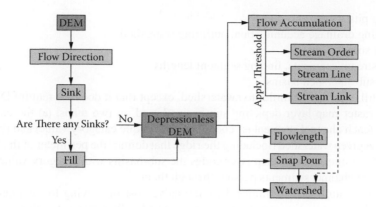

FIGURE 5.7 Flow chart of GIS operations for extracting watershed information from DEMs with the ArcGIS software. (*Source:* http://webhelp.esri.com/arcgisdesktop/9.2/.)

surface drainage. The emphasis here is on the use of raster DEMs because the USGS data are the most available source of digital elevation data.

As an example, the ArcGIS® software package includes the hydrology toolset, which provides many of the watershed delineation functions. The following lists the GRID tools in the hydrology toolset followed by a brief description of each. The hydrology tools can be applied individually or used in sequence to create a stream network or delineate watersheds. Figure 5.7 illustrates the process of operations for extracting watershed information.

- Basin: creates a raster delineating all drainage basins
- Fill: fills sinks in a surface raster to remove small imperfections in the data
- Flow Accumulation: creates a raster of accumulated flow to each cell
- Flow Direction: direction from each cell to its steepest downslope neighbor
- Flow Length: calculates distance, or weighted distance, along a flow path
- Sink: creates a raster identifying all sinks or areas of internal drainage
- Snap Pour Points: snaps pour points to the cell of highest flow accumulation within a specified distance
- Stream Link: assigns unique values to sections of a raster linear network between intersections
- Stream Order: assigns a numeric order to segments of a raster representing branches of a linear network
- Stream to Feature: converts a raster representing a linear network to features representing the linear network
- Watershed: determines the contributing area above a set of cells in a raster

These functions can be accessed through the ArcGIS Spatial Analyst® interface, or called directly using various programming options. The Arc Macro Language (AML) is another powerful tool within ARC/INFO which allows a user to build customized applications, including graphical user interfaces (GUIs). Also, in recent versions of the software, the GRID tools have been made accessible, through ActiveX controls, to languages such as Visual Basic on the Windows NT operating system.

There are open-source options that provide similar watershed processing functions. The GRASS GIS program (http://grass.itc.it/index.php) has a number of terrain-processing routines that can be integrated into computer programs:

- r.watershed, a popular series of programs for delineating and segmenting of channel networks and for identifying a watershed boundary
- Filtering of elevation data

- Locating pits
- Calculating drainage accumulation, outlining watershed
- Creating stream network
- Coding stream segments, finding segment lengths
- Finding subwatershed basins
- r.basins.fill, a program similar to r.watershed, except that it does not require DEM; it generates a raster map layer depicting sub-basins based on two maps: (a) the coded stream network (each channel segment is "coded" with a unique category value) and (b) the ridges within the given watershed (including the ridge that defines the perimeter of the watershed); the resulting output raster map layer codes the sub-basins with category values matching those of the channel segments passing through them
- r.cost, a program that determines the cumulative cost of moving to each cell on a cost surface (the input raster map) from other specified cell(s) whose locations are specified by their coordinates; r.drain, a program that traces a flow through a least-cost path in an elevation model; results for cells are similar to those obtained when using the seg version of the r.watershed program

A variety of preprocessed products have been developed for wide distribution. The Elevation Derivatives for National Applications (EDNA), a multilayered Web-based database (http://edna. usgs.gov), is a derivative of the NED that has been combined with the National Hydrography Dataset (NHD) to provide seamless 30-m resolution raster and vector data (Figure 5.8). EDNA was developed to improve drainage basin boundaries and provide modelers with consistent DEM-derived layers for use in basin characterization. EDNA layers currently available include filled DEMs, aspect, contours, flow accumulation, flow direction, reach catchments, shaded relief, slope, and synthetic

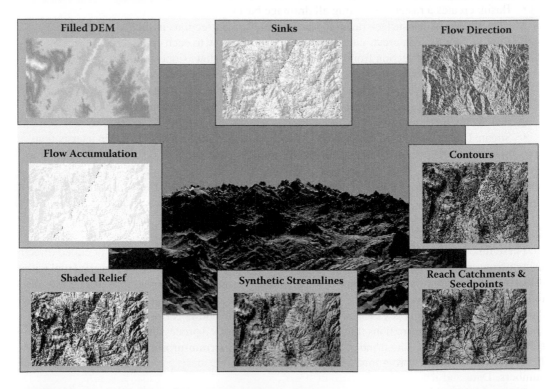

FIGURE 5.8 (See color insert following page 136.)   EDNA is a multilayered seamless database derived from a version of NED that has been conditioned for improved hydrologic flow representation. (*Source:* http://edna.usgs.gov.)

streamlines. Using EDNA data sets, upstream and downstream drainage areas can be derived from any location for purposes of flood analysis, pollution studies, and hydropower generation projects.

Procedures for watershed DEM processing were summarized by DeBarry (1999). The D8 method is the most widely used raster DEM processing method. The method identifies the steepest downslope flow path between each cell of a raster DEM and its eight neighbors (hence the name D8 method), and defines this path as the only flow path leaving the raster cell. The method also accumulates catchment area downslope along the flow paths connecting adjacent cells. The drainage network is identified by selecting a threshold catchment area at the bottom of which a source channel originates, and then classifying all cells with a greater catchment area as part of the drainage network. This drainage network identification approach is simple and directly generates connected networks. A widely used version is that of Jenson and Domingue (1988). Shortcomings of the early watershed DEM processing routines included difficulties of drainage definition in the presence of depressions, flat areas, and flow blockages. These features are often the result of data noise, interpolation errors, and systematic production errors in DEM elevation values. The topographic parameterization model TOPAZ (Garbrecht and Martz 1999) addressed these problems by introducing a trend surface from higher to lower terrain. This approach results in a convergent flow direction pattern over the flat surface that is also consistent with the topography surrounding the flat surface. The TOPAZ is also able to identify inland catchments that do not drain to the ocean.

An issue that is often of concern with drainage networks extracted from DEMs is the precise positioning of channel links in the digital landscape. Comparison with actual maps or aerial photos often shows discrepancies, particularly in low-relief landscapes. A primary reason for this discrepancy is the approximate nature of digital landscapes that cannot capture important topographic information that is below the DEM resolution. Though the channel position in the digital landscape is consistent with the digital topography, it may not reflect the actual drainage path in the field. From a practical point of view, this dilemma can be overcome by "burning in" the path of the channels along predigitized pathways. This is achieved by artificially lowering the elevation of the DEM cells along digitized lines or raising the entire DEM except along stream paths (Maidment 2002a). However, caution is advised with this method because it may produce flow paths that are not consistent with the digital topography. There is also an issue with "artifacts" in the DEM that originate from their development; older DEMs have this problem. Older DEMs produced by methods that are now obsolete have been filtered during the NED assembly process to minimize artifacts that are commonly found in data produced by these methods. Artifact removal greatly improves the quality of the slope, shaded-relief, and synthetic drainage information that can be derived from the elevation data. Figure 5.9 illustrates the results of this artifact removal filtering. NED processing also includes steps to adjust values where adjacent DEMs do not match well and to fill areas of missing data between DEMs. These processing steps ensure that the NED has no void areas and that artificial discontinuities have been minimized.

Implementation of watershed DEM processing for the EDNA products involved a staged implementation of DEM processing and error checking (Franken et al. 2002; Kost and Kelly 2002). EDNA data development occurred in three stages, the first of which uses semi-automated techniques to create preliminary derivative data from the unconditioned NED elevation data. The second stage uses the stage 1 data, within a set of ArcView tools, to create preliminary watersheds and subwatersheds and to identify and flag discrepancies between the derivative data and existing data sets that portray true hydrologic features. The stage 2 tools enable the user to select an outlet catchment and automatically aggregate the upstream catchments to form watersheds or subwatersheds. This automatic aggregation is implemented by use of a Pfafstetter coding scheme (Verdin and Verdin 1999; Pfafstetter 1989). Figure 5.10 shows an example of watersheds with their names and 10-digit numbers. Data and information created in stage 2 are used in stage 3 to develop a hydrologically conditioned NED data set. These data are then used to create a more hydrologically correct derivative database. This final step in stage 3 processing results in vertical integration of the final EDNA data with the NHD and other data sets.

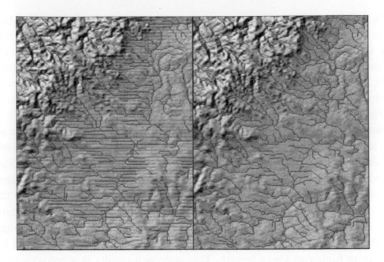

FIGURE 5.9   Shaded-relief representation of the Rocky Point, Wyo., 7.5-min DEM is shown on the left. The same area is shown on the right after NED artifact filtering. The superimposed lines are synthetic drainage networks derived from each elevation data set. (*Source:* http://ned.usgs.gov/.)

FIGURE 5.10   EDNA watersheds with Pfafstetter index numbering. (From Kost and Kelly 2002.)

5.3.3   Arc Hydro Data Model and Tools

Arc Hydro is an ArcGIS-based system developed by the Center for Research in Water Resources (CRWR) of the University of Texas at Austin and the Environmental Systems Research Institute, Inc. (Maidment 2002b). The Arc Hydro data model provides a template for the creation and manipulation of a wide variety of hydrologic- and water-resources-related objects and their associated attributes, including the appropriate rules governing their topological interaction. The basic Arc Hydro data model is described in Chapter 3 (GIS Data and Databases). Of interest here are the Arc Hydro tools.

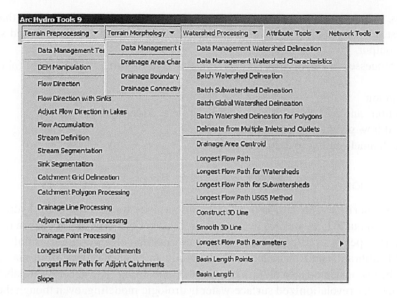

FIGURE 5.11   Arc Hydro tools provide functions for watershed processing. (From Maidment 2002a. With permission.)

The Arc Hydro tools are utilities based on the Arc Hydro data model for performing many of the aforementioned tasks of deriving hydrography data from DEMs, extracting drainage networks, delineating watersheds, and deriving geometric networks of drainage features for flow tracing. The tools provide raster, vector, and time-series functionality, and many of them populate the attributes of Arc Hydro features. Arc Hydro provides a basic functionality that can be expanded by adding database structures and tools for a wide variety of applications in hydrology and water resources. Figure 5.11 displays the toolbar for Arc Hydro tools as an extension registered in ArcMap®, showing the extensive functionality available in Arc Hydro. Additional description of the Arc Hydro data model is provided in Chapter 3.

5.3.4 SURFACE-WATER HYDROLOGIC MODEL MODULES

Hydrologic models can be generally categorized in terms of how they deal with time, randomness, and the level of spatial detail (HEC 2000). Distinctions include: (a) event or continuous, (b) lumped or distributed, (c) empirical (system theoretic) or conceptual, (d) deterministic or stochastic, and (e) measured or fitted parameter. An event model simulates a single storm. The duration of the storm may range from a few hours to a few days. A continuous model simulates a longer period, predicting watershed response both during and between precipitation events. A distributed model is one in which the spatial (geographic) variations of characteristics and processes are considered explicitly, while in a lumped model these spatial variations are averaged or ignored. A collection of watersheds where each is represented as "lumped" may be considered as "distributed" overall. An empirical model is built upon observations of input and output, without seeking to represent explicitly the process of conversion (e.g., unit hydrograph). Conceptual models may be physically based (e.g., Green-Ampt infiltration dynamics) or conceptual only (e.g., soil-moisture partitioning represented by linear reservoir dynamics). There is a trend toward physics-based models, although the abstract component storage models perform well when calibrated. A stochastic model describes the random variation and incorporates the description in the predictions of output; a deterministic model assumes that all input, parameters, and processes are free of random variation and known with certainty. Stochastic models are increasingly being used to characterize the uncertainty bounds of model outputs (i.e., risk assessment). A measured-parameter model involves the direct measurement of parameters. A fitted-parameter model, on the other hand, includes parameters that cannot be measured. Instead, the parameters must be found by fitting the model with observed values of the input and the output.

The basic processes of a hydrologic model, illustrated in Figure 5.1, include the following:

- Precipitation
- Abstractions and infiltration losses, soil-moisture accounting
- Overland flow generation and routing
- Stream channel routing hydraulics

5.3.5 PRECIPITATION

Precipitation, primarily as rainfall, is of primary importance for surface-water hydrology, as it is the main driving input to land-surface hydrologic processes. Certainly, confidence in the hydrologic modeling effort depends to a large extent on the availability of high-quality rainfall and runoff data for model calibration and verification. Traditionally, rainfall estimated from sparse rain-gauge networks has been considered a weak link in watershed modeling. The general availability of radar-rainfall products has revolutionized surface-water hydrologic modeling, even though these products are not perfectly accurate. In general, precipitation varies on the scale of kilometers, while rain gauges in the United States are spaced 10s to 100s of kilometers apart; because of this, the likeli-hood of obtaining rain-gauge data within a particular watershed is small. For this reason, it is usu-ally required that spatial interpolation procedures be applied to develop rainfall estimates in the catchment of interest, as discussed in the following section.

5.3.5.1 Rain-Gauge Data Spatial Interpolation

The common problem of obtaining rainfall data for the watershed of interest using point rain gauges is addressed using spatial interpolation procedures. Interpolation methods are appropriate when an attribute measured at sample points is a spatially continuous field variable. Usually, the interpola-tion process involves estimating the rainfall values onto a regular grid. Alternatively, contours may be fitted to the grid, and the data represented as vector objects with labels or as polygon objects having the contours as boundaries. The interpolated surface may also be represented as a TIN. There are a wide variety of procedures for interpolation and supporting literature (e.g., Issacs and Srivastava 1989; HEC 2000). For this discussion, five methods are described: (a) nearest neighbor, (b) isohyetal, (c) triangulation, (d) distance weighting, and (e) kriging. A concern with interpolation using sparse network data is that different interpolation methods may yield differing results.

> *Nearest neighbor*: The nearest-neighbor method simply assigns a value to the grid equal-ing the value of the nearest data point. For a scattered spacing of gauges, the method results in Thiessen polygons, also known as Voronoi polygons of influence. As illustrated in Figure 5.12a, the gauge nearest each point in the watershed may be found graphically by connecting the gauges and constructing perpendicular bisecting lines; these form the boundaries of polygons surrounding each gauge. The area within each polygon is nearest the enclosed gauge, so the weight assigned to the gauge is the fraction of the total area that the polygon represents. The nearest-neighbor method produces stair-step surfaces with sharp discontinuities at the edges. The procedure is also applied in the conversion of vector contours to raster images, where grid values are assigned the value of the nearest contour pixel.
>
> *Isohyetal*: This too is an area-based weighting scheme. Contour lines of equal precipitation are estimated from the point measurements, as illustrated by Figure 5.12b. This allows a user to exercise judgment and knowledge of a basin while constructing the contour map. MAP is estimated by finding the average precipitation depth between each pair of contours

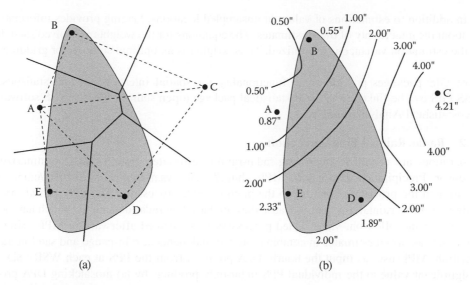

FIGURE 5.12 Illustration of mean areal-precipitation-depth computation schemes: (a) Thiessen polygon method, (b) isohyetal method. (From HEC 2000.)

(rather than precipitation at individual gauges), and then weighting these depths by the fraction of total area enclosed by the pair of contours.

Triangulation: The triangulation method involves the joining of adjacent data points by a line to form a lattice of triangles. A TIN is formed with each triangle treated as a planar surface, with the surface maintaining the original data points. The equation of each triangular surface can be exactly determined from the point values, and values at any intermediate point on the surface can be computed through trigonometry. Delaunay triangulation produces triangles that are as close to equilateral as possible, which permits inclusion of break lines into the surface where discontinuities occur (e.g., at drainage divides).

Distance weighting: Distance weighting is a moving-average procedure using points within a specified zone of influence. One of the most commonly used techniques is one that computes the weights as inversely proportional to the square of the distance between the point of interest and each of the data points. Powers other than 2 may be used to change the rate of decay of the weighting function. Inverse distance weighting produces a surface that usually does not pass through the original data points, and selection of the distance exponent is somewhat arbitrary. They do have the advantages of being easy to use and produce smooth surfaces.

Kriging: Kriging is a more analytical approach to the point-interpolation problem. Although similar to the inverse distance-weighting method in that it is also based on a weighted sum of the points within a zone of influence, the weights in kriging are determined from a set of n simultaneous linear equations, where n is the number of points used for the estimation. The procedure, described by Issacs and Srivastava (1989), enforces the phenomenon that points closer together tend to be strongly correlated, whereas those far apart lack correlation. The spatial correlation, expressed as the covariance between pairs of sample points and the sample points being estimated, is first computed and the weights then determined. Over the field of interest, the variogram is formed as an inverse plot of the autocorrelogram, the covariance versus distance. Variograms provide a measure of the spatial continuity of the interpolated surface. Generalizations of the variogram to speed up computations may be modeled as linear, spherical, exponential, and Gaussian, and may exhibit a "nugget" effect as well as "range" and "sill" values. The nugget effect represents some residual variance that may exist at zero distance. The range is a distance beyond which the covariance is not influenced by distance, and the covariance reaches a background value called the sill.

In addition to estimations of values at unsampled locations, kriging provides information about the uncertainty of those estimates. The equations for the weights are solved such that the estimation variance is minimized. Thus, kriging is an optimal solution for gridding.

Most GIS packages provide some functionality for spatial interpolation and statistics. The GRASS GIS can be linked with the R statistical package (open source). The ArcGIS software has the Geostatistical Analyst® module.

5.3.5.2   Radar-Rainfall Bias Correction

GIS operations are central to improving radar-rainfall estimates. Fulton (2002) summarized the Multisensor Precipitation Estimator (MPE; http://www.nws.noaa.gov/ohd/hrl/hag/empe_mpn/) algorithm and the procedures built upon the previous PPS processing. The PPS rainfall products are only first-guess rainfall products because they are based on radar data alone, which have known biases. Important follow-on, value-added processing is performed afterward in MPE using these PPS products as initial estimates in combination with independent rain-gauge and satellite rainfall information. MPE uses as input the hourly DPA products from the PPS at each WSR-88D, but it adds significant value to the individual PPS radar-only products by (a) mosaicking DPA products from adjacent regional radars to create larger regional gridded hourly rainfall maps; (b) quantitatively incorporating real-time rain-gauge data to calibrate the radar-rainfall estimates and reduce biases; (c) incorporating satellite-based rainfall estimates to add additional rainfall information to the radar and rain-gauge rainfall estimates; and (d) permitting manual forecaster interaction, editing, and quality control of the input data and products.

In overlapping regions of adjacent radars, the lowest-to-ground, unblocked radar estimate is used to create the mosaic so that rainfall is estimated as close to the Earth's surface as possible. Rain-gauge data are then incorporated in a mean, radar-wide sense to remove possible spatially uniform biases between radar and rain gauges each hour (Seo et al. 1999). This is necessary since WSR-88D rain-rate estimates can sometimes have uniform hourly biases (e.g., overestimation by 30%) due to radar reflectivity miscalibration problems. The calibration problems are uniform across the radar and are therefore very amenable to removal through mean field bias correction. The next step is to merge the radar and rain-gauge data together using geostatistical optimal estimation procedures to make local rainfall corrections based on the rain-gauge data (Seo 1998). Figure 5.13 illustrates the results of a rain-gauge bias correction. Independent satellite-based rainfall estimates produced by NOAA's National Environmental Satellite Data Information Service (NESDIS) are then substituted in regions where radar beam blockages by terrain prevent reasonable estimation by radar alone. Additional processing capability allows computation of a local-bias-adjusted multisensor field using the rain-gauge data together with the radar data (Seo and Breidenbach 2002).

5.3.6   Abstractions, Infiltration, and Soil Moisture

Soil-moisture accounting is a fundamental component of hydrologic models. Effective precipitation is the amount remaining for overland flow after abstractions for ponding, infiltration, and evaporation. Since the soil–water balance is a physical process that takes place mainly vertically, abstractions can be analyzed at a specific location without involving the surrounding area. In terms of GIS, this characteristic implies that evaluating the abstractions can be accomplished for grid cells that do not interact with each other, so no linkage with adjacent cells is necessary. With each of the soil-infiltration models, precipitation loss is found for each computation time interval, and this is subtracted from the precipitation depth for that interval. The remaining depth is the precipitation excess; this depth represents the volume available as runoff, which may be routed in various ways.

The most common soil-moisture modeling approaches involve computation of soil infiltration and soil-profile moisture dynamics. Infiltration models have been successfully applied that involve (a) empirical equations based on field observations and (b) equations based on the mechanics of

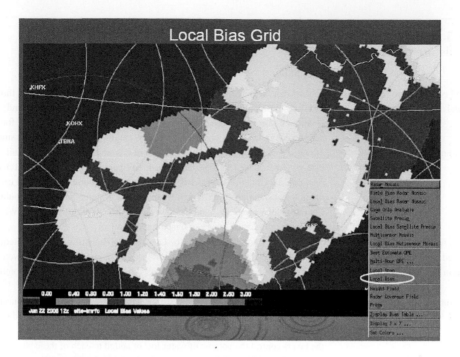

FIGURE 5.13   Example of radar-rainfall bias correction using rain-gauge data. (From Fulton 2005.)

unsaturated and saturated flow in porous media. Table 5.3 provides summaries for the commonly used models, most of which are applied for rainfall events. Details on the equations and procedures for estimating parameters for these models are provided by the HEC (2000).

Continuous simulations require that the soil-moisture dynamics be modeled for periods between rainfall events. Thus, the recovery of soil moisture due to evapotranspiration (ET) needs to be addressed. ET is the loss of water from the canopy interception, surface depression, and soil-profile storages. In the SMA model, potential ET demand currently is computed from monthly pan-evaporation depths, multiplied by monthly varying pan-correction coefficients and scaled to the time interval. The potential ET volume is satisfied first from canopy interception, then from surface interception, and finally from the soil profile. Within the soil profile, potential ET is first fulfilled from the upper zone, then the tension zone. If potential ET is not completely satisfied from one storage in a time interval, the unsatisfied potential ET volume is filled from the next available storage.

A soil–water-balance modeling system (SWBMS) was developed by Reed (1998) to evaluate alternative methods for using digital soil maps in rainfall-runoff modeling. The two soil databases of interest are the State Soil Geographic (STATSGO), at 1:250,000 scale, and the Soil Survey Geographic (SSURGO), at 1:24,000 scale. Simple conceptual models (e.g., Green-Ampt) were used to describe the infiltration, percolation, and evaporation processes. Representative soil hydraulic properties were assigned based on USDA texture class. Distributed soil properties were used in conjunction with distributed NEXRAD Stage III rainfall data to make rainfall excess calculations. Calibration and validation of the SWBMS model yielded interesting insights. In the two-layer conceptual soil model, the top layer controls direct runoff, but the bottom layer serves as an important control on how much water evaporates and how much water percolates to the groundwater reservoir. When the spatial variability of surface texture was reduced through resampling of the soil properties, model performance decreased. Although the 1:250,000-scale soil map contains considerably less spatial detail than the 1:24,000-scale map, the use of each data set as input in model validation runs produced similar results.

TABLE 5.3
Soil-Moisture and Infiltration Models

Model	Description
Initial and constant-rate loss model Deficit and constant-rate model	Assumes that the maximum potential rate of precipitation loss is constant throughout an event. An initial loss is added to the model to represent interception and depression storage. Simple and widely used, but no direct physical basis for parameters. Deficit and constant-rate models can be used for continuous simulation.
SCS curve number (CN) loss model	Estimates precipitation excess as a function of cumulative precipitation, soil cover, land use, and antecedent moisture. A basic soil parameter is the potential maximum retention. Precipitation excess will be zero until the accumulated rainfall exceeds the initial abstraction. The CN represents the runoff potential; there are published CN tables for various soils and land uses (see HEC 2000, Appendix A). Widely used, but not based on theory of unsaturated flow.
Green and Ampt (G-A) loss model (simplified form is Horton's infiltration model)	Based on an unsaturated flow form of Darcy's law (Richard's equation). Requires estimates of soil saturated hydraulic conductivity, moisture deficit, and wetting front suction; these parameters can be based on soil characteristics. Becoming more common, but less experience with use.
Soil-moisture accounting (SMA) model	Continuous simulation as a series of storage layers for canopy interception, surface depression, soil profile, and groundwater. The soil-profile zone is divided into an upper zone and a tension zone. Rates of inflow to, outflow from, and capacities of the layers control the volume of water lost or added to each of these storage components. SMA model parameters must be determined by calibration with observed data.

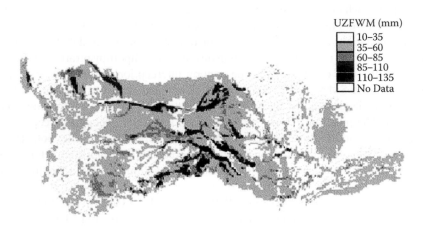

UZFWM (mm)
- 10–35
- 35–60
- 60–85
- 85–110
- 110–135
- No Data

FIGURE 5.14 Upper-zone free-water maximum capacity for the Arkansas River basin. (From Reed et al. 2002.)

Koren et al. (2000) developed a set of equations that can be used to derive 11 of the SAC-SMA (Sacramento Soil-Moisture Accounting Model) parameters from the Soil Conservation Service (SCS) curve number (McCuen 1982), properties that can be inferred from soil texture (e.g., porosity, field capacity, wilting point, and saturated hydraulic conductivity) and soil depth. These equations were developed based on both physical reasoning and empirical relationships. Using 1-km soil property grids derived from STATSGO data (Miller and White 1999; Koren et al. 2000) produced a priori SAC-SMA parameter grids covering the conterminous United States. The a priori values of the UZFWM (upper-zone free water maximum) parameter for the Arkansas River basin are shown in Figure 5.14.

Comparisons of soil-moisture accounting methods with measured runoff responses indicate that a wide variability of results can be obtained, depending on the model and parameter selections. A study by Schaake et al. (2004) compared four land-surface models (LSMs) of the North American Land Data Assimilation System (NLDAS) with each other and with available observations. The NLDAS project is designed to provide enhanced soil and temperature initial conditions for numerical weather/climate prediction models. The models include (a) Noah, (b) Mosaic, (c) VIC, and (d) Sacramento (SAC). All of the models attempt to represent the layered structure of the soil column and the root zone. The Noah and Mosaic models use a numeric solution of the Richard's equation to simulate the soil-moisture dynamics; the VIC and SAC represent soil water storage as conceptual water storage components (e.g., tension and free water). For the Schaake et al. (2004) study, soil-moisture fields were compared with soil-moisture observations in Illinois. There, total water storage from each of the models was highly correlated with the observations. Simulated values from both the SAC and Noah models agreed well with the measured values. In contrast, the absolute values of water storage from Mosaic and VIC did not agree well with the observed values. Values of the average 2-year range of total water storage for 27 river basins in the Arkansas–Red River basin were compared with the 2-year range of total water storage simulated by each of the four LSMs. The SAC model reproduced the best storage ranges for the 27 basins, which are located in a wide range of climate regimes.

5.3.7 EVAPORATION AND EVAPOTRANSPIRATION

Evaporation is the process of water transfer from land and water surfaces back to the atmosphere. Evapotranspiration (ET) involves transpiration through plant leaves as well as evaporation. Combined accounting for evaporation and transpiration is called consumptive use. Continuous hydrological models must account for evaporation, ET, and soil moisture during periods of no precipitation. Procedures for estimating evaporation and ET incorporate considerations of energy supply (solar radiation), vapor transport (wind velocity and humidity gradient), and moisture supply. Potential ET is that which would occur from a vegetative surface when moisture supply is not limited. Actual ET is less than potential ET when the soil dries out. Methods for ET estimation are described in various texts on hydrology (e.g., Chow et al. 1988; Viessman et al. 1989). Common methods for estimating evaporation include water and energy budgets, mass-transfer equations, and pan evaporation (direct measurement). Methods for determining ET are based on various physics-based theories, analyses of water and energy budget, or empirical approaches. Popular methods include the Thornwaite (for potential ET), Blaney-Criddle, Priestly-Taylor, and Penman-Montieth methods, in order of increasing complexity and data requirements (Jensen et al. 1990).

GIS procedures for computing evaporation and ET include interpolation of the input meteorological data from weather stations to the location of interest. These interpolations may be accomplished in a manner similar to those used for rain-gauge data. Computations may be accomplished using map-algebra techniques for regular grid structures across the landscape, taking account of differences in temperature, precipitation, elevation, soils, vegetative cover, and other variables as required. Chapter 7 (GIS for Water Supply and Irrigation Systems) presents GIS procedures for estimating ET from irrigated lands.

The hydrology component of the Groundwater Loading Effects of Agricultural Management Systems (GLEAMS) model is illustrative of procedures to estimate initial soil water content for hydrologic model simulations (Knisel et al. 1993; Knisel and Williams 1995). The GLEAMS ET routine was used by Skahill and Johnson (2000) as part of a 2-D hydrologic model application. The GLEAMS model operates on a daily time step and computes potential evapotranspiration using either the Priestly-Taylor or the Penman-Monteith method. Raster GIS data layers describing elevation, hydrologic soil group, land use and land cover, percent clay, percent silt, and soil texture were used together with published guidelines (Chow et al. 1988; Knisel et al. 1993) to estimate parameter values. The raster GIS data layers describing percent clay, percent silt, and soil texture were used to determine the number of computational soil layers. Monthly averages for dew-point temperature, solar radiation, and wind movement are based on values provided in the GLEAMS model supplemental database. Water content can be interpolated to model grid cells using the GRASS GIS command s.surf.tps, which utilizes spline with tension.

5.3.8   RUNOFF MODELS

Runoff is generated from excess precipitation that has not infiltrated or been stored on the land surface. This direct runoff is translated from its location on the ground to the nearest stream channel (overland flow) in the watershed outlet. Various empirical and conceptual methods are used to accomplish this runoff operation (HEC 2000):

> *Empirical models* (also referred to as system theoretical models): These are the traditional unit hydrograph (UH) models. The system theoretical models attempt to establish a causal linkage between runoff and excess precipitation without detailed consideration of the internal processes.
>
> *Conceptual models*: A common conceptual model is the kinematic-wave model of overland flow. It represents the physical mechanisms that govern the movement of the excess precipitation over the watershed land surface and in small collector channels in the watershed.

5.3.8.1   Unit Hydrograph Methods

The unit hydrograph (UH) is a well-known, commonly used empirical model of the relationship of direct runoff to excess precipitation (HEC 2000). As originally proposed by Sherman (1932), it is "the basin outflow resulting from one unit of direct runoff generated uniformly over the drainage area at a uniform rainfall rate during a specified period of rainfall duration." The UH can be viewed as the characteristic "signature" of runoff from a watershed that implicitly represents the influence of watershed area, shape, slope, time of travel, and other watershed characteristics. The underlying concept of the UH is that the runoff process is linear, so the runoff from greater or less than one unit is simply a multiple of the unit runoff hydrograph. The UH approach is sometimes termed a "lumped" approach because the method's inputs are averaged across the watershed or represented by just a few parameters, and the prediction is for the outlet point only.

Various versions of the UH have been developed, including: user-specified UH, Clark's UH, Snyder's UH, the SCS UH, and ModClark UH. These are options incorporated into the HEC-HMS software package (HEC 2000). The user-defined UH can be derived from gauged rainfall and flow records (Chow et al. 1988), although records to support this procedure are seldom available at the location of interest. The other methods are referred to as synthetic UH methods and are based on determination of watershed geometric and flow time characteristics. For example, the Snyder UH is based on determination of the basin lag time, which is related to a basin coefficient (determined empirically), the length of the main stream from outlet to the divide, and the length from the outlet point nearest the basin centroid. The SCS UH method estimates lag time using Manning's equation to determine flow velocity for sheet and channel flow paths from the divide to the outlet. Physical parameters for these computations include channel cross-sections, slopes, lengths, and roughness.

The Clark UH model accounts explicitly for variations in travel time to the watershed outlet from all regions of a watershed; this results in the so-called time-area map (Figure 5.15). The time-area map is used to formulate the time-area histogram, which is then used to define the UH. The influence of basin storage may be incorporated by a linear reservoir routing operation (Clark 1945).

The time-area map can be used directly to translate nonuniform excess runoff from the various grid locations in the watershed into the outlet runoff hydrograph; this is referred to as the modified Clark (ModClark) method. Maidment (1993) presented one of the first models of this type based on the time-area map. Kull and Feldman (1998) assumed that travel time for each cell in the watershed

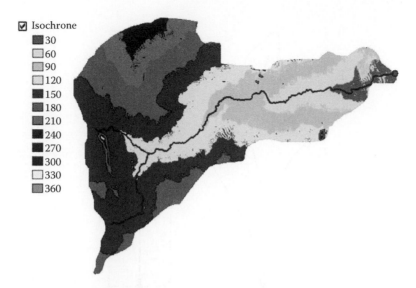

Isochrone
- 30
- 60
- 90
- 120
- 150
- 180
- 210
- 240
- 270
- 300
- 330
- 360

FIGURE 5.15 ModClark time-area method was used to generate 30-min isochrones for Spring Creek watershed, Fort Collins, CO.

was simply proportional to the time of concentration scaled by the travel length of the cell over the maximum travel length. A kinematic-wave travel-time scheme was employed by Saghafian et al. (2002). The ModClark approach has been implemented in the HEC-HMS hydrologic model package (HEC 2003). GeoHMS uses ArcView and Spatial Analyst® for the development of grid-based data for linear quasi-distributed runoff transformation.

5.3.8.2 Flow Routing

Routing of flows through the channel network is accomplished using a variety of mathematical representations of the fundamental equations of channel flow hydraulics. For example, the HEC-HMS model package supports the lag, Muskingum, modified Puls, Muskingum Cunge, and kinematic-wave models (HEC 2000). More-complex model formulations are applied where unsteady flow conditions occur.

For the HEC-HMS model (HEC 2000), a watershed can be represented conceptually as a wide, open channel, with inflow to the channel equal to the excess precipitation (Figure 5.16). It then solves the equations that simulate unsteady shallow water flow for the overland flow planes and the open channel to obtain the watershed-runoff hydrograph. The governing partial differential equation is approximated with a finite-difference scheme. To estimate runoff with the kinematic-wave model, the watershed is described as a set of elements that include: (a) overland flow planes, (b) sub-collector channels, (c) collector channels, and (d) the main channel. Data required for overland flow planes include the typical length, slope, and area; these can be derived from geometric operations of a GIS. Data on roughness and losses can be related to vegetation and soil types. Channels require information on cross-sectional shape as well as roughness and slope. Instability of the kinematic-wave computation with respect to time and distance steps has been a concern with this method. The HEC representation of this method provides for automatic stability checks of the time and distance intervals until the computation remains stable.

Regardless of the channel-routing method, GIS representations of the network as a binary-tree collection of reaches and junctions are central to coordinating the logical sequencing of computations from an upstream to downstream direction. The network database includes the topological relations to identify upstream and downstream nodes as well as junction characteristics. Sometimes the indexing code generated by the stream network derivation (e.g., Pfafstetter code) is used to schedule the sequencing of computations. For example, the computational sequence for performing channel-

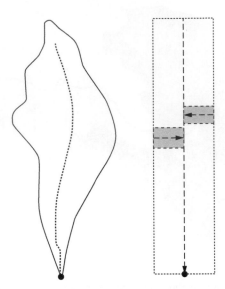

FIGURE 5.16 HEC-HMS simple watershed model with kinematic-wave representation. (From HEC 2000.)

routing computations can be based on a tabular file provided by the TOPAZ model of Garbrecht and Martz (1999), which is based on stream order. The file is modified to include additional physical information describing channel geometry and Manning's *n* values for each individual channel link. A boundary condition for the channel routing is the upstream inflow hydrograph, which, after routing through the channel reach, becomes the downstream outflow. Outflow hydrographs at junctions are additively combined, taking account of any time offset. The GIS relational database contains the alphanumeric data on channel shape, roughness, slope, and other relevant data. Manning's *n* values for channels can be estimated on the basis of field observations and published guidelines (e.g., Chow et al. 1988).

Rainfall-runoff modeling generates flow hydrographs at selected locations in a drainage network; this is the direct, quick runoff of precipitation. There is another component of streamflow that is sustained during "fair weather" periods of no precipitation; this is called the *base flow*. Base flow is drainage of runoff from prior precipitation that has been stored temporarily in the watershed, plus the delayed subsurface runoff from the current storm. Groundwater drainage can be a major source of base flow. Base flow is sometimes represented as a monthly constant value or as an exponentially decreasing value with time. Some conceptual models account explicitly for this storage and for the subsurface movement within the watershed. The base flow is added to the direct runoff to obtain the total channel flow for routing purposes.

5.3.8.3 Distributed Runoff Modeling

The increasing availability of high-resolution DEMs and land-surface data provides a foundation for distributed models of the watershed. The ModClark time-area method described above is an example of a semidistributed model. A large basin can be represented as a collection of sub-basins, and the overall system representation can be considered to be distributed. For fully distributed models, the excess precipitation amounts computed for various locations in the watershed, usually on a grid structure, are routed overland and through channels to the basin outlet. The distributed approach allows flow predictions at many points internal to the watershed.

The NWS Research Modeling System (HL-RMS) is an example of a fully distributed approach (Smith et al. 2004). The HL-RMS has been developed to support proof-of-concept research comparing distributed modeling approaches with the more traditional lumped models based on the UH. Some of the main features of the current HL-RMS are

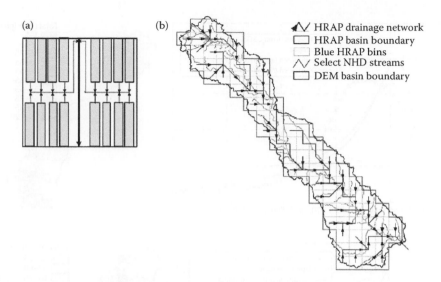

FIGURE 5.17 HL-RMS distributed model representations: (a) conceptual hillslopes within a 4-km HRAP grid, (b) cell-to-cell drainage network. (From Smith et al. 2004.)

- Ingests gridded NEXRAD-based products.
- Basic modeling unit is the NEXRAD grid cell (≈4 km).
- Rainfall-runoff calculations are done independently for each grid cell.
- Runoff is routed over hillslopes within a model cell.
- Channel routing is done from cell to cell.
- Rainfall-runoff calculations can be done using lumped or distributed rainfall and lumped or distributed parameters.
- Uses the Sacramento Soil Moisture Accounting Model (SAC-SMA) (Burnash et al. 1973).
- Uses the kinematic method for both hillslope and channel routing.
- Writes output parameter, state, or forcing grids that can be displayed in ArcView GIS.

In HL-RMS, the impervious, surface, and direct-runoff components are routed over conceptual hillslopes within each NEXRAD cell to a conceptual channel. Because of the relatively large size of the 4-km model cells, the cells are subdivided into conceptual hillslopes to make overland flow distances physically realistic. A drainage-density parameter in the model is used to subdivide a cell into equally sized overland flow planes (Figure 5.17a). These hillslopes drain to a conceptual channel segment within the same cell. Cell-to-cell channel routing is done using flow-direction networks like that illustrated in Figure 5.17b. Three parameters are defined in each cell for kinematic overland flow routing: hillslope slope, hillslope roughness, and drainage density. Representative hillslope slopes are estimated using DEM data (initially with 30-m DEM data for basin-scale applications and 400-m DEM data for regional-scale applications) by first computing the local slope of each DEM cell in the study domain using the ArcInfo® slope function, and then averaging all of the DEM cell slopes in each 4-km model cell. A kinematic routing scheme is applied for the channels; parameters are based on stage discharge, channel cross section, and other geomorphic data. The HL-RMS was applied to the Blue River basin in Oklahoma. Figure 5.18 illustrates simulated results.

The HL-RMS was applied as part of a Distributed Model Intercomparison Project (Smith et al. 2004), where a number of distributed models were applied and compared. Results of Phase I of that project include the following:

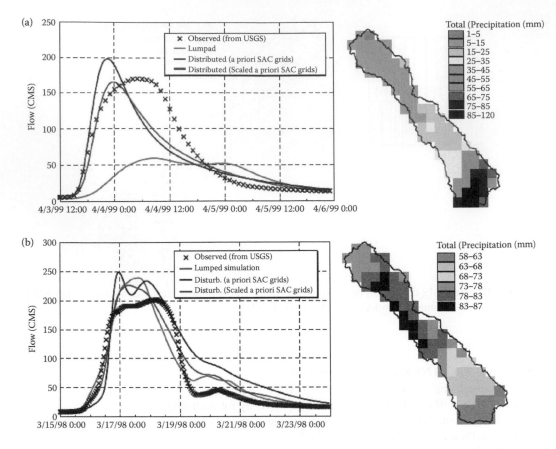

FIGURE 5.18 Modeled results for the Blue River basin in Oklahoma: (a) case with highly nonuniform rainfall, (b) case with relatively uniform rainfall. (From Smith et al. 2004.)

- Although the lumped model outperformed distributed models in more cases than distributed models outperformed the lumped model, some calibrated distributed models can perform at a level comparable to or better than a calibrated lumped model (the current operational standard).
- The wide range of accuracies among model results suggests that factors such as model formulation, parameterization, and the skill of the modeler can have a bigger impact on simulation accuracy than simply whether or not the model is lumped or distributed.
- Clear gains in distributed-model performance can be achieved through some type of model calibration. On average, calibrated models outperformed uncalibrated models during both the calibration and validation periods.
- Among calibrated results, models that combine techniques of conceptual rainfall-runoff and physically based distributed routing consistently showed the best performance in all but the smallest basin.
- Simulations for smaller interior basins, where no explicit calibration was done, exhibited reasonable performance in many cases, although not as good statistically as results for larger, parent basins.

5.4 SURFACE-WATER HYDROLOGY MODELS

There are a number of surface-water hydrologic models having integrated GIS interfaces and databases. Table 5.4 lists some of the more popular modeling packages. Some other hydrologic models

TABLE 5.4
Surface-Water Hydrologic Models

Product/Model	Description	Company	Web Site
HEC-HMS GEO-HMS	Various options for flood event simulation; hydrographs at different locations along streams; HMS can use gridded rainfall data	U.S. Army Corps of Engineers	http://www.hec.usace.army. mil/software/hec-hms/
HSPF	Extended period simulation of hydrologic processes and water quality	U.S. EPA and USGS	http://water.usgs.gov/ software/lists/ surface_water/
MIKE 11 UHM MIKE 11 RR	MIKE 11 is event runoff simulator using unit hydrograph techniques; MIKE 11 RR is a lumped-parameter continuous simulator; links to ArcMap	DHI Water & Environment	http://www.dhigroup.com/ Software.aspx
NWS Hydrology Lab Research Modeling System (HL-RMS)	Modeling framework for testing lumped, semi-distributed, and fully distributed hydrologic modeling approaches	NWS Hydrologic Research Lab	http://www.weather.gov/ ohd/hrl/distmodel/ distmod.htm#tools
Precipitation-Runoff Modeling System (PRMS), with Modular Modeling System (MMS)	Modular, deterministic, distributed-parameter hydrologic modeling system based on hydrologic response units (HRUs); MMS user interface	U.S. Geological Survey	http://water.usgs.gov/ software/lists/ surface_water/
SWMM 5 SWMM (RUNOFF) 4.30	Dry- and wet-weather flow simulator for urban areas; Windows interface	U.S. Environmental Protection Agency	http://www.epa.gov/ ednnrmrl/models/swmm/
TOPMODEL	Rainfall-runoff model that bases its distributed predictions on an analysis of catchment topography	Integrated into GRASS GIS simulation models as r.topmodel	http://grass.itc.it/intro/ modelintegration.html
TR-20 Win TR-55 Win	Storm-event watershed model; generates, routes, and combines hydrographs at points within a watershed; TR-55 for urban apps.; Windows interface	USDA, Natural Resources Conservation Service	http://www.wsi.nrcs.usda. gov/products/W2Q/H&H/ H&H_home.html
TREX CASC2D	2-D physically based, event watershed model that simulates rainfall and runoff; TREX version simulates erosion	Dr. Pierre Julienne, Dept. Civil Eng., Colorado State Univ.	http://www.engr.colostate. edu/~pierre/; http://grass.itc.it/intro/ modelintegration.html
Watershed Modeling System (WMS)	Comprehensive graphical modeling environment for all phases of watershed hydrology and hydraulics	U.S. Army Engineer R&D Center, Coastal and Hydraulics Laboratory	http://chl.erdc.usace.army. mil/wms
WATFLOOD	2-D hydrologic model that interfaces with radar-rainfall and remote-sensing data sets	Dr. Nicholas Kouwen, Dept. Civil Eng. Univ. Waterloo, CA	http://www.civil.uwaterloo. ca/Watflood/

that focus on urban stormwater runoff are summarized in Chapter 8 (GIS for Wastewater and Stormwater Systems). A large listing of hydrologic models can be found at the Hydrologic Modeling Inventory Web site (http://hydrologicmodels.tamu.edu/).

There are several models that are categorized as open source, and these are available via the GRASS GIS. Models incorporated into GRASS to date include r.topmodel (Bevin 2001) and r.hydro. CASC2D (Downer et al. 2002). However, the present trend in watershed-model coupling with GIS is the development of interfaces that serve to facilitate the creation of watershed-model input data sets. These interfaces read data from GIS systems such as ArcInfo and GRASS, allow the user to process the data into a form required by the hydrologic model, write input files, and run the model as an external process. Many intermediate interfaces also allow postprocessing of model output and potential transfer of model output back to the GIS.

REFERENCES

Bevin, K. 2001. *Rainfall-runoff modelling: The primer.* New York: John Wiley.

Bitner, D., T. Carroll, D. Cline, and P. Romanov. 2002. An assessment of the differences between three satellite snow cover mapping techniques. Paper presented at the 70th Annual Western Snow Conference. http://www.nohrsc.noaa.gov/technology/pdf/wscproceedings.pdf.

Burnash, R. J., R. L. Ferral, and R. A. McGuire. 1973. A generalized streamflow simulation system: Conceptual modeling for digital computers. Technical report. Sacramento, Calif.: Joint Fed. and State River Forecast Cent.

Chow, V. T., D. R. Maidment, and L. W. Mays. 1988. *Applied Hydrology.* New York: McGraw-Hill.

Clark, C. D. 1945. Storage and the unit hydrograph. *ASCE Trans.* 110: 1419–1446.

Cline, D., and T. Carroll. 1998. Operational automated production of daily, high-resolution, cloud-free snow cover maps of the continental U.S. Core science demonstration for the GEWEX Continental-Scale International Project (GCIP) Mississippi River Climate Conference. St. Louis, Mo. http://www.nohrsc.noaa.gov/html/papers/gcip98/gcip98.htm.

DeBarry, P., ed. 1999. *GIS modules and distributed models of the watershed.* Reston, Va.: ASCE.

Downer, C. W., F. L. Ogden, W. D. Martin, and R. S. Harmon. 2002. Theory, development, and applicability of the surface water hydrologic model CASC2D. *Hydrol. Process.* 16: 255–275.

Franken, S. K., D. J. Tyler, and K. L. Verdin. 2002. Development of a national seamless database of topography and hydrologic derivatives. Sioux Falls, S.D.: USGS EROS Data Center. http://edna.usgs.gov/papers.asp.

Fulton, R. A. 1999. Sensitivity of WSR-88D rainfall estimates to the rain-rate threshold and rain gauge adjustment: A flash flood study. *Weather Forecasting* 14: 604–624.

Fulton, R. A. 2002. Activities to improve WSR-88D radar rainfall estimation in the National Weather Service. In *Proc. 2nd Federal Interagency Hydrologic Modeling Conference.* Las Vegas, Nev. http://www.nws.noaa.gov/oh/hrl/papers/.

Fulton, R. A. 2005. Multisensor Precipitation Estimator (MPE) workshop. Advanced Hydrologic Applications course. National Weather Service Training Center. http://www.nws.noaa.gov/oh/hrl/papers/papers.htm.

Fulton, R. A., J. P. Breidenbach, D.-J. Seo, D. A. Miller, and T. O'Bannon. 1998. The WSR-88D rainfall algorithm. *Weather Forecasting* 13: 377–395.

Garbrecht, J., and L. W. Martz. 1999. TOPAZ: An automated digital landscape analysis tool for topographic evaluation, drainage identification, watershed segmentation and subcatchment parameterization; TOPAZ overview. ARS publication No. GRL 99-1. U.S. Department of Agriculture, Agricultural Research Service, Grazinglands Research Laboratory, El Reno, Okla.

Grimes, D. I. F., E. Pardo-Igu´zquiza, and R. Bonifacio. 1999. Optimal areal rainfall estimation using rain gauges and satellite data. *J. Hydrol.* 222: 93–108.

Huffman, G. J., R. F. Adler, P. A. Arkin, A. Chang, R. Ferraro, A. Gruber, J. J. Janowiak, R. J. Joyce, A. McNab, B. Rudolf, U. Schneider, and P. Xie. 1997. The Global Precipitation Climatology Project (GPCP) combined precipitation data set. *Bull. Am. Meteorol. Soc.* 78: 5–20.

Huffman, G. J., R. F. Adler, M. M. Morrissey, S. Curtis, R. J. Joyce, B. McGavock, and J. Susskind. 2001. Global precipitation at one degree daily resolution from multi-satellite observations. *J. Hydrometeorol.* 2: 36–50.

HEC (Hydrologic Engineering Center). 2000. HEC-HMS hydrologic modeling system. Technical reference manual. Davis, Calif.: U.S. Army Corps of Engineers, USACE-HEC. http://www.hec.usace.army.mil/software/hec-hms/.

HEC (Hydrologic Engineering Center). 2003. Geospatial hydrologic modeling extension (HEC-GeoHMS). Davis, Calif.: U.S. Army Corps of Engineers, USACE-HEC. http://www.hec.usace.army.mil/software/hec-geohms/.

Issacs, E. H., and R. M. Srivastava. 1989. *An introduction to applied geostatistics.* New York: Oxford University Press.

Jackson, T. J. 1999. Remote sensing of soil moisture in the southern Great Plains hydrology experiment. In *Proc. International Geoscience and Remote Sensing Symposium.* Piscataway, N.J.: IEEE.

Jackson, T. J., and A. Y. Hsu. 2001. Soil moisture and TRMM microwave imager relationships in the southern Great Plains 1999 (SGP99) experiment. *IEEE Trans. Geosci. Remote Sensing* 39: 1632–1642.

Jackson, T. J., A. Gasiewski, A. Oldak, M. Klein, E. Njoku, A. Yevgrafov, S. Christiani, and R. Bindlish. 2002a. Soil moisture retrieval using the C-band polarimetric scanning radiometer during the southern Great Plains 1999 experiment. *IEEE Trans. Geosci. Remote Sensing* 40: 2151–2161.

Jackson, T. J., A. Y. Hsu, A. M. Shutko, Y. Tishchenko, B. Petrenko, B. Kutuza, and N. Armand. 2002b. Priroda microwave radiometer observations in the SGP97 hydrology experiment. *Int. J. Remote Sensing* 23: 231–248.

Jackson, T. J., D. M. Le Vine, C. T. Swift, T. J. Schmugge, and F. R. Schiebe. 1995. Large area mapping of soil moisture using the ESTAR passive microwave radiometer in Washita '92. *Remote Sensing Environ.* 53: 27–37.

Jensen, M. E., R. D. Burman, and R. G. Allen, eds. 1990. Evapotranspiration and irrigation water requirements. In *ASCE manuals and reports on engineering practice, No. 70*. Reston, Va.: ASCE.

Jenson, S. K., and J. O. Domingue. 1988. Extracting topographic structure from digital elevation data for geographic information system analysis. *Photogrammetric Eng. Remote Sensing* 54: 1593–1600.

Knisel, W. G., and J. R. Williams. 1995. *Hydrology components of CREAMS and GLEAMS models: Computer models of watershed hydrology*, ed. V. P. Singh, 1069–1114. Highlands Ranch, Colo.: Water Resources Publications.

Knisel, W. G., R. A. Leonard, F. M. Davis, and A. D. Nicks. 1993. GLEAMS Version 2.10, Part III. User manual. U.S. Department of Agriculture, Agricultural Research Service, Conservation Research Report series.

Koren, V. I., M. Smith, D. Wang, and Z. Zhang. 2000. Use of soil property data in the derivation of conceptual rainfall-runoff model parameters. In *Proc. 15th Conference on Hydrology*, 103–106. Long Beach, Calif.: American Meteorology Society.

Kost, J. R., and G. G. Kelly. 2002. Watershed delineation using the National Elevation Dataset and semiautomated techniques. USGS paper 0421. http://edna.usgs.gov/Edna/pubs/p0421/p0421.html.

Maidment, D. R. 1993. Developing a spatially distributed unit hydrograph by using GIS. In *Proc. HydroGIS'93*. IAHS publ. No. 211. http://www.cig.ensmp.fr/~iahs/redbooks/211.htm.

Maidment, D. R., ed. 2002a. *Arc Hydro: GIS for water resources*. Redlands, Calif.: ESRI Press.

Maidment, D. R. 2002b. National Hydro Data Programs. PowerPoint presentation. University of Texas, Austin: Center for Research in Water Resources.

McCuen, R. H. 1982. *A guide to hydrologic analysis using SCS methods*. Englewood Cliffs, N.J.: Prentice-Hall.

Miller, D. A., and R. A. White. 1999. A conterminous United States multi-layer soil characteristics data set for regional climate and hydrology modeling. *Earth Interactions* 2. http://earthInteractions.org.

NRC (National Research Council). 2007. *Integrating multiscale observations of U.S. waters*. Committee on Integrated Observations for Hydrologic and Related Sciences. http://www.nap.edu/catalog/12060.html.

OFCM (Office of the Federal Coordinator for Meteorological Services and Supporting Research). 2006. Meteorological handbook No. 11: Doppler radar meteorological observations. Part C, WSR-88D products and algorithms. FCM-H11C-2006. http://www.ncdc.noaa.gov/oa/radar/radarresources.html.

Pfafstetter, O. 1989. Classification of hydrographic basins: Coding methodology. Unpublished manuscript. DNOS, Rio de Janeiro. Translated by J. P. Verdin, U.S. Bureau of Reclamation, Brasilia, Brazil.

Rango, A. 1993. Snow hydrology processes and remote sensing. *Hydrological Proc.* 7: 121–138.

Reed, S. M. 1998. Use of digital soil maps in a rainfall-runoff model. CRWR online report 98-8. http://www.crwr.utexas.edu/reports/1998/rpt98-8.shtml.

Reed, S., M. Smith, V. Koren, Z. Zhang, D. Seo, F. Moreda, V. Kuzmin, Z. Cui, and R. Anderson. 2002. Distributed modeling for improved NWS river forecasts. NOAA technical report. http://www.weather.gov/ohd/hrl/.

Saghafian, B., P. Y. Julien, and H. Rajaie. 2002. Runoff hydrograph simulation based on time variable isochrone technique. *J. Hydrol.* 261: 193–203.

Santanello, J. A., C. D. Peters-Lidard, M. E. Garcia, D. M. Mocko, M. A. Tischler, M. S. Moran, and D. P. Thoma. 2007. Using remotely sensed estimates of soil moisture to infer soil texture and hydraulic properties across a semi-arid watershed. *Remote Sensing Environ.* 110: 79–97.

Schaake, J. C., et al. 2004. An intercomparison of soil moisture fields in the North American Land Data Assimilation System (NLDAS). *J. Geophys. Res.* 109: D01S90. doi:10.1029/2002JD003309.

Seo, D.-J. 1998. Real-time estimation of rainfall fields using radar rainfall and rain gauge data. *J. Hydrol.* 208: 37–52.

Seo, D.-J., and J. Breidenbach. 2002. Real-time correction of spatially nonuniform bias in radar rainfall data using rain gauge measurements. *J. Hydrometeorol.* 3 (2): 93–111.

Seo, D.-J., J. Breidenbach, and E. Johnson. 1999. Real-time estimation of mean field bias in radar rainfall data. *J. Hydrol.* 223: 131–147.

Sherman, L. K. 1932. Stream-flow from rainfall by the unit-graph method. *Eng. News Records* 108: 501–505.

Skahill, B., and L. Johnson. 2000. F2D: A kinematic distributed watershed rainfall-runoff. NOAA technical memorandum ERL FSL-24. Boulder, Colo.: Forecast Systems Lab.

Smith, M., V. Koren, Z. Zhang, S. Reed, D. Seo, F. Moreda, V. Kuzmin, Z. Cui, and R. Anderson. 2004. NOAA NWS distributed hydrologic modeling research and development. NOAA technical report NWS 45. http://www.nws.noaa.gov/oh/hrl/distmodel/NOAA_TR45.pdf.

Sorooshian, S., K. Hsu, X. Gao, H. V. Gupta, B. Imam, and D. Braithwaite. 2000. Evaluation of PERSIANN system satellite-based estimates of tropical rainfall. *Bull. Am. Meteorol. Soc.* 81: 2035–2046.

Susskind, J., P. Piraino, L. Rokke, L. Iredell, and A. Mehta. 1997. Characteristics of the TOVS pathfinder path A dataset. *Bull. Am. Meteorol. Soc.* 78: 1449–1472.

Thorne, V., P. Coakley, D. Grimes, and G. Dugdale. 2001. Comparsion of TAMSAT and CPC rainfall estimates with rainfall, for southern Africa. *Int. J. Remote Sens.* 22 (10): 1951–1974.

Verdin, K. L., and J. P. Verdin. 1999. A topological system for delineation and codification of the Earth's river basins. *J. Hydrol.* 218: 1–12.

Vicente, G., R. A. Scofield, and W. P. Menzel. 1998. The operational GOES infrared rainfall estimation technique. *Bull. Am. Meteor. Soc.* 79: 1883–1898.

Viessman, W., Jr., G. L. Lewis, and J. W. Knapp. 1989. *Introduction to hydrology*. New York: Harper & Row.

Vieux, B. E. 2004. *Distributed hydrologic modeling using GIS*. 2nd ed. New York: Springer.

Wang, J. R., J. C. Shiue, T. J. Schmugge, and E. T. Engman. 1989. Mapping surface soil moisture with L-band radiometric measurements. *Remote Sensing Environ.* 27 (3): 305–312.

6 GIS for Groundwater Hydrology

6.1 OVERVIEW

Groundwater is sometimes referred to as the hidden source of water supply because it resides in the subsurface. It is of vital importance in areas where dry summers and extended droughts cause surface supplies to disappear. More than 1.5 billion people worldwide and more than 50% of the population of the United States rely on groundwater for their primary source of drinking water (Alley et al. 2002). Threats to groundwater quality have risen in importance, given the increased dependence on groundwater supplies and the long times required for clearance of contamination.

GIS has found extensive application for groundwater assessments, as there are many types and large amounts of data involved. Proper evaluation of groundwater resources requires thorough hydrologic, geologic, and hydraulic investigations. The spatial scope may be quite local for a specific pumping well, or it may range in size from a few hundred hectares to entire basins and even countries. Use of simulation and management models is widespread in such studies, and GIS has become a primary technology for coordinating the data management and providing the interface for groundwater model development.

This chapter reviews the application of GIS for groundwater assessment and modeling. Firstly, background information on groundwater hydrology and management concepts is provided, and groundwater data and models are summarized. Groundwater *quantity* data management and modeling concepts are then reviewed, and case studies of GIS applications are described. The chapter concludes with descriptions of groundwater *quality* assessment models.

6.2 GROUNDWATER HYDROLOGY AND MANAGEMENT

Sound management of groundwater involves assessing the role of the groundwater system in the hydrologic balance, prediction of the long-term pumping capacity and effects, and evaluation of water-quality conditions.

Groundwater availability in any location is quite site specific, given the combination of natural hydrologic and geologic conditions as well as the human-induced changes that might exist. Quantitative methods have been developed for characterizing groundwater flows and quality, and these provide an acceptable basis for management decisions. However, these methods require adequate data pertaining to hydrogeologic conditions, and these data are notoriously difficult to obtain. GIS data-management and analysis procedures help considerably in groundwater management practice.

Groundwater is an integral part of the hydrologic cycle, being recharged through infiltration of rainwater, snowmelt, irrigation return flows and streams at the land surface, and discharging water to streams and wetlands during low-flow periods. The hydrogeologic setting is a key to whether a deposit can yield water in economic quantities and its susceptibility to contamination. Although the chemical and biological character of groundwater is acceptable for most uses, the quality of some sources, particularly shallow groundwater, is often changing due to human activities. Figure 6.1 illustrates the dynamic aspects of groundwater flow systems, their recharge, interactions with surface water, and the different time scales involved.

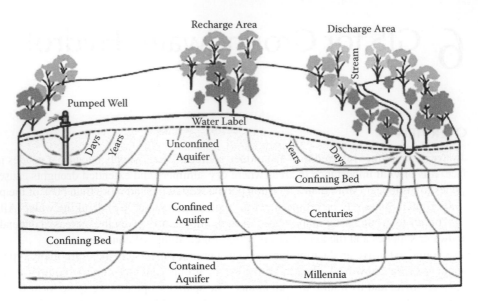

FIGURE 6.1 Groundwater flow paths vary greatly in length, depth, and travel time from points of recharge to points of discharge in the groundwater system. (From Winter et al. 1998.)

Groundwater-bearing formations that are sufficiently permeable to yield usable quantities of water are called aquifers. An aquifer overlain by a relatively impermeable layer is said to be confined; if the formation is exposed to the surface, it is unconfined. The hydraulics of porous media flow in the two situations is distinctly different. Water flow in a confined aquifer is analogous to that in pipes; in an unconfined aquifer, flow is analogous to that in an open channel. The height that water rises to in a confined aquifer is a measure of the aquifer pressure; it is called the piezometric level, and such a well is called artesian. A confined aquifer may become unconfined due to pumping when the piezometric surface drops below the confining layer.

Streams either gain water from inflow of groundwater or lose water by outflow to groundwater. Some streams do both, gaining in some reaches and losing in other reaches. Under natural conditions, groundwater makes some contribution to stream flow in most physiographic and climatic settings. A pumping well can change the quantity and direction of flow between an aquifer and stream in response to different rates of pumping. The adjustments to pumping of an actual hydrologic system may take place over many years, depending upon the physical characteristics of the aquifer, the degree of hydraulic connection between the stream and aquifer, and locations and pumping histories of wells.

Management of groundwater resources is directed to the functions that aquifers play in the water resources of an area. These may include the roles for water supply, storage, pipeline, mining, filtering, energy, and natural hydrologic balance (Fetter 1994). Water supply from wells is the most obvious function and commands our attentions for assessment of yield potential and facilities development. Supply management is guided by the concept of "safe yield," which has been defined as the amount of water that could be pumped on a continuing basis without depleting the storage reserve; this is sometimes called the "sustainable yield." Here, the water removed is balanced by the amount of recharge that occurs. The pipeline concept relates to the transmission of recharge water from elsewhere; these recharge zones become the focus of land-management practices that minimize the threat of contamination. The safe-yield concept has been augmented to incorporate economic and legal considerations. For example, it may be warranted to mine the groundwater at rates that exceed the sustainable yield and thereby deplete the storage reserve. Such actions take account of the initial investment potential of development so that sustainable supplies might be developed at a later time. Artificial recharge and storage is another approach that seeks to replace aquifer waters during periods of surplus so that groundwater supplies are available during periods of drought.

6.3   GROUNDWATER DATA

The foundation of a usable groundwater model is the availability of high-quality data. GISs typically are an integral part of the database system to assist in organizing, storing, and displaying the substantial array of needed information. Principal types of data commonly required are listed in Table 6.1. Some, such as precipitation data, are generally available and relatively easy to obtain at the time of a hydrologic analysis. Other data and information, such as geologic and hydrogeologic maps, are difficult and expensive to obtain and can require years to develop. Still other data, such as a history of water levels in different parts of an aquifer system, require foresight in order to obtain measurements

TABLE 6.1
Data Required for Analysis of Groundwater Systems

Hydrogeology

Topographic data on the stream drainage network, surface-water bodies, landforms, cultural features, and locations of structures and activities related to water

Geologic maps of surficial deposits and bedrock

Hydrogeologic maps showing extent and boundaries of aquifers and confining units

Maps of tops and bottoms of aquifers and confining units

Saturated-thickness maps of unconfined (water table) and confined aquifers

Average hydraulic conductivity maps for aquifers and confining units and transmissivity maps for aquifers

Maps showing variations in storage coefficient for aquifers

Estimates of age of groundwater at selected locations in aquifers

Hydrology

Locations of active pumps, amounts, and time patterns (current and historical)

Well hydrographs and historical head (water level) maps for aquifers

Precipitation data

Evaporation data

Stream-flow data, including measurements of gain and loss of stream flow between gauging stations

Maps of the stream drainage network showing extent of normally perennial flow, normally dry channels, and normally seasonal flow

Estimates of total groundwater discharge to streams

Measurements of spring discharge

Measurements of surface-water diversions and return flows

Quantities and locations of interbasin diversions

Amount of groundwater consumed for each type of use and spatial distribution of return flows

Location of recharge areas (areal recharge from precipitation, losing streams, irrigated areas, recharge basins, and recharge wells) and estimates of recharge

Water quality

Geochemical characteristics of Earth materials and naturally occurring groundwater in aquifers and confining units

Spatial distribution of water quality in aquifers, both areally and with depth

Temporal changes in water quality, particularly for contaminated or potentially vulnerable unconfined aquifers

Sources and types of potential contaminants

Chemical characteristics of artificially introduced waters or waste liquids

Maps of land cover–land use at different scales, depending on study needs

Stream-flow quality (water-quality sampling in space and time), particularly during periods of low flow

Administrative data

Jurisdictions and administrative boundaries

Ownership boundaries

Water rights categorization

Source: Adapted from Winter et al. (1998).

over time, if they are to be available at all. As examples, these data would include depths and thicknesses of hydrogeologic units from lithologic and geophysical well logs, water-level measurements to help define predevelopment water-level maps for major aquifers as well as water-level maps at various times during development, groundwater sampling to document pre- and postdevelopment water quality, and simultaneous measurements of stream flow and stream quality during low flows to indicate possible contributions of discharging groundwater to surface-water quality.

6.4 GROUNDWATER MODELS

6.4.1 Overview

Quantitative methods for assessing groundwater flow and contamination migration originate with Darcy's studies in the 1850s. He determined that the velocity of flow is directly related to the hydraulic gradient and aquifer permeability, as seen in equation 6.1. Natural hydraulic gradients seldom exceed 0.2% to 0.4%, and velocities at these gradients range from less than 100 m/year in sandstones to 100 m/day in coarse gravels.

$$v = -K\frac{\partial h}{\partial l} = -Ks \tag{6.1}$$

where v = the apparent, or Darcy, velocity (not the actual velocity through the interstices of the soil); h = drop in phreatic surface- or groundwater table between two points; l = horizontal distance between the points; s = hydraulic gradient; and K = Darcy's coefficient, dependent on soil material and fluid characteristics, also called the *hydraulic conductivity*, with dimensions L/T. Darcy's coefficient is most often determined by field pumping tests, which require considerable effort and expense. Given assumptions on the homogeneous character of aquifer materials, horizontal flow, and an aquifer of infinite extent, then Darcy's equation can be solved analytically to provide equations useful for estimating aquifer responses to pumping. However, these conditions seldom occur, so the conventional approach is to use advanced numerical models (e.g., finite-difference type) to represent the dynamics of groundwater hydraulics and solute transport.

6.4.2 Finite-Difference Model MODFLOW

Groundwater modeling tools are used to represent an approximation of the field data and to assess the behavior of the groundwater system under varying climatic conditions (drought conditions) or changes in water consumption, population growth, or changes in land use. The most popular computer model of the numerical type is the modular finite-difference groundwater flow model (MODFLOW) developed by the U.S. Geological Survey (McDonald and Harbaugh 1988; Harbaugh 2005). MODFLOW was designed to simulate aquifer systems in which (a) saturated-flow conditions exist, (b) Darcy's law applies, (c) the density of groundwater is constant, and (d) the principal directions of horizontal hydraulic conductivity or transmissivity do not vary within the system. These conditions are met for many aquifer systems for which there is an interest in analysis of groundwater flow and contaminant movement. For these systems, MODFLOW can simulate a wide variety of hydrologic features and processes. Steady-state and transient flows can be simulated in unconfined aquifers, confined aquifers, and confining units. A variety of features and processes such as rivers, streams, drains, springs, reservoirs, wells, evapotranspiration, and recharge from precipitation and irrigation also can be simulated (Figure 6.2).

MODFLOW simulates groundwater flow in aquifer systems using the finite-difference method. In this method, an aquifer system is divided into rectangular blocks by a grid (Figure 6.3). The grid of blocks is organized by rows, columns, and layers, and each block is commonly called a *cell*. For each cell within the volume of the aquifer system, the user must specify aquifer properties. The user

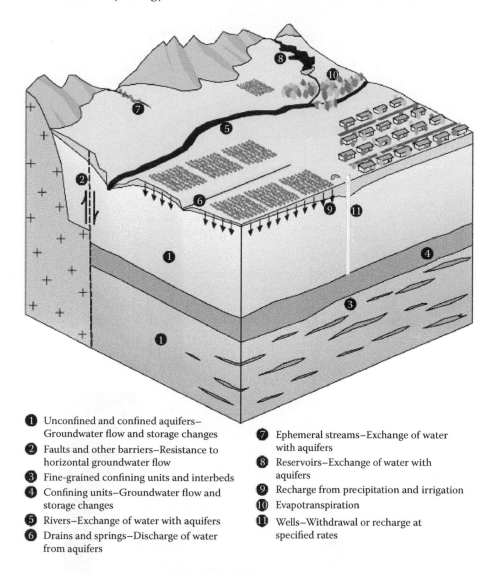

1. Unconfined and confined aquifers–
   Groundwater flow and storage changes
2. Faults and other barriers–Resistance to
   horizontal groundwater flow
3. Fine-grained confining units and interbeds
4. Confining units–Groundwater flow and
   storage changes
5. Rivers–Exchange of water with aquifers
6. Drains and springs–Discharge of water
   from aquifers
7. Ephemeral streams–Exchange of water
   with aquifers
8. Reservoirs–Exchange of water with
   aquifers
9. Recharge from precipitation and irrigation
10. Evapotranspiration
11. Wells–Withdrawal or recharge at
    specified rates

FIGURE 6.2   Features of an aquifer that can be simulated by MODFLOW. (From USGS 1997.)

also specifies information relating to wells, rivers, and other inflow and outflow features for cells corresponding to the locations of the features. MODFLOW uses the input to construct and solve equations of groundwater flow in the aquifer system. The solution consists of head (groundwater level) at every cell in the aquifer system (except for cells where head was specified as known in the input data sets) at intervals called *time steps*.

Many studies require information such as the average rate of movement of groundwater and contaminants. Also, information often is needed on the recharge or capture areas for water discharging to wells, springs, stream reaches, and other features. Although MODFLOW does not compute this information directly, simulation with MODFLOW provides basic information needed for such analyses. The particle-tracking program, MODPATH, is a postprocessing program for MODFLOW to estimate flow paths and times of travel in groundwater systems. An accompanying program, MODPATH-PLOT, displays particle paths, contours, and model features. MODPATH can be used for studies of steady-state and transient flows. Common applications include studies of paths and time of travel of contaminant movement (Figure 6.4) and source (recharge) areas of wells, springs, rivers, and other features.

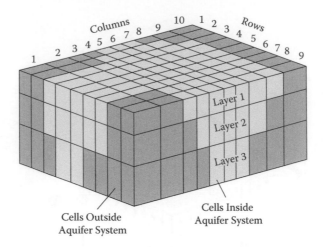

FIGURE 6.3 Example of model grid for simulating three-dimensional groundwater flow. (From USGS 1997.)

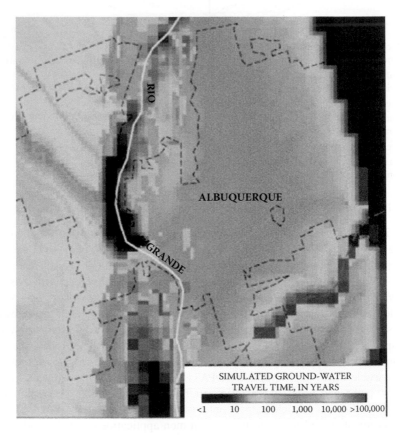

FIGURE 6.4 (See color insert following page 136.) Application of particle tracking to estimate groundwater travel time. (From USGS 1997.)

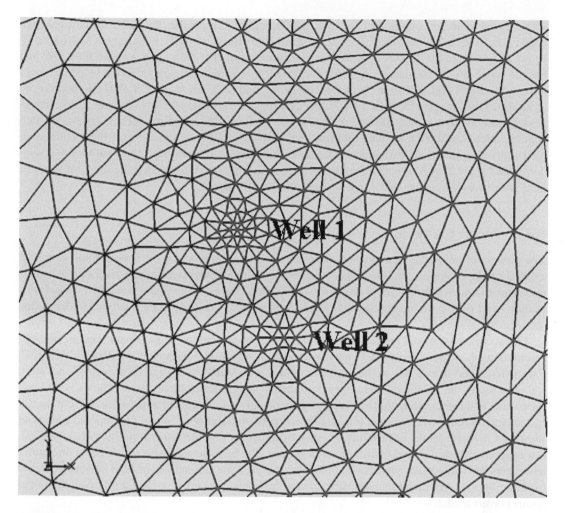

FIGURE 6.5   Two-dimensional finite-element mesh showing fine resolution of the mesh for modeling wells. (From Tracy et al. 2007.)

6.4.3   FINITE-ELEMENT MODELS

In contrast to the rectangular format of the finite-difference models, there is another geometric construct, called *finite element* (FE), for representing the spatial variability of aquifer properties. The finite-element approach uses triangular elements, in contrast to the block elements most finite-difference (FD) models use (Figure 6.5). The FE methods are not as popular as the FD due to the complexity of the solution algorithms. However, there are some programs available, such as SUTRA, a two-dimensional (2-D) or three-dimensional (3-D) density-dependent flow model by the USGS; Hydrus, a commercial unsaturated flow model; FEFLOW, a commercial modeling environment for subsurface flow as well as solute- and heat-transport processes; and COMSOL Multiphysics (FEMLAB), a commercial general modeling environment.

6.4.4   GROUNDWATER QUALITY MODELING

To evaluate or remediate contaminated aquifers, it is important to be able to compute the concentration of dissolved chemicals. Hydrologists and water managers need to know how contaminants will spread in the future, or how the concentration distributions will respond to changes in the flow field that might arise from new water-supply stresses, remediation efforts, or other management activities.

As an example, MOC3D is a solute-transport program that is integrated with MODFLOW and has the capability to calculate changes in concentration of a single solute as affected by processes of advection, dispersion, diffusion, fluid sources, decay, and retardation (USGS 1997). In addition to input data requirements for MODFLOW, the user must also define porosity, thickness, dispersiveness, initial concentrations, and source-fluid concentrations. Model output includes the calculated concentration distribution in space and time. MOC3D is based on a particle-tracking method coupled with an explicit finite-difference procedure. This numerical technique is optimal for advection-dominated problems. Alternative numerical methods are planned to broaden the range of types of field problems for accurate and efficient application of the model.

6.4.5 MODEL CALIBRATION

The process of adjusting the model input values to reduce the model error is referred to as model calibration. Hydrologists commonly use water levels from a model layer to construct contour maps for comparison with similar maps drawn from field data. They also compare computed water levels at individual cells with measured water levels from wells at corresponding locations to determine model error.

The accuracy of model predictions is limited by the correctness of the model (i.e., proper representation of relevant processes) and the uncertainty in model parameters (Alley et al. 2002). Parameter uncertainty is due to the limited accuracy with which values can be measured and, more importantly, to the substantial heterogeneity inherent in aquifer characteristics. The difficulties in measuring and representing this heterogeneity adequately is a fundamental problem in groundwater hydrology, and this will continue to place limits on the reliability of model predictions, even with improved models. Model uncertainty can also be dependent on the type of questions being asked. For example, hydraulic head distributions can reasonably be estimated with limited data on aquifer heterogeneity. Conversely, predictions of solute transport and concentrations can be very sensitive to minor uncertainty in the spatial distribution of hydraulic properties, even for relatively homogeneous porous media.

In conventional or "forward modeling," model parameters (such as aquifer properties) are specified, and water levels and flow quantities are computed (USGS 1997). For most aquifer systems, however, more information is available from field data on water levels, flows, and advective transport or groundwater age than on input parameters. Typically, input parameters are adjusted during model calibration using a trial-and-error process. This calibration process can yield acceptable agreement between computed model results and field data, but it is time consuming, may not produce parameter values that result in the best fit of field data, and does not result in quantitative estimates of uncertainty in model results and estimated parameter values.

Inverse modeling is a more formal approach to model calibration that includes automatic parameter adjustment to match field data. Program MODFLOWP is the USGS version of MODFLOW that includes automatic parameter estimation (USGS 1997). MODFLOWP uses a weighted least-squares objective function as a measure of how well model results agree with field measurements. Weights are used to reflect the reliability of individual measurements. Parameters that can be estimated by MODFLOWP include transmissivity; hydraulic conductivity; storage coefficient; vertical leakage; vertical and horizontal anisotropy; hydraulic conductance between aquifer systems and rivers, drains, and other features; areal recharge; maximum evapotranspiration; pumping; and water levels at constant-head boundaries. The resulting parameter values are "best-fit," in that they provide the closest match between measured and simulated heads and flows, as measured by the objective function. The task of the modeler is to evaluate how well these calculated values represent the physical system being simulated.

TABLE 6.2
GIS Operations Supportive of Groundwater Modeling

Data management

Database of spatial data keyed to location and depth

Capture of archived site inventory data and conversion to GIS formats

Collation of aquifer attribute data on hydrogeologic factors, hydrology, and quality

Design of model FD grids or FE mesh

Automatic formulation of model input data

Visualization of input data for error checking and consistency

Statistical interpolations to assign field data to aquifer extent

Groundwater system modeling

Establishment of aquifer model boundary conditions

Systematic assignment of model parameters

Interactive model simulations

Sensitivity analyses aided by GIS-based parameter changes

Model output review

Display of model outputs in color-coded map-oriented formats (with animations)

Map and graphical comparisons of aquifer simulation results with field calibration data

Model reporting and archive

6.5 GIS FOR GROUNDWATER MODELING

6.5.1 OVERVIEW

As listed in Table 6.1, groundwater models require a number of disparate and large data sets that are difficult to manage. GIS can help with the modeling process by coordinating data collection, providing comprehensive database operations, supporting systematic model parameter assignments, conducting spatial analysis (e.g., spatial statistics) functions, and displaying model results in understandable color-map formats. Table 6.2 lists a variety of these GIS operations supportive of groundwater modeling.

6.5.1.1 Model Data Development

Groundwater systems are often represented using gridded data; grids are used to efficiently create and visualize spatial distributions for pre- and postprocessing of the model (Radin 2006). Grid functions make it easy to compare and modify input data. Identifying attribute values of concern in large data files is made easier with GIS when they can be visualized in map formats. Most data are gathered at points, which are then interpolated using geostatistical techniques into surfaces of elevations (land surface, piezometric). GIS grids, coverages, and shapefiles are used to create the majority of the input data sets for MODFLOW, including hydrogeology and stratigraphy, hydrogeologic parameters, boundary conditions, and initial conditions. Coverages and shapefiles are used to represent rivers, drains, and wells. Geoprocessing is used to create polygons with unique soils, precipitation, evapotranspiration, and land use, which can then be combined to generate recharge and evapotranspiration arrays.

The California Integrated Water Flow Model (IWFM) uses an FE approach to simulate groundwater flow, surface-water flow, and surface-water–groundwater interaction (Dogrul 2007). Figure 6.6 shows an example of the application of the IWFM for generating input data to the triangular mesh from land-use data. Agricultural and urban water demands can be prespecified or calculated internally based on different land-use types. Water reuse is also modeled as well as tile drains and lakes or open-water areas. A main feature of IWFM is a "zone budget"-type of postprocessor that includes subsurface flow computations across element faces.

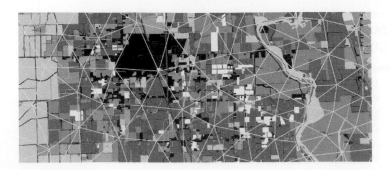

FIGURE 6.6 (See color insert following page 136.)   Land use with finite-element model grid overlay for integrated water-flow model (IWFM). (From Heywood et al. 2008. With permission.)

With the aid of GIS spatial analyses, grids of the modeled groundwater levels and drawdowns are created to demonstrate model calibration, verification, and assessment management actions on the model. Groundwater levels in various model layers and runs are easily compared using a raster calculator. Grids make it easy to create parameter changes for calibration trials. Drawdown maps are routinely developed to visualize the impacts of management options.

6.5.1.2   Model Integration

Groundwater-model developers often face choices on the degree to which GIS functions are integrated with the groundwater model. Various degrees of GIS and groundwater-model integration range from:

1. Linking a GIS to a groundwater model through data-transfer programs
2. Integrating a model with a GIS database
3. Embedding modeling capabilities within a GIS

The advantages and disadvantages of these three interface methods were described by Watkins et al. (1996). For the linked GIS model (option 1), the groundwater model and GIS are separate, and computer routines are used to convert data from the GIS into formatted files that are readable by the model code. Similar routines are used to convert model outputs into GIS formats. This option requires a minimum of programming, but it does require repetitive manual operations to run and review model simulations, and such efforts will accrue over time. Integrating the groundwater model with the GIS (option 2) involves a stronger linkage through a single database and a user interface to facilitate the data exchanges. This option removes database redundancy and makes modeling operations easier and more efficient. However, it does require more programming. Embedding the model in the GIS (option 3) involves a single database with no data conversions, where the GIS intrinsic functions are fully available through a customized user interface. This approach supports full interactivity for model development, simulations, and reviewing of results. Raster-based GISs allow changing the resolution of a finite-difference model grid, and spatial statistical functions allow interpolations to be easily accomplished. The time and effort needed to develop a fully embedded GIS groundwater-modeling system are quite large. Such effort would only be warranted for high-value aquifer systems requiring continuing model support for management. Modern groundwater-modeling packages have been developed to represent the third level of GIS-model embedding, thereby greatly reducing the customization effort.

6.5.1.3   GIS Databases

In concert with the surface-water domain, there has been a movement toward a geodatabase approach for groundwater data and modeling support. Zeiler (1999) described the ESRI geodatabase data model as an object-oriented model introduced with the ArcGIS® software. The basic form of the model contains a geodatabase, feature data sets, feature classes, and object classes. An example of a feature data set would be a group of monitoring stations, where a feature class would be the locations of the observation wells and the object class would be the water level data collected at all the stations. The model is packaged in a geodatabase, which contains the feature data sets. Feature data sets, in turn, contain all the feature classes in a model and the relationships among them within a common coordinate system. An object class is a nonspatial entity, like a data table, and feature classes are objects plus spatial coordinates.

The Arc Hydro data model (Maidment 2002) defines a "hydrologic information system" that is a blend of geospatial and temporal data supporting hydrologic analysis and modeling. The incorporation of groundwater elements is necessary to provide a wider array of applications to Arc Hydro users as well as to provide a data model that more accurately reflects surface-water and groundwater interactions within the hydrologic cycle. Maidment et al. (2004) extended the Arc Hydro geodatabase concepts to include groundwater. Groundwater applications range from regional studies, which usually describe the flow in aquifers as two-dimensional, to site investigations that model the three-dimensional nature of the flow through the aquifer architecture. Other common applications use the interaction between surface water and groundwater as part of surface-water modeling. The groundwater geodatabase is focused on a multidimensional representation of groundwater in the ArcGIS application. The groundwater geodatabase consists of four main components: (a) Hydrogeology, (b) GeoMap, (c) Modeling (for posting modeled results), and (d) Surfaces (rasters and triangulated irregular networks [TINs]). The Hydrogeology feature data set is a set of vector objects (points, lines, polygons, and multipatches) that represent hydrogeologic features used in groundwater studies. The Modeling feature data set is a set of vector feature classes that can represent common modeling objects such as cells and elements. This feature data set is primarily used to post the results of a modeling study within a GIS. The Surfaces group includes both rasters and TINs and is used to define elevations or spatially variable aquifer parameters.

The 2-D data model is useful for studying subjects such as aquifer recharge from streams, water balances, and contaminant transport between surface- and groundwater systems. The focus of this data model is on the points of river networks and groundwater interaction. The edge feature class already exists in the Arc Hydro data model (HydroEdge); thus only the aquifer and well feature classes need to be added to the data model. In addition to the relationship between river edges and aquifers, the connection between well features and aquifers can also be defined. Wells are related to aquifers through an aquifer identifier (AquiferID). The aquifer–well relationship describes the aquifer in which the well is screened. Although the well may be drilled through several layers and aquifers, the actual connection is through the well's screen. Once the relationships between river network edges, aquifers, and wells are established, applications such as water balances, recharge/discharge estimations, and solute transport between the surface- and groundwater systems can be represented.

A 3-D representation of the geologic framework, aquifer architecture, and boreholes within groundwater systems provides advanced capabilities for regional groundwater studies as well as site investigations. The 3-D data model enables an interface with groundwater modeling software such as MODFLOW. Figure 6.7 illustrates the classes and relationships of the 3-D data model.

Boreholes provide a portholelike window into the subsurface and are the primary source for information regarding the state of an aquifer at any given time. The majority of data describing aquifer characteristics and temporal groundwater conditions are related to boreholes; therefore, the development of a borehole feature is important to accurately describe the groundwater system. A borehole can be viewed as a point feature, with attributes defining measures in the vertical dimension, or as a line feature (vertical line), with events on the line giving the same information. In both cases, the well contains information in three dimensions. Information in the vertical dimension can be measured from the land-surface elevation or from a benchmark, which is a known location with a measured elevation. Figure 6.8 shows a geodatabase model developed for a spill-remediation site (Fleetwood 2008).

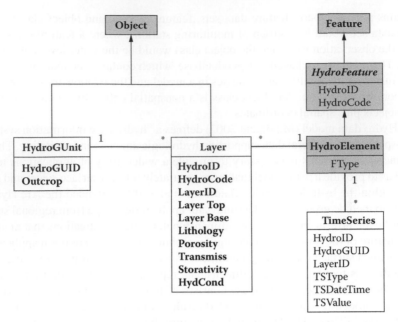

FIGURE 6.7 Classes and relationships of the three-dimensional data model. (From Maidment et al. 2004. With permission.)

FIGURE 6.8 Geodatabase model for a hazardous-material spill-remediation site. (From Fleetwood 2008. With permission.)

The following case studies illustrate GIS methods and techniques to expand the analysis of groundwater systems. By using a standard GIS interface such as ArcView, the study of groundwater systems can be more straightforward and intuitive.

6.5.2 CASE STUDIES

6.5.2.1 Cherry Creek Well Field

An example of a MODFLOW groundwater model supported by GIS was demonstrated by Rindahl (1996) and Inbau and Rindahl (1998) for the Cherry Creek alluvium upstream of Cherry Creek Reservoir near Denver, Colorado. The original model consisted of a 72-column × 244-row grid of 200-ft squares aligned along the axis of Cherry Creek. Input parameters to the model for each cell are tabulated in input files read by the MODFLOW program. Additional parameters such as stream flows, well pumping, and time periods are also input to the program. Outputs are typically groundwater elevation, drawdowns from the starting conditions, and stream flows in either printed form or binary files. Printed output was quite voluminous and requires a through knowledge of both the particular model and groundwater hydraulics.

The project used ArcView GIS for preparing groundwater models and analyzing simulation results. Input files are difficult and time consuming to construct, and once the input files are written, future modifications pose a challenge. Typically, displaying the results of MODFLOW simulations required postprocessing software, which usually does not display model or information in a georeferenced format. A GIS was seen as a possible solution to the interface challenges with the MODFLOW model. ArcView, with Spatial Analyst®, was used as an efficient tool for both pre- and postprocessing of MODFLOW data. It was demonstrated that ArcView could display drawdowns, stream flows, and aquifer elevations from MODFLOW simulations (Figure 6.9). Using ArcView's ability to join GIS coverages and text tables, an easy-to-use interface for displaying and analyzing complex model simulation results was developed. The next step in interfacing ArcView with MODFLOW was to use ArcView as a method to develop MODFLOW input files. To achieve this, routines were developed for both the well (WEL) and stream-flow routing (STR) packages.

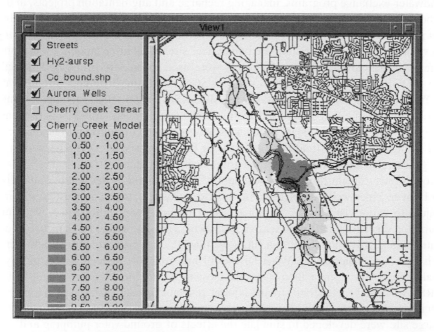

FIGURE 6.9 Cherry Creek well-field drawdowns generated from MODFLOW simulation. (From Rindahl 1996. With permission.)

For the Cherry Creek groundwater project, the modeling effort preceded the GIS coverage and development work. Subsequently, the MODFLOW and GIS were integrated. Originally, the MODFLOW grid and other features were designed and laid out using AutoCAD®. Once GIS was recognized as a tool to assist the modeling effort, the grid CAD file was converted into an ArcInfo® polygon coverage, and the TRANSFORM function was used to georectify the coverage. A copy of the model grid was created using only the cells that were used in the tributary aquifer modeling. Both pumping-well and monitoring-well coverages were generated from CAD and survey data. The stream-flow cell coverage was created from the model grid coverage using only the cells where stream flow is modeled. Additional GIS layers were used for reference purposes only. The line stream layer in the study area was developed from 1:24,000-scale USGS DLG (digital line graph) data. Other GIS layers such as streets were developed by the city's public works department as part of the city's base map.

ArcView was successfully used to create MODFLOW input files for both the well- and stream-flow-routing packages. This tool allowed MODFLOW users to create and manipulate complex MODFLOW input files in an easy and efficient way. Avenue scripts allowed flexibility in modifying well-pumping and stream-flow scenarios for various "what if" scenarios. One main advantage of using ArcView to input data into the model was to simplify repetitive and difficult tasks while creating or modifying large MODFLOW input text files. To aid in analyzing the MODFLOW model results, an ArcView application was developed to provide a clear, georeferenced view of the model results. By using MODFLOW coverages in a GIS, various ancillary analyses could be conducted, such as distance measurements, nearness, and overlays. Animations of stress-period variables were also demonstrated.

6.5.2.2 Conjunctive Stream–Aquifer Model

A stream–aquifer management decision-support system (DSS) was developed by Fredericks et al. (1998) for the South Platte River below Denver; it was also applied to the Lower Snake River (Shannon et al. 2000) and several other groundwater systems. The DSS is intended to support improved regional strategies for daily water administration, drought contingency planning, evaluating groundwater-exchange programs, managing recharge and augmentation projects, and resolving conflicts between urban, agricultural, and environmental concerns. Impacts of a groundwater-augmentation plan on river and tributary flows over time and space were assessed, including a comparative evaluation of the use of simplified analytical stream–aquifer models vs. a more realistic finite-difference groundwater model. GIS tools were used to prepare grid-based spatial data for input into MODRSP, a modified version of the USGS 3-D finite-difference groundwater model MODFLOW. Response functions generated by MODRSP are provided to support simulation of spatially varied and time-lagged return/depletion flows from stream–aquifer interactions. Results of the case study indicated significant differences between using groundwater response coefficients developed from preassigned stream depletion factor (SDF) values, as currently used in the basin, and those generated using a finite-difference groundwater model.

The groundwater–stream-flow model was applied to a segment of the South Platte River downstream of Denver. Since wells are included in the priority system, pumping must be administered to prevent injury to senior water rights. To protect senior surface-water rights, and prevent interruption of well pumping during the irrigation season, well owners are allowed to implement a groundwater-augmentation plan. This allows replacement of well-depletion flows through groundwater recharge, water exchanges, and water trades. Water District #1 has over 500 decreed large-capacity wells, 20 conditional or approved augmentation plans, and close to 100 monitored groundwater-recharge sites.

MODRSP was used to generate stream–aquifer response coefficients for each of the 196 wells and 32 recharge sites identified in the Bijou Augmentation Plan. Sets of coefficients for each well and recharge site were developed to simulate the effects of groundwater pumping and recharge on the South Platte River and its major tributaries. The Lower South Platte River was divided into 29 reaches with 11 separate tributaries.

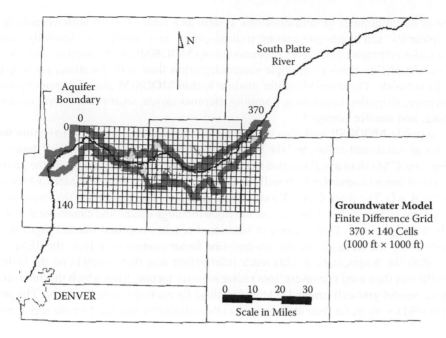

FIGURE 6.10 Groundwater model finite-difference grid for Lower South Platte study. (From Fredericks et al. 1998. With permission.)

The network for the MODRSP finite-difference groundwater model was constructed to cover all of Water District #1 located within the South Platte alluvial aquifer. A 370 × 140 groundwater grid network, with each cell having dimensions of 1000 × 1000 ft, was developed using GIS techniques (Figure 6.10). MODRSP transmissivity, boundary condition, river, and well data input files were developed using GIS and database-management procedures. Hydrography for the study area was read into AutoCAD® (Autodesk, Inc.) from U.S. Census Bureau TIGER files, with data edited into single AutoCAD polylines. Various federal- and state-agency maps were used to identify all river reaches, drains, canals, and reservoirs. All maps and spatial data prepared for the numerical finite-difference groundwater model were geocoded with a common UTM (Universal Transverse Mercator) reference system, requiring the conversion of several map sources from Lat-Long references.

Aquifer transmissivity data were digitized from transmissivity maps for the Greeley, Weldonna, and Brush reaches of the South Platte River. The GIS package IDRISI® (Eastman 1992) was used to develop a raster grid file from the contour data, with cells outside the aquifer assigned a transmissivity value of zero. Data in units of 1000 gal/day/ft were converted to ft^2/s for input into MODRSP. The boundary file was also developed using IDRISI, with cells located within the aquifer assigned a value of 1 and cells outside the aquifer assigned a value of 0 to represent no flow conditions. Reservoirs were specified as constant head boundaries and given a value of −1.

Since the governing groundwater flow equation is linear and time invariant, linear system theory can be applied via the principle of superposition (Bear 1979), allowing individual excitation events to be calculated independently and their responses linearly combined. Applying linear system theory to the groundwater flow equation allows use of Green's function to solve the resulting nonhomogeneous boundary value problem (Maddock 1972). Response of the groundwater system to external excitations such as pumping, recharge, or infiltration at any point in space and time can be expressed as a set of unit coefficients independent of the magnitude of the excitation.

Maddock and Lacher (1991) developed MODRSP as a modified version of MODFLOW for calculating kernel or response functions for stream–aquifer interactions. MODRSP is an appropriate numerical model for determining response-function coefficients, since it allows modeling of a multiaquifer groundwater flow system as a linear system with irregularly shaped areal boundaries and nonhomogeneous transmissivity and storativity. Spatially distributed stream–aquifer response coefficients generated using MODRSP can be used to allocate groundwater return/depletion flows to multiple return/depletion flow node locations anywhere in the river basin network. The stream–aquifer module within MODSIM allows consideration of reservoir seepage, irrigation infiltration, pumping, channel losses, return flows, river depletion due to pumping, and aquifer storage.

Three separate MODRSP well files were prepared representing the Bijou irrigation wells, the Bijou recharge canals and drains, and the Bijou recharge ponds and reservoirs. Well data were digitized using AutoCAD from a well location map. The associated grid cell, along with the groundwater grid row and column number for each well, was directly calculated from the well x,y locations provided in AutoCAD. The Bijou Canal, Kiowa Creek, and Bijou Creek recharge sites were subdivided into 26 separate recharge sites. Locations of the Bijou recharge drains and canals are available from TIGER hydrography data. Demarcation of individual reaches required manual digitizing of this information using the recharge site stream-depletion factor contour map from the Bijou Irrigation Company Plan for Augmentation. This reach information was then overlain on the hydrography data. IDRISI was then used to convert from vector to raster format, from which the finite-difference groundwater model grid-cell row and column locations for each site were produced. The process of identifying grid locations for river and tributary cells is similar to that used for the recharge channel and drain well file, although somewhat more complicated.

With transmissivities available for each grid cell as an IDRISI® raster file, saturated thickness data were digitized into AutoCAD from USGS saturated thickness maps for the Greeley, Weldonna, and Brush reaches of the South Platte River. The resulting contour map was transferred to an IDRISI vector file format, from which IDRISI was then used to develop a raster grid file from the saturated thickness contour data. Cells outside the aquifer were assigned a saturated thickness value of zero. The IDRISI transmissivity and saturated thickness files were imported into an attribute database, and the data required for river and tributary grid cells were extracted. Reach-width data were assigned as a polyline width to each river reach and tributary in AutoCAD and extracted as grid-cell data using AutoLISP programs and spreadsheet software. Width data were then linked by cell to the river and tributary cell data in DBASE IV. Tributary width was derived using data from the Colorado Highway Department Bridge Division database. The South Platte River width was set at 150 ft (50 m) based on previous South Platte River studies. A separate AutoLISP program was developed to determine river reach lengths in each grid cell, and spreadsheet software was used to assign a cell location to each grid and its attribute width. This information was then linked with the river and tributary data in the attribute database. Riverbed conductance values were also calculated, and the final results were output from the attribute database to an ASCII text file for use by MODRSP.

MODRSP was set up to generate response coefficients for a 120-month period, with specific yield set at a constant value of 0.16. The response data file output from MODRSP was processed using the attribute database, with cells accumulated by recharge and well site. Well, recharge, river reaches, and tributary site numbers were assigned actual MODSIM node values. The results were exported to an ASCII text file and input into MODSIM as 1079 sets of coefficient data representing monthly response data for a 10-year period for 193 wells, 30 recharge sites, 13 river reaches, and 4 tributaries.

6.5.2.3 Rio Grande Valley Groundwater Model

The Rio Grande basin within Colorado is located in south-central Colorado and encompasses approximately 7500 square miles. The primary feature of the basin is an open, almost treeless, relatively flat valley floor (known as the San Luis Valley) surrounded by mountains. Agricultural activities account for more than 85% of basin water consumption, with an estimated 638,000 acres under irrigation. The primary crops are potatoes, carrots, small grains, and alfalfa.

The basin has been the focus of a major groundwater modeling effort directed to assessing the amount of recoverable water (CWCB 2005). There has been interest by Colorado in meeting its commitments to New Mexico and Texas under the Rio Grande Compact and to assist the United States in meeting its commitments to Mexico under the Treaty of 1906. Various federal and state legislation and court decrees have been put forth to authorize pumping from the basin at various amounts to meet downstream obligations. There are also requirements for water rights administration for both surface- and groundwater in the Colorado portion of the Rio Grande basin. The Rio Grande Decision Support System (RGDSS 2005) program has been developed to provide the tools and information appropriate for water management of the region.

Saturated sedimentary and volcanic rock layers comprise the aquifers of the valley, from which numerous wells draw water. Groundwater continues to recharge the pore spaces in the aquifers by percolation from surface streams, leakage through canals, deep percolation from irrigation, and recharge from the mountains surrounding the valley. This complex, interconnected aquifer system is, in many areas, in hydrologic connection with the surface-water system. The aquifer system is generally composed of a shallow, unconfined aquifer and a deeper, confined aquifer. The uppermost water-saturated layer of sand and gravel, down to a depth of about 100 ft across most of the valley, is the unconfined aquifer. Below the unconfined aquifer, in the central part of the valley, are a number of clay layers that serve to separate, although not totally disconnect, the unconfined aquifer from deeper water-bearing layers of sand, gravel, and fractured volcanic rocks. The deeper layers, of which there are many, together make up the confined aquifer because of the overlying and confining clays. Water flows from many wells completed in the confined aquifer due to natural artesian pressure.

The RGDSS was directed to addressing these groundwater issues by activities for groundwater modeling, consumptive use modeling, new data collection, and DSS integration. The groundwater simulation modeling effort required new data on stream flows, piezometric pressures, consumptive uses, stratigraphy, topography, and wells. GIS databases and tools of various types provided a primary means for accomplishing the project. The coverages developed for the RGDSS groundwater model included rivers/streams, canals, drains, surface-irrigated lands, groundwater-irrigated lands, wells, nonirrigated lands, soils, rim inflows, diversion locations, and gauge locations. All GIS coverages were developed on a common datum and with consistent units.

The groundwater computer modeling package MODFLOW was used as the primary simulation tool to analyze the movement and impact of pumping wells on the surface-water system in the Rio Grande basin. Procedures for developing input files for MODFLOW were described by Rindahl and Bennett (2000). The groundwater modeling interface system (GMS) involved integration of GIS coverages, relational databases, and consumptive-use model results into a coordinated package for generating input files required by the following MODFLOW packages: (a) basic package, (b) block centered flow package, (c) general head boundary package, (d) stream or river package, (e) output control, (f) solver package, and (g) drain package. GMS also exports the model cell-grid system so that it can be processed by the ArcView data analysis.

The GIS coverages and the model grid file were then used by an ArcView Avenue application to compute model inputs on a cell-by-cell basis. This application determines, for each structure, the cells that are associated with a particular groundwater flux term (deep percolation, pumping, canal recharge, evapotranspiration, and pumping). For example, a ditch system or structure may provide groundwater recharge via canal leakage to one or more groundwater cells. Similarly, a structure may provide groundwater recharge via deep percolation from irrigated lands to one or more model cells. Other groundwater flux terms that have a spatial component include well pumping, precipitation recharge on nonirrigated lands, and rim inflows. To associate the irrigated lands with one or more wells serving them, a spatial comparison between the irrigated lands and the well coverage was performed. A parcel was estimated to be served by groundwater if the parcel was irrigated by sprinklers or had an irrigation well located within or in proximity to the parcel. Figure 6.11 illustrates the well-to-parcel assignment procedure.

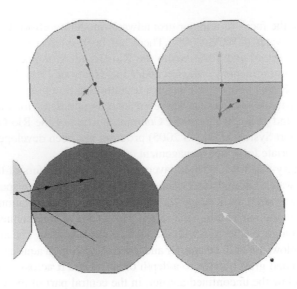

FIGURE 6.11 Well-to-parcel assignments are illustrated to show which parcels are served by a particular well. (From RGDSS 2005. With permission.)

The state's Consumptive-Use Model, StateCU, performs a water supply analysis of consumptive use (CU) by structure. As part of the CU analysis, StateCU provides canal loss, surface water applied, and capacity-limited pumping by structure. In addition to the parametric CU data (crop coefficients, root zone depth, etc.), the CU model uses historic climate (temperature, precipitation, and frost data) and diversion records obtained from the relational database, HydroBase. It also uses the irrigated lands, soil types, and water supply (surface water only, groundwater only, or both) from the spatial data. The four output files resulting from the State CU model include time series of: (a) surface water diverted and applied for each structure, (b) total canal losses for each structure, (c) well pumping for each structure, (d) deep percolation from irrigation, and (e) precipitation recharge for various geographic zones in the model.

Similar spatial-analysis procedures were applied in assigning aquifer recharges from canal leakage, surface-water irrigation, groundwater irrigation, rim inflows, and precipitation on nonirrigated lands. For example, recharge from canal leakage for each cell containing a canal or lateral is generated using the following formula:

$$\text{TotLeakage}_{ijk} = \frac{\left(\text{Length}_{ijk} \times \text{Weight}_{ijk} \times \text{Leakage}_k\right)}{\displaystyle\sum_{i,j}\left(\text{Length}_{ijk} \times \text{Weight}_{ijk}\right)} \tag{6.2}$$

where
i	= model cell row
j	= model cell column
k	= structure number
Length_{ijk}	= length of structure k in cell i,j
Weight_{ijk}	= weight to be applied from structure k in cell i,j
Leakage_k	= canal loss from structure k
TotLeakage	= total recharge from canal loss to cell i,j

Canal length (Length_{ijk}) and weight (Weight_{ijk}) are obtained from the ArcView data analysis. The weight typically assigned to each cell is 1.0, although a user may adjust the weight to represent canal segments that carry a greater volume of water, different soil types, etc. The monthly canal loss by structure (Leakage$_k$) is obtained from the StateCU analysis. The preprocessor executes the above formula to estimate total canal leakage by cell (TotLeakage) at any desired time step (monthly, annual, average annual, etc.) within the study period.

In addition to assigning irrigation wells to groundwater-irrigated parcels, the GIS is used to estimate the distribution of pumping for each well by model layer. The layers defining the bottom elevation of each aquifer are established as grids in an Excel spreadsheet, with every bottom elevation by layer for each cell. These tabular data are imported as an event theme in ArcView. The event theme has points at each model cell center, with the bottom elevations for each layer as attributes. Using the Spatial Analyst extension to ArcView, a surface is created as an ArcView grid that passes through each elevation point. These grids are then used for analysis. The first layer is the ground elevation in an ArcView grid format for the model area, and each subsequent layer is the bottom elevation for each aquifer in ArcView grid format. Attributes for each well obtained from the relational database are used to define the percentage of well capacity and pumping from each well layer.

Calibration of the Rio Grande basin groundwater model was performed for steady-state (1990–1998) and average monthly (1970–2002) study periods (RGDSS 2005). It was determined to be adequate when the differences between observed and simulated flow values and heads were minimized while maintaining aquifer parameters within a reasonable range. Calibration included evaluation of: (a) groundwater budget; (b) change in storage; (c) stream flow, diversions, and gain-loss; (d) observation wells; (e) total evapotranspiration; and (f) other nonnumeric data (e.g., dry cells, flooded cells). The steady state is the simplest comparison, but it does not include any storage change estimate or the impact of seasonal water level changes on head-dependent terms like stream gain, flowing wells, and general heads. The monthly is the most complicated comparison because it includes storage changes and includes the impact of monthly water level changes on head-dependent terms.

Figure 6.12 presents a residuals plot that shows the differences between simulated heads and observed heads for the 903 wells where observed data were compared to simulated values; the map display is for layer 1 of the model (the topmost layer of the four-layer model). The residual plots indicate the model reproduces observation wells fairly accurately, with better results in the center of the valley than the boundaries. Simulated heads in the Costilla Plain (southeast portion of the basin) match observed heads relatively poorly. Many of the large residuals occur near the model perimeter, where a very steep gradient of the ground surface and substantial subsurface faulting exist. Observation wells commonly only partially penetrate a model layer. MODFLOW calculates the vertical gradient between layers but not within layers. Therefore, near the model perimeter, where a significant vertical gradient occurs, a difference between the simulated and observed groundwater level values is to be expected.

6.5.3 GROUNDWATER QUALITY AND MODELING

Groundwater and surface-water systems are interconnected through recharge of surface water into the subsurface and by discharge of groundwater back to surface streams and wetlands. The natural balance between surface water and groundwater usually leads to groundwater having a high-quality level free of particulates and bacteria, and requiring little or no treatment. This high quality is often cited as an advantage of groundwater over surface-water supply sources. However, polluted surface water increasingly finds its way into the subsurface, resulting in deteriorated quality that is difficult to remove. There are many sources of groundwater contamination, including septic tanks and other wastewater discharges, leaking underground storage tanks, landfills, mine drainage, road salting, and urban runoff. Agricultural activities for farm animal wastes, excessive fertilizer and pesticide applications, and irrigation return flows are a major concern for groundwater pollution potential. Historic discharges of hazardous chemicals onto the land and in buried disposal sites have resulted in extensive groundwater contamination and prompted legislation requiring site remediation. Novotny (2003) provides a comprehensive review of groundwater pollution sources and mechanisms.

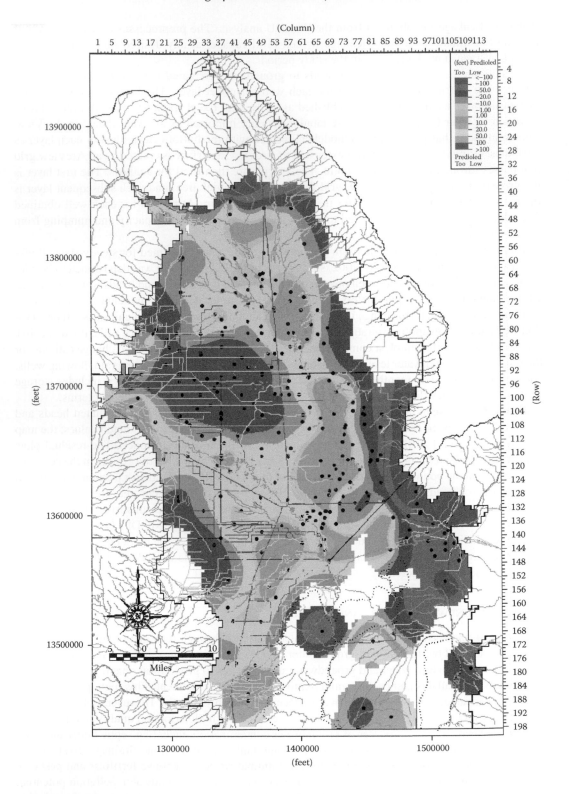

FIGURE 6.12 (See color insert following page 136.) Residual head for steady-state simulation for layer 1 of the Rio Grande groundwater model. (From RGDSS 2005. With permission.)

The transport and spread of contamination in groundwater is governed in part by the hydraulics of porous media. Advective transport of contaminants occurs with the bulk of water and is related to the average linear velocity as defined by Darcy's law. However, groundwater velocity fields are typically not uniform due to heterogeneity and anisotropy of aquifer materials. There is a tendency for contaminated water to progress preferentially through the higher conductivity paths of an aquifer, thus negating predictions based on average conditions. There results a spreading of the contaminants in the longitudinal and transverse directions; this is called *hydrodynamic dispersion*. Transverse dispersion is perpendicular to the primary flow direction and is smaller than longitudinal dispersion, which is along the direction of flow. There is also molecular diffusion, but this is usually small in comparison. Furthermore, the contaminant may interact with the aquifer materials and be retarded through adsorption; or the contaminant, such as bacteria, may biodegrade. Biodegradation can actually be a method of remediation or treatment.

Models of groundwater contamination are many and varied. Various procedures for accounting for pollutant loading conditions on aquifer recharge areas have been developed to support land-management strategies that protect groundwater quality. The DRASTIC (see next section) model is a popular example of this type. Classical solutions to the hydrodispersive equations have been applied with reasonable results to contamination situations where steady-state hydraulics and instantaneous or continuous release conditions can be assumed (e.g., Fetter 1994). These equations are generally represented as "plume" models. Numerical extensions of the classical approaches have been developed that take better account of the heterogeneity of aquifer materials and complex boundary conditions. Examples of these groundwater quality modeling approaches are described in the following section.

6.5.4   DRASTIC

The DRASTIC method has been used to characterize groundwater vulnerability to contamination at many locations (e.g., Rundquist et al. 1991; Rupert 1999). Groundwater vulnerability maps are designed to show areas with the greatest potential for groundwater contamination on the basis of hydrogeologic and human factors. The maps are developed by using GIS to combine data layers such as land use, soils, and depth to water. Typically, groundwater vulnerability is determined by assigning point ratings to the individual data layers and then adding the point ratings together when those layers are combined into a vulnerability map. DRASTIC is named for the seven factors considered in the method: Depth to water, net Recharge, Aquifer media, Soil media, Topography, Impact of vadose zone media, and hydraulic Conductivity of the aquifer (Aller et al. 1985):

- D—Depth to water table: Shallow water tables pose a greater chance for the contaminant to reach the groundwater surface as opposed to deep water tables.
- R—Recharge (net): Net recharge is the amount of water per unit area of the soil that percolates to the aquifer. This is the principal vehicle that transports the contaminant to the groundwater. The more the recharge, the greater the chances of the contaminant to be transported to the groundwater table.
- A—Aquifer media: The material of the aquifer determines the mobility of the contaminant through it. An increase in the time of travel of the pollutant through the aquifer results in more attenuation of the contaminant.
- S—Soil media: Soil media is the uppermost portion of the unsaturated/vadose zone characterized by significant biological activity. This, along with the aquifer media, will determine the amount of percolating water that reaches the groundwater surface. Soils with clays and silts have larger water-holding capacity and thus increase the travel time of the contaminant through the root zone.
- T—Topography (slope): The higher the slope, the lower is the pollution potential due to higher runoff and erosion rates. These include the pollutants that infiltrate into the soil.

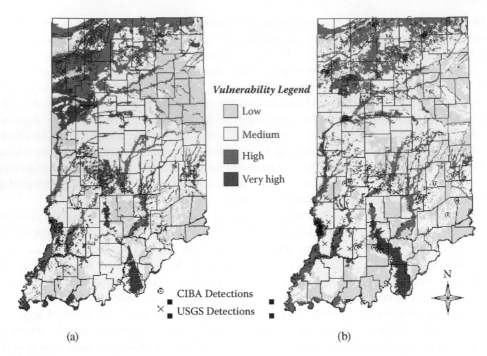

FIGURE 6.13 DRASTIC ratings for Indiana: (a) pesticides, (b) nitrates. (From Cooper et al. 2007. With permission.)

- I—Impact of vadose zone: The unsaturated zone above the water table is referred to as the vadose zone. The texture of the vadose zone determines how long the contaminant will travel through it. The layer that most restricts the flow of water will be used.
- C—Conductivity (hydraulic): Hydraulic conductivity of the soil media determines the amount of water percolating to the groundwater through the aquifer. For highly permeable soils, the pollutant travel time is decreased within the aquifer.

Each factor is assigned a rating for different ranges of the values. Each factor is also assigned a weight based on its relative significance in affecting the pollution potential. The typical ratings ranges are from 1 to 10, and the weights are from 1 to 5. The DRASTIC Index is computed by summation of the products of rating and weights for each factor as follows:

$$\text{DRASTIC Index} = D_r D_w + R_r R_w + A_r A_w + S_r S_w + T_r T_w + I_r I_w + C_r C_w \tag{6.3}$$

where D_r, R_r, A_r, S_r, T_r, I_r, and C_r are the ratings, and D_w, R_w, A_w, S_w, T_w, I_w, and C_w are the weights. A high numerical index is indicative of a geographic area that is likely to be susceptible to groundwater pollution. Figure 6.13 shows DRASTIC results for the state of Indiana.

There has been some concern with the DRASTIC approach because it is often applied without linkage to sampled data. Rupert (1999) demonstrated an improvement to the basic DRASTIC concept in Idaho by calibrating the point rating scheme to measured nitrite plus nitrate as nitrogen concentrations in groundwater. Statistical correlations were performed between the nitrogen concentrations and land use, soils, and depth-to-water data; nonparametric statistics and a GIS were used to quantify the relations. On the basis of the relations, a point rating scheme was developed that classifies areas according to their potential for groundwater contamination by nitrogen. That point rating scheme then was entered into the GIS, and the probability map was produced. The Wilcoxon rank-sum nonparametric statistical test (Ott 1993) was used to correlate nitrogen concentrations with land use, soils, and depth-to-water data. This test determines whether differences in nitrogen concentrations are statistically

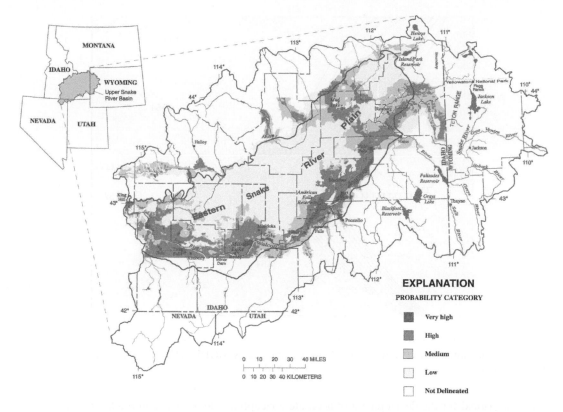

FIGURE 6.14 Probability of groundwater contamination by nitrogen, eastern Snake River Plain, Idaho. (From Rupert 1999.)

significant between the various data groups, for example, whether nitrogen concentrations in groundwater in irrigated agricultural areas are statistically different from concentrations in rangeland areas. The test calculates a p-value; if the resulting p-value is less than 0.05, then the data sets are significantly different at the 95% confidence level. For that report, the 95% confidence level was used as the cutoff value for determining whether differences between data sets were statistically significant. Figure 6.14 shows the potential for elevated nitrogen concentrations that was developed.

6.5.5 CONTAMINANT PLUME MODELING

Demonstration of the use of GIS tools for monitoring and modeling oil-spill distributions through time has been accomplished for a number of contaminant plume situations. Experience indicates that, even for a small site, if the appropriate data are available, then the GIS can be a valuable tool in understanding hydrocarbon plume migration and can help in tracking site treatment and remediation progress. Linking surface waters to groundwater can be done if you have the high-resolution stream data required for small spill sites. Tracking analyses and forward modeling of plume migration are now standard practice (Figure 6.15). Building 3-D geological models and combing them with porosity and permeability distributions across the site are now becoming available; such visualizations help in providing the 3-D view of the site with different surfaces and lithologies. The combined results provide managers with distribution profiles for spill containment and treatment that result in better resource management and huge time and money savings. The U.S. EPA distributes various public-domain groundwater models as well as links to commercial products (http://www.epa.gov/ada/csmos/models.html).

GIS databases are invaluable for studying basin-scale groundwater problems with complex lithology and decades of groundwater elevation and quality data. Compilation and efficient retrieval of

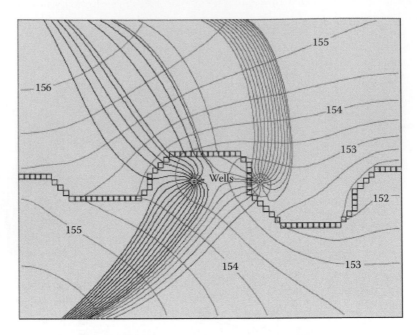

FIGURE 6.15   Flow paths of particles discharging at two wells. Pathlines, which cross the contours of water levels (denoted by feet above sea level), were computed by MODPATH. Squares are river cells. (From USGS 1997.)

data are critical for the development of accurate conceptual site models and tracking dynamic site conditions. Fleetwood (2008) demonstrated the development of a geodatabase and geoprocessing application for implementation in a GIS to aid in environmental data management and analysis. The GIS compiled for the project encompassed 20 years of historical environmental site data for a former petroleum refinery. The geodatabase houses tabular analytical data as well as geologic and groundwater well data, along with spatial data layers of well and boring collection sites, parcel data, roads, and historical site features. The main purpose of the geoprocessing application is to re-create a time series of hydrocarbon thickness maps dating back to 1986. To manage tables and feature classes, the geodatabase was constructed using ArcCatalog®. The analysis tool was built using ArcMap/ArcObjects® using Visual Basic for Applications. The geoprocessing tool used the ArcGIS Spatial Analyst® extension to create a raster surface for each time period representing the hydrocarbon thickness. The application was written to obtain data from the geodatabase, create the raster surfaces, symbolize by layer file, name the raster, and return the output to the geodatabase. Various spatial interpolation methods were applied to map plume location and thickness based on the data obtained from the sampled observation well. The interpolation methods included natural neighbors, inverse distance weighted, and ordinary kriging; the natural-neighbor method was chosen to best represent the contaminant distribution (Figure 6.16).

6.6   VISUALIZATION

The role of visualization has become increasingly important in order to convey site information to stakeholders who are not technical experts. A demonstration of visualization techniques applied to groundwater was provided by Rogoff et al. (2008). They combined GIS databases with the Environmental Visualization System (EVS) Software® by Ctech Corporation. EVS was used to develop 3-D visualizations integrating GIS information, database queries, geologic information, and groundwater model output. Visualizations of several basin-scale sites combining GIS and EVS were developed using the Ctech 4DIM Player®. This application allows unlimited zoom, rotation, and

FIGURE 6.16   Example of the chosen natural-neighbor interpolation method. (From Fleetwood 2008. With permission.)

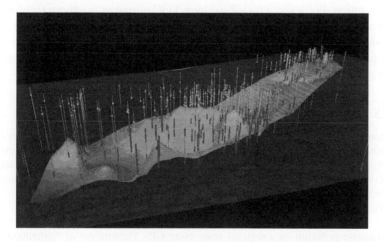

FIGURE 6.17 (See color insert following page 136.)   Well posts and screen intervals of wells, and the bedrock surface topology. A transparent georeferenced aerial photograph allows viewers to remain oriented as the model is rotated interactively using the EVS 4DIM Player. (From Rogoff et al. 2008. With permission.)

panning, resulting in unparalleled abilities to evaluate any desired area of the site in detail. Videos of groundwater flow model output were also demonstrated. One case study was for a Superfund site in the southwest. A GIS database with lithologic information from over 400 borings, groundwater elevations, and analytical results obtained over two decades was assembled using ArcView. The lithologic information was used to prepare shapefiles of isoelevation contours of hydrostratigraphic subunits in the complex underlying alluvium and bedrock. Shapefiles of quarterly groundwater isoelevations and VOC (volatile organic compounds) isocontour maps were prepared for wells screened in the subunits. The GIS shapefiles were visualized in 3-D in EVS, and animations of groundwater flow and VOC plumes through time were prepared. Figure 6.17 shows well posts and screen intervals of wells, and the bedrock surface topology.

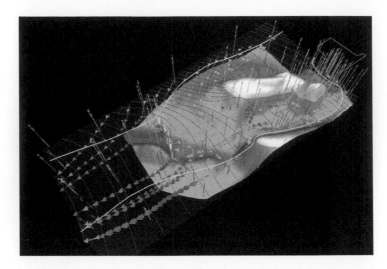

FIGURE 6.18 (See color insert following page 136.) Groundwater flow and equipotential lines. Note the influence of the bedrock ridges, visible as white surfaces protruding above the water table, on groundwater flow. Capture zones of three large extraction wells and stagnation areas are readily apparent in animations. (From Rogoff et al. 2008. With permission.)

Another case study was conducted by Rogoff et al. (2008) to portray the subsurface character of contaminant migration. For a site with a 2-mi-long VOC plume in Southern California, GIS information was combined with output from a finite-element groundwater flow model with 42,000 cells and 7 layers. Modeled outputs for various extraction well scenarios were combined with the GIS information to prepare animations of particle tracks from four regions of interest. Figure 6.18 shows a visualization of groundwater flow paths.

REFERENCES

Aller, L., T. Bennett, J. H. Lehr, and R. J. Petty. 1985. DRASTIC: A standardized system for evaluating groundwater pollution potential using hydrogeologic settings. EPA/600/2–85/018. U.S. Environmental Protection Agency, Robert S. Kerr Environmental Research Laboratory, Office of Research and Development.

Alley, W. M., R. W. Healy, J. W. LaBaugh, and T. E. Reilly. 2002. Flow and storage in groundwater systems. *Science* 296 (June): 1985–1990.

Bear, J. 1979. *Hydraulics of groundwater*. New York: McGraw-Hill.

Cooper, B. S., K. C. Navular, B. A. Engel, and L. Hahn. 2007. Groundwater vulnerability evaluation to pesticide and nitrate pollution on a regional scale using GIS. Department of Agricultural and Biological Engineering, Purdue University, W. Lafayette, Ind.

CWCB. 2005. Rio Grande Decision Support System. Colorado Water Conservation Board. http://cdss.state.co.us/DNN/.

Dogrul, E. C. 2007. Integrated Water Flow Model (IWFM v3.0). Theoretical documentation. California Dept. Water Resources, Hydrology Development Unit, Modeling Support Branch, Bay-Delta Office.

Eastman, J. R. 1992. IDRISI: A grid-based geographical analysis system. Grad. School of Geography, Clark Univ., Worcester, Mass.

Fetter, C. W. 1994. *Applied hydrogeology*. 3rd ed. New York: Macmillan.

Fleetwood, J. 2008. Development of a geodatabase and geoprocessing tool for environmental data management. Masters report, Dept. Civil Engineering, University of Colorado at Denver.

Fredericks, J., J. Labadie, and J. Altenhofen. 1998. Decision support system for conjunctive stream–aquifer management. *J. Water Resour. Plan. Manage. ASCE* 124 (2): 69–78.

Harbaugh, A. W. 2005. MODFLOW-2005: The U.S. Geological Survey modular ground-water model—the Ground-Water Flow Process. U.S. Geological Survey Techniques and Methods 6-A16. Washington, D.C.: USGS.

Heywood, B. J., K. J. Heisen, and K. E. Hard. 2008. Utilizing the power of GIS for surface water/groundwater modeling. Paper presented at AWRA 2008 Spring Specialty Conference, San Mateo, Calif.

Maddock, T., III. 1972. Algebraic technological functions from a simulation model. *Water Resour. Res.* 8 (1): 129–134.

Maddock, T., III, and L. Lacher. 1991. MODRSP: A program to calculate drawdown, velocity, storage and capture response functions for multi-reservoir systems. HWR Rep. No. 91-020. Dept. Hydrol. and Water Res., Univ. Arizona, Tucson.

Maidment, D. R., ed. 2002. *Arc Hydro: GIS for water resources.* Redlands, Calif.: ESRI.

Maidment, D. R., N. L. Jones, and G. Strassberg. 2004. Arc Hydro groundwater geodatabase. Center for Research in Water Resources, Univ. Texas, Austin. https://webspace.utexas.edu/gstras/MyWebsite/publications/ArcHydroGWGeoDBFinalForReview.doc.

McDonald, M. G., and A. W. Harbaugh. 1988. A modular three-dimensional finite-difference ground-water flow model. In *Techniques of water resources investigations of the U.S. Geological Survey.* Book 6, chap. A1. Denver, Colo.: USGS.

Novotny, V. 2003. *Water quality: Diffuse pollution and watershed management.* 2nd ed. New York: John Wiley and Sons.

Ott, R. L. 1993. *An introduction to statistical methods and data analysis.* Belmont, Calif.: Duxbury Press, Wadsworth Publishing.

Radin, H. A. 2006. Using ArcGIS for preprocessing and postprocessing a Modflow-96 groundwater model. Paper presented at GIS and Water Resources IV, AWRA Spring Specialty Conference.

RGDSS (Rio Grande Decision Support System). 2005. Calibration of the Rio Grande basin groundwater model. Report to the Colorado Water Conservation Board, Denver, Colo.

Rindahl, B., 1996. Groundwater modeling and analysis using the USGS MODFLOW program and ArcView. In *Proc. ESRI 16th Annual Users Conference.* Redlands, Calif.: ESRI.

Rindahl, B., and R. Bennett. 2000. Design and implementation of a data-centered groundwater modeling system. In *Proc. ESRI Annual Users Conference.* Redlands, Calif.: ESRI. http://gis.esri.com/library/userconf/proc00/professional/papers/PAP443/p443.htm.

Rogoff, E., J. Kivett, and C. Hickman. 2008. Integration of GIS and EVS for three-dimensional hydrogeologic visualizations. Paper presented at AWRA 2008 Spring Specialty Conference, San Mateo, Calif.

Rundquist, D. C., A. J. Peters, L. Di, D. A. Rodekohr, R. L. Ehrman, and G. Murray. 1991. Statewide groundwater-vulnerability assessment in Nebraska using the DRASTIC/GIS model. *Geocarto Int.* 6 (2): 51–58.

Rupert, M. G. 1999. Improvements to the DRASTIC ground-water vulnerability mapping method. U.S. Geological Survey water-resources fact sheet FS-066-99. http://id.water.usgs.gov/nawqa/PUBLICATIONS.html.

Shannon, T., J. W. Labadie, M. Baldo, and R. Larson. 2000. Integration of GIS and river basin network flow modeling. Paper presented at ESRI International Users Conference, San Diego, Calif. http://gis.esri.com/library/userconf/proc00/professional/papers/PAP628/p628.htm.

Tracy, F. T., T. C. Oppe, and S. Gavali. 2007. Testing Parallel Linear Iterative Solvers for Finite Element Groundwater Flow Problems. NAS Technical Report; NAS-07-007. September.

USGS (U.S. Geological Survey). 1997. Modeling ground-water flow with MODFLOW and related programs. USGS fact sheet FS-121-97. http://pubs.usgs.gov/fs/FS-121-97/.

Watkins, D. W., D. C. McKinney, D. R. Maidment, and M.-D. Lin. 1996. Use of geographic information systems in groundwater flow modeling. *J. Water Resour. Plan. Manage.* 122 (2): 88–96.

Winter, T. C., J. W. Harvey, O. L. Franke, and W. M. Alley. 1998. Ground water and surface water: A single resource. U.S. Geological Survey circular 1139. Washington, D.C.: USGS.

Zeiler, M. 1999. *Modeling our world: The ESRI guide to geodatabase design.* Redlands, Calif.: ESRI.

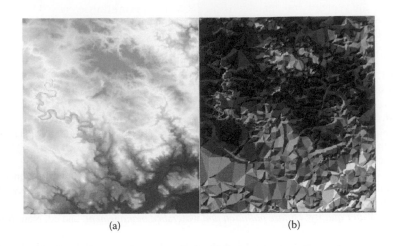

(a)   (b)

FIGURE 3.7   DEM represented as a) grid, and b) TIN. (From Close 2003)

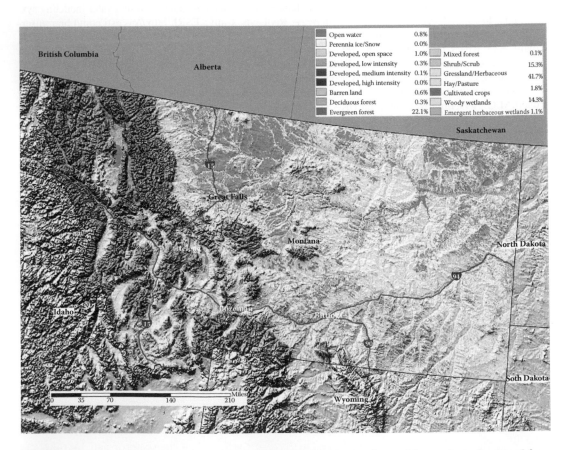

Open water	0.8%			
Perennia ice/Snow	0.0%			
Developed, open space	1.0%	Mixed forest	0.1%
Developed, low intensity	0.3%	Shrub/Scrub	15.3%
Developed, medium intensity	0.1%	Gressland/Herbaceous	41.7%
Developed, high intensity	0.0%	Hay/Pasture	1.8%
Barren land	0.6%	Cultivated crops	
Deciduous forest	0.3%	Woody wetlands	14.3%
Evergreen forest	22.1%	Emergent herbaceous wetlands 1.1%

FIGURE 3.15   Land-use–land-cover map for Montana is an example of one of the products developed from LandSat 7 imagery for all states in the United States (Source: http://landcover.usgs.gov/.)

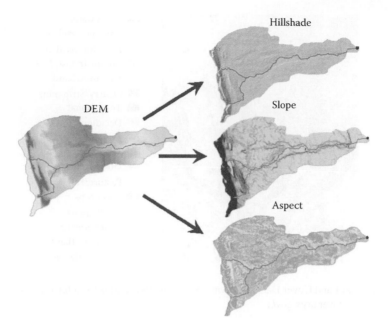

FIGURE 4.6 Slope, aspect, and hillshade can be derived from an elevation raster layer.

Marshes ■ Water ■ Farm land ■ Aquatic bed ■ Wet meadow

FIGURE 4.13 Image-processing techniques are used to classify land characteristics. Example identifies wetland areas in mixed agricultural landscape (Hsu and Johnson 2007).

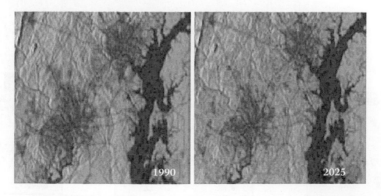

FIGURE 4.17 Land-use-change projections for 2025, made with a land-use-change model, show likely areas of new urban growth in yellow (high probability) and in greens (light green is moderate probability; dark green is low probability). (Source: USGS 1999.)

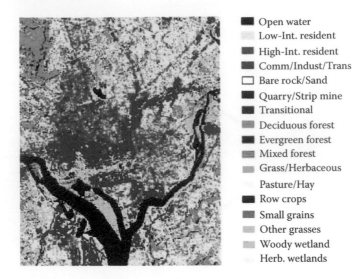

■	Open water
	Low-Int. resident
■	High-Int. resident
■	Comm/Indust/Trans
□	Bare rock/Sand
■	Quarry/Strip mine
■	Transitional
■	Deciduous forest
■	Evergreen forest
■	Mixed forest
■	Grass/Herbaceous
	Pasture/Hay
■	Row crops
■	Small grains
■	Other grasses
■	Woody wetland
	Herb. wetlands

FIGURE 5.4 National Land Cover Dataset is available nationwide and is used for watershed modeling studies. (Source: http://landcover.usgs.gov/.)

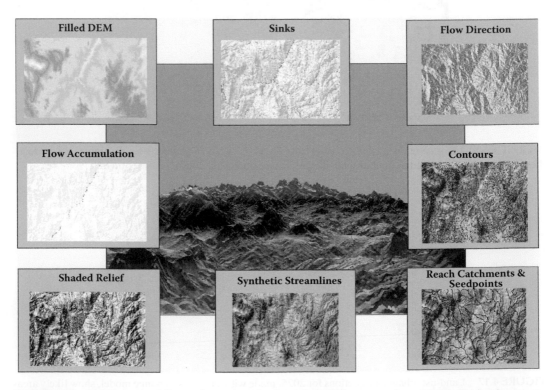

FIGURE 5.8 EDNA is a multilayered seamless database derived from a version of NED that has been conditioned for improved hydrologic flow representation. (Source: http://edna.usgs.gov.)

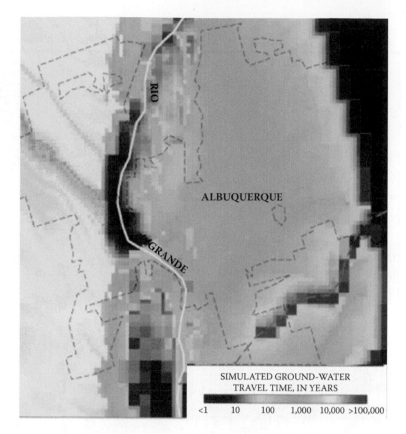

FIGURE 6.4 Application of particle tracking to estimate groundwater travel time. (From USGS 1997.)

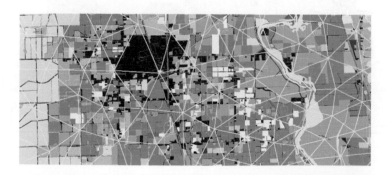

FIGURE 6.6 Land use with finite-element model grid overlay for integrated water-flow model (IWFM). (From Heywood et al. 2008. With permission.)

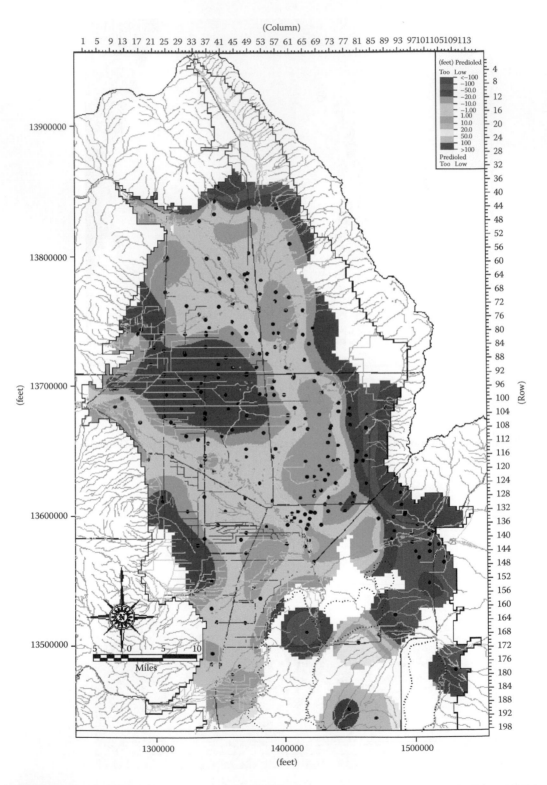

FIGURE 6.12 Residual head for steady-state simulation for layer 1 of the Rio Grande groundwater model. (From RGDSS 2005. With permission.)

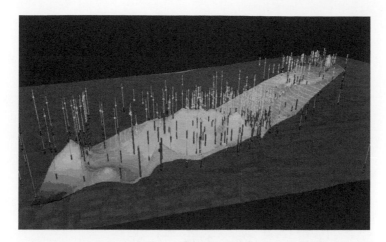

FIGURE 6.17 Well posts and screen intervals of wells, and the bedrock surface topology. A transparent georeferenced aerial photograph allows viewers to remain oriented as the model is rotated interactively using the EVS 4DIM Player. (From Rogoff et al. 2008. With permission.)

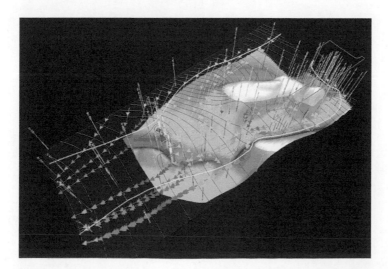

FIGURE 6.18 Groundwater flow and equipotential lines. Note the influence of the bedrock ridges, visible as white surfaces protruding above the water table, on groundwater flow. Capture zones of three large extraction wells and stagnation areas are readily apparent in animations. (From Rogoff et al. 2008. With permission.)

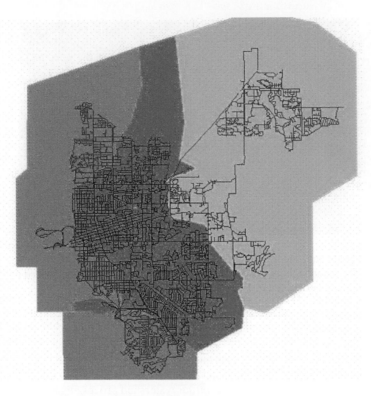

FIGURE 7.4   Pipe-network model completed with interconnections to pumps and tanks, and assignments to pressure zones. (From Szana 2006. With permission.)

FIGURE 8.6   Impervious surface areas (shown in yellow) extracted by image processing. (From Washburn et al. 2003. With permission.)

Proposed Floodplain

EXPLANATION OF ZONE DESIGNATIONS

ZONE | EXPLANATION

AE Zone — Defined as areas of special flood hazard inside the 100-year floodplain where water surface elevations have been determined. A 100-year flood event is a flood that has a one-percent chance of being equaled or exceeded in any given year.

X Shaded — Defined as areas of moderate flood hazard that are determined to be within the 500-year floodplain where there is a 0.2-percent chance of being equaled or exceeded in any given year, areas in the 100-year floodplain where average depths are less than one foot; and areas protected by levees from the 100-year flood.

X Unshaded — Areas determined to be outside 500-year floodplain

0   2,500   5,000   7,500   10,000
Feet

Date: August 2007

FIGURE 9.7 Proposed South Boulder Creek 100-year floodplain zone designation map (provisional, not fully approved). (Courtesy City of Boulder, Colo.; http://www.southbouldercreek.com/.)

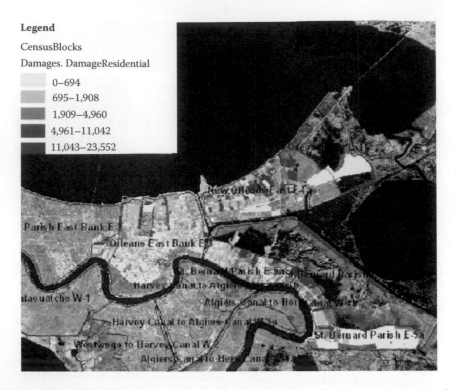

FIGURE 9.9 Distribution of Hurricane Katrina–generated residential direct property damages by census block; damages in thousands of dollars. (From IPET 2007. With permission.)

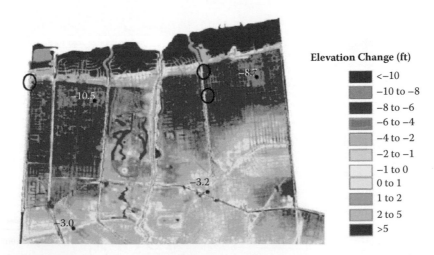

FIGURE 9.10 Changes in ground surface elevation due to land subsidence between 1895 and 2001 (URS 2006). Comparison is based on 1895 historic topographic map and 1999–2001 digital elevation models (DEMs). (From IPET 2007. With permission.)

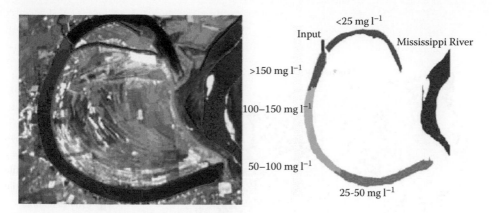

FIGURE 10.2 Landsat TM image of Lake Chicot, Arkansas (left), and a derived image (right) showing categories of suspended sediments mapped in Lake Chicot based on the radiance in the TM image. (From Ritchie et al. 2003. With permission.)

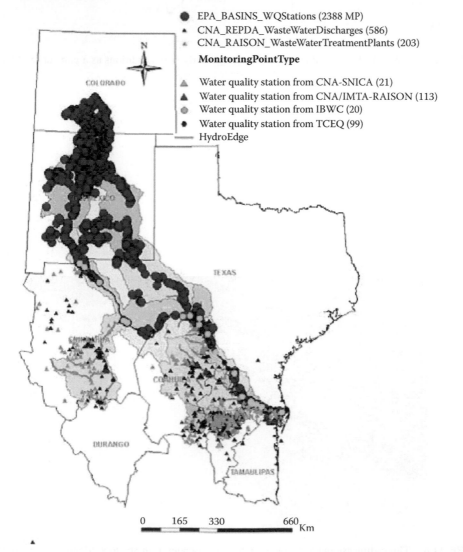

FIGURE 10.9 Monitoring points included in the WQDM geodatabase. (From McKinney and Patino-Gomez 2006. With permission.)

FIGURE 11.3 NWIS Web retrieval portrays river-gauge daily flow conditions as a percentile of the period of record flows for the selected day. (Source: http://waterdata.usgs.gov/nwis/.)

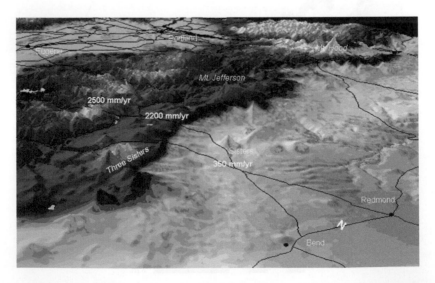

FIGURE 11.4 Three-dimensional view of the Oregon Cascades rain shadow. Mean annual precipitation drops from 2200 mm/year at the crest of the Cascades to only 350 mm/year just down the hill to the east. (From Daly et al. 1994. With permission.)

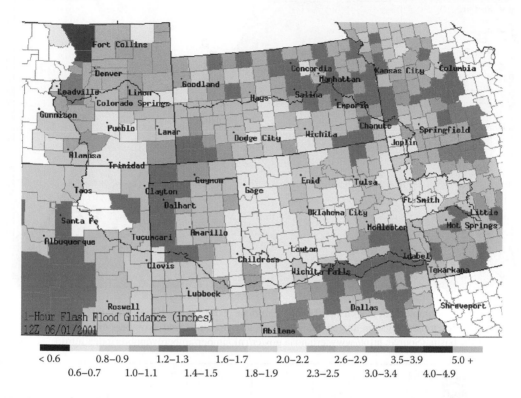

FIGURE 11.9 Flash-flood guidance product. (Source: http://www.srh.noaa.gov/abrfc/.)

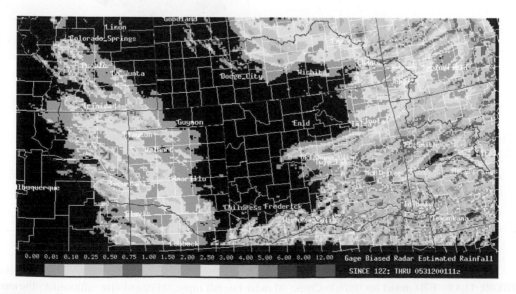

FIGURE 11.10 NEXRAD radar-rainfall display. (Source: http://www.srh.noaa.gov/abrfc/.)

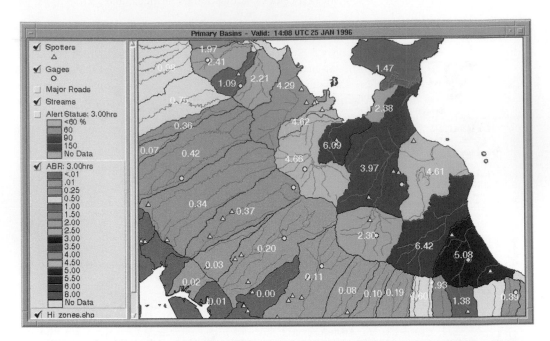

FIGURE 11.12 Sample ArcView GIS AMBER basin display. Basins are color coded by the 3-h ABR value for the basin. Streams and gauge and spotter locations also are displayed. Numbers are the actual ABR in inches for the basin. (Source: http://www.erh.noaa.gov/er/rnk/amber/amberov/ambrerindex.html.)

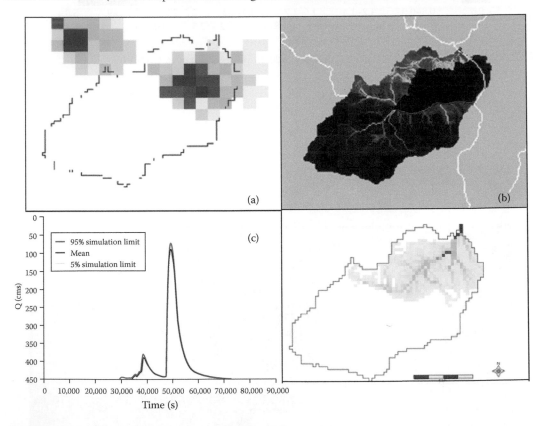

FIGURE 11.13 F2D model for Buffalo Creek: (a) radar-rainfall input, (b) cumulative infiltration, (c) runoff hydrograph, and (d) runoff flow distribution. (From Skahill and Johnson 2000. With permission.)

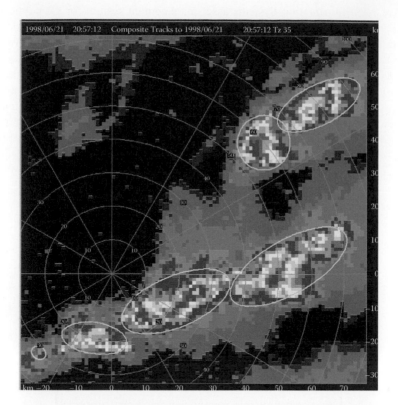

FIGURE 11.14 TITAN identifies storms as ellipses (or polygons) where radar reflectivity exceeds some threshold level. (Source: http://www.ral.ucar.edu/projects/titan/home/index.php.)

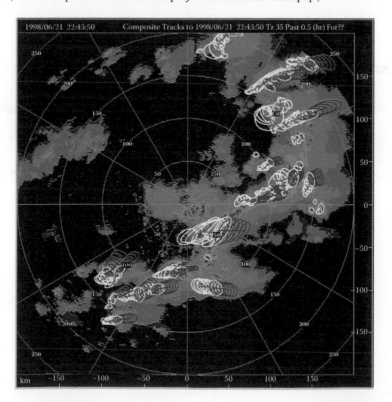

FIGURE 11.16 TITAN forecast display shows the current storm location in cyan, the recent (30-min) past in yellow, and the 30-min forecast in red. (Source: http://www.ral.ucar.edu/projects/titan/.)

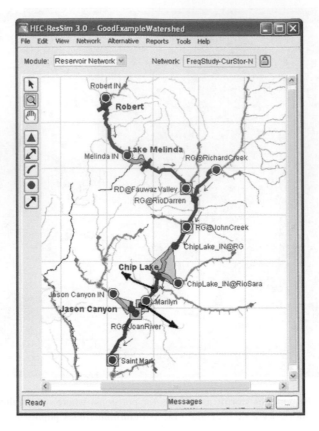

FIGURE 12.2 HEC-ResSim is a general-purpose river basin simulation package. (Source: Klipsch and Hurst 2007; http://www.hec.usace.army.mil/software/hec-ressim/.)

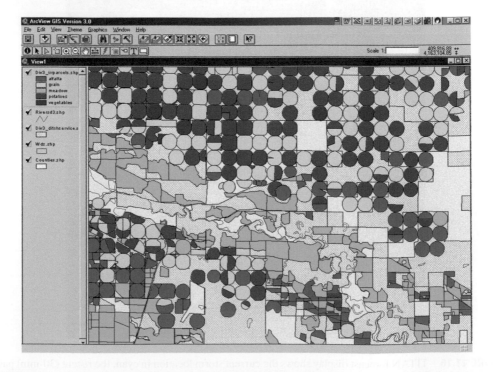

FIGURE 12.7 Display of irrigated acreage obtained through interactive query. (Source: http://cdss.state.co.us.)

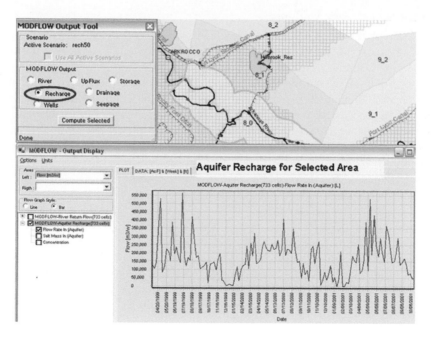

FIGURE 12.19 Geo-MODFLOW tool in ArcMap for integrating MODFLOW output with the Geo-MODSIM surface-water network. (From Gates et al. 2006. With permission.)

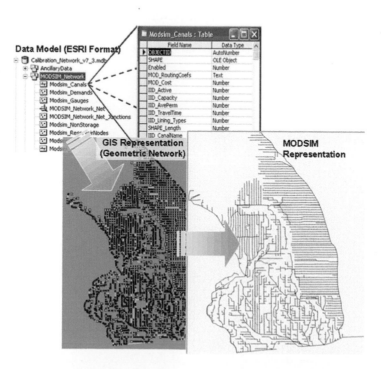

FIGURE 12.21 Generation of MODSIM network from Geometric Network developed from Geo-MODSIM data model. (From Triana and Labadie 2007. With permission.)

7 GIS for Water-Supply and Irrigation Systems

7.1 OVERVIEW

Water-supply and irrigation systems are fundamental infrastructure components sustaining public health and agricultural productivity. In any specific locale, there are distinctive circumstances of climate, topography, and geology that control the amounts of water available and provide the backdrop on opportunities for its capture and distribution. Water-supply system designs are also strongly influenced by the prevailing legal, administrative, and political (i.e., jurisdiction boundary) factors, which may limit the range of choice.

This chapter reviews water-supply systems and components, aspects of their design, and GIS concepts and tools that support these activities. Potable-water and irrigation supply systems are shown to be similar in some respects, although potable supplies require a great degree of positive control over quality. GIS tools are applied for assessment and evaluation of supply-facility alternatives. Estimation of demand is a fundamental GIS task that drives capacity requirements. Pipe networks and canal distribution systems are seen to be basic networks for which sophisticated simulation and optimization algorithms are applied. Selected case studies are reviewed.

7.2 WATER-SUPPLY AND IRRIGATION SYSTEMS PLANNING AND DESIGN

Water-supply systems are composed of various components for raw water collection, storage, treatment as needed, and distribution. Figure 7.1 illustrates components of a typical urban water-supply system and their interconnections. The water-supply subsystem comprises source, storage, transmission, treatment, and distribution through primary and secondary mains. Design of the surface-water-supply system concerns the locations and capacities of diversion works and storage, as well as the operations of these to meet multiple purposes and objectives. This topic is addressed in Chapter 5 (GIS for Surface-Water Hydrology) and Chapter 12 (GIS for River Basin Planning and Management). Groundwater supplies are addressed in Chapter 6 (GIS for Groundwater Hydrology). The supply distribution system design is based on network models that simulate pressurized pipe flows to nodal demand points and service storage tanks that maintain adequate water pressure and supply water for local demands and fire flows. Customer service connections are mainly tapped into distribution pipes and rarely into primary mains. Separate pressure zones are typically established in areas of topographic relief to maintain delivery pressures within prescribed limits. The design of a water-supply system requires forecasting of demands and planning for capacity additions so that demands may be met with minimal risk of supply shortfalls. Not every water-supply system has all components; some water-supply systems depend only on groundwater that serves as source storage and may not require treatment.

The primary elements of an irrigation system are similar to those for municipal water supplies, except raw-water treatment is typically not necessary. Source supplies arise from surface water, which is captured in reservoirs, and groundwater. Conveyance of supplies from the source is made through canals, ditches, and pipelines to the points of delivery. Diversion structures split canal flows into portions for the various irrigation projects. Other hydraulic equipment, including pumps, valves, and siphons, may be required to apply water to the fields. A portable pipe may be used to carry water from a well to a field. Considerable labor is involved in operating an irrigation system,

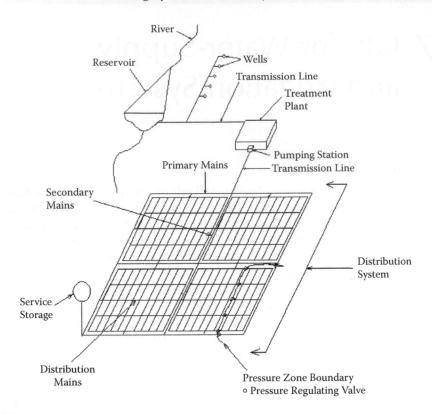

FIGURE 7.1 Water-supply system components and interrelations.

so efforts are made toward systems automation to reduce costs. Table 7.1 lists components and data for water-supply and irrigation systems.

7.3   WATER-SUPPLY SYSTEM DESIGN

7.3.1   ESTIMATION OF WATER-SUPPLY DEMANDS

Water-supply and irrigation system planning and design begin with demand estimation. Current and forecast services for the connected population and associated residential, commercial, and industrial flow demands establish the need for system capacity. These demand estimates are often based on current and forecast land-use patterns, which in turn are based on comprehensive plans, zoning maps, land-use mappings by imagery, and land-change models.

Water use varies from city to city, depending on climate, population characteristics, industrialization, and extent of conservation practices. Municipal uses for water may be divided into various categories such as domestic, commercial and industrial, and public uses (Table 7.2). Typically, in the United States, an average of 600 L/capita-day (160 gal/capita-day) is assumed, but this figure may differ considerably in a given locality, so records of actual water use and land-use maps are used to refine demand estimates. Domestic water use is that water used by private residences and apartment houses, etc., for drinking, bathing, lawn and garden watering, and sanitary and other purposes. Lawn and garden watering during warm weather periods often causes peak demands for the domestic sector.

Site-specific analyses of demands supported by GIS are typically used in lieu of the general-purpose demand factors cited here. Also, in any city, demand will vary from season to season, day to day, and hour to hour. Therefore, it is required that the average demands as well as the time variability be estimated. Table 7.2 shows peaking factors to be applied to average-day demands for

TABLE 7.1
Water-Supply and Irrigation Systems Data

Terrain
  DEM
  Slopes
  Drainage

Hydrological boundaries
  Watersheds and sub-basins

Supply sources
  Dams and reservoirs
  Groundwater and wells
  Recycling and re-use
  Transbasin diversions

Water availability
  Restrictions on water availability
    (time periods and water volumes)
  Diversion and application records
  Groundwater contribution
  Water quality (bacterial, salinity)
  Conservation practices
  Customer delivery records

Conveyance
  Canals
  Transmission lines
  Distribution canals
  Diversion structures

Land use
  Land uses
  Population
  Industries
  Zoning
  Master plan
  TAZ
  Forecast land use

Water-supply infrastructure
  Treatment plants
  Primary mains
  Secondary mains
  Distribution network
  Pumps
  Service storage tanks
  Pressure zones
  Valves
  Meters

Administrative boundaries
  Political boundaries
  Water-supply districts
  Irrigation districts
  Farm parcels

Irrigation infrastructure
  Water application facilities (sprinklers, furrows, etc.)
  Diversion structures
  Drains
  Flow gages

Agroclimate
  Precipitation (effective)
  Number of rainfall events
  Temperature (max. and min.)
  Wind speed
  Relative humidity
  Reference evapotranspiration

Soils (multilayered soil)
  Layer depths
  Soil textural percentages
  Soil water at field capacity
  Soil water at wilting point
  Initial soil water content
  Soil salinity

Crops
  Crops and cropping schedules
  Root depths and water-yield response factors
  Crop planting or season initiation date
  Dates of the crop development stages
  Crop coefficients
  Root depths
  Soil water depletion fraction for no stress
  Seasonal water-yield response factors

Irrigation practices
  Type of irrigation
  Irrigation thresholds
  Options to define application depths
  Fraction of soil surface wetted
  Number and depth of irrigation events
  Conservation practices

various demand categories. Capacity and costs of pipes, storage reservoirs, pumps, and other facilities are determined for alternative designs using forecasts of demands and flows, scheduled over design lifetimes. In a given locale, estimates based on recorded records in the city under study or data from similar cities are the best guide in selecting values for water-supply demands for planning

TABLE 7.2
Example Water-Use Factors

Land Use	Unit Flow Factors			Peaking Factors	
	Avg. Day	Max Day	Max Hour	MD/AD [a]	MH/AD [a]
	(L/min/ha)				
Multifamily	47			2.50	4.90
Commercial & office	17			2.00	3.98
High-rise office	116			1.89	2.48
High-rise industrial	20			1.53	1.66
Warehouses	4			1.53	1.66
Public	7			2.47	3.13
High-demand parks	7			2.5	4.00
	(L/min/tap)				
Single-family homes		6.0	7.5		

Source: Adapted from Hubly (1998).

[a] Average day (AD), maximum day (MD), or maximum hour (MH).

purposes. Given analyses of per capita usage and population forecasts out to a design period, total demands are computed.

Demands can be estimated based on maps of the land uses and various unit flow factors. The *land-use-area approach* computes average day (AD) demand as the land-use area × unit flow factor, where the unit flow factor is for the average day (AD). Maximum day (MD) or maximum hour (MH) demands are computed as the AD demand × peaking factor, where the peaking factor is the ratio of MD or MH to the AD demand rate. For a service area with several land uses, the total demand is calculated by summing the estimated water demands for each land use. The *number of users approach* computes demand as the number of users × unit flow factor, where the unit flow factor is the AD, MD, or MH water demand rate per user. Peak MD and MH flows may also be computed as AD demand × peaking factor. If the data on the number of users are not available, then they can be estimated based on the number of users per unit area. Sometimes more than one of these methods is used to estimate demand for a mixed-use area.

An example is illustrated in Figure 7.2 and Table 7.3 (adapted from Hubly 1998). The task is to find the maximum day (MD) and maximum hour (MH) demand for the area shown in Figure 7.3. Land use and area are shown for each parcel. The single-family area is zoned for 2.5 residences per ha. Table 7.3 includes MD and MH unit flow factors for single-family land use, so peaking factors are not required. However, if such unit flow factors were not available for the MD and MH demands, the calculation would use an AD unit flow factor and peaking factors as illustrated for the other land uses in this example. Note that this example estimates demands for the entire service area; for network modeling, these demands would be disaggregated to specific nodes in the network.

The unit flow and peaking factors shown above are samples of the broad spectrum of values commonly used. The factors will vary substantially among the geographic areas. The values included herein are useful for educational purposes, but they should not be used in practice without confirmation that these values truly represent the demand patterns of the subject service area. An important task in designing water-distribution systems is the selection of representative unit flow and peaking factors. When not specified by jurisdictional design specifications, a good approach to selecting these factors is to find existing service areas with available data that have populations, climates, topography, irrigation practices, economic conditions, and demographic patterns similar to the subject service area. Peaking factors will vary with the size of the service area; large service areas will have lower peaking factors.

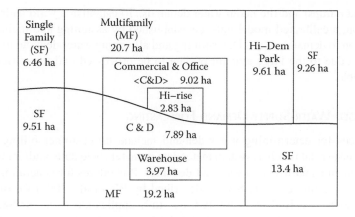

FIGURE 7.2 Example land-use plan used to illustrate the land-use method for computing water demands.

TABLE 7.3
Example of Peak-Flow Estimation Using Land-Use Method

Land Use	Area (ha)	Units per ha	Unit Flow Factors (L/min) AD	MD	MH	Peaking Factors MD/AD	MH/AD	Demands (L/min) MD	MH
Single family	38.63	2.5		6	7.5			579	917
Multifamily	37.92		47			2.50	4.90	4,691	9,194
Commercial & office	16.91		17			2.00	3.98	575	1,144
High-rise office	2.83		116			1.89	2.48	620	814
Warehouses	3.97		4			1.53	1.66	24	26
Irrigated park	7.61		7			2.05	4.00	138	269
							Total	6,628	12,365

Source: Adapted from Hubly (1998).
Note: Average day (AD), maximum day (MD), or maximum hour (MH).

Most municipal water-distribution systems in the United States are designed to provide water for fighting fires. The adequacy of the water-supply system in providing fire flow is a factor in setting fire insurance rates, so an inferior design will have an economic effect on the service area. The fire-fighting requirement consists of (a) a minimum flow rate, called the *needed fire flow* (NFF), which is provided at a pressure at or above a minimum level, and (b) a storage volume in each service storage tank that is large enough to provide all the water needed to fight the worst expected fire in the service area of the tank. The fire-fighting requirement is the largest NFF and fire storage required for any structure in the tank service area. So, the estimation of fire demand for a service area consists of two steps, estimation of (a) NFF and (b) fire duration for the worst fire condition within the service area. Most insurance companies and fire departments use the Insurance Services Office (ISO) method for calculating NFF and duration; the ISO and other methods are presented in an AWWA manual (AWWA 1998). Ratings depend on building construction, size, and contents. These data may be obtained from assessors' records and field inspections. An NFF network scenario generally drives the sizing of primary mains in a distribution system.

In contrast to the system-level demand estimates illustrated above, the development of a water-distribution model requires that the network be assigned to the service area, that it be topologic (i.e., connected), and that details be defined for the geometrical properties of hydraulic elements (pipes, valves, tanks, and pumps) and the nodal water demands. To establish a valid hydraulic and water-quality simulation, a calibrated model must be established by assigning and maintaining accurate water consumption to demand nodes. Determining and allocating water demand to a node are often tedious processes due to the large amount of data to be processed and the variety of formats in which it is available.

7.3.2 GIS-BASED WATER-SUPPLY DEMAND FORECASTING

Major data sources for determining water demand include (a) customer billing meter data, (b) regional/area water-use data such as system meter data or meter route data, and/or (c) GIS-based data such as population and land-use data with unit demands. Procedures for generating water demands for pipe-network modeling using GIS were described by Wu et al. (2003) for the WaterGEMS® software and by Prins and Bodeaux (2000). Case study applications, such as Szana (2006), also illustrate the basic ideas of GIS functionality for database development and modeling. In general, water demands are based on land-use maps augmented by a relational database, or geodatabase, which incorporates attribute data such as customer ID, land-use category, water-use records, and per-unit planning factors.

Customer billing records can be used to determine water demands given historic records. Tax assessors' data may also be used to identify customers, although there would not be metered flow data from this source; here, water-demand estimates may be developed using the unit-flow-factor approach illustrated above (Tables 7.2 and 7.3). Customer billing data must be exported and processed to calculate the representative demands such as average-day, peak-day, and maximum-day demand. Due to the simplified model representation of a real water system, many customers may be supplied by one node in a model. Geocoding, a standard GIS process to establish the location of customers based on their address, can be used to assign customer demands to a model using one of the following methods: (a) nearest node, (b) nearest pipe, or (c) meter aggregation (Wu et al. 2003). The meter aggregation method involves the development of a polygon around each demand junction that includes all the customer meters that are assigned to that junction. The Thiessen polygon method is an efficient and well-accepted approach for creating these polygons; however, it should be applied with consideration of pressure zones in the distribution system. Figure 7.3 illustrates an example of demand projection by land use.

Regional water-use data may be used where water consumption may not be metered for every individual water customer but metered for a region or a large area. Regional demands are distributed to pressure zones, traffic analysis zones (TAZ), or neighborhoods using an (a) equal flow distribution, (b) proportional to area, and/or (c) proportional to population. Unit flow factors by customer type may be incorporated into the nodal demand assignments.

Future water demands are usually based on the land uses defined by the land-use plan. Several data sources that can be used in the development of the land-use plan include the (a) comprehensive plan, (b) land-use zoning, (c) parcel data, (d) traffic zone data, and/or (e) planning zone data. A municipality's comprehensive plan documents the long-term development plan for the current incorporated areas and the areas within its planning jurisdiction. General land-use categories such as low-, medium-, and high-density residential areas are often used (Table 7.2). The comprehensive plan's land-use definitions for undeveloped areas include open space, parks, schools, public services, rights-of-ways, and commercial and retail development. The comprehensive plan may also include land-use and population projections.

Land-use zoning and the associated zoning codes provide guidelines to developers for planning development within the agency's jurisdiction. Typically, zoning is assigned to parcels with specific zoning designations for parks, open spaces, and public facilities. If defined in GIS, zoned areas exclude rights-of-way for parks and open space, neighborhood streets, and major thoroughfares. A municipal agency's zoning codes are often the best data source for establishing target development densities for each land-use category.

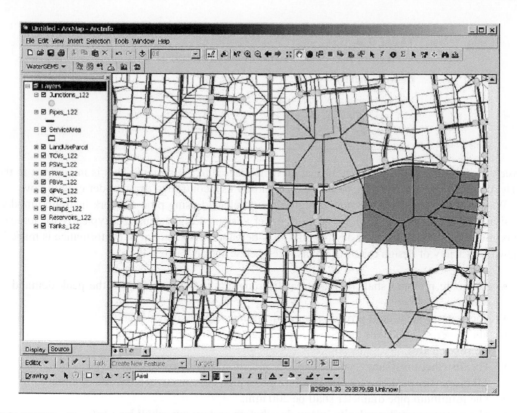

FIGURE 7.3 Example of demand projection by land use. (From Wu et al. 2003. With permission.)

Parcel data are often available from the assessor's office. These data include a parcel identification number (PIN) along with the parcel's development status. Field inventory surveys or high-resolution aerial photography may also be used to establish current land uses. Existing land use can then be extracted by linking the inventory table to the parcel coverage using a GIS Join function with the relational database.

Most cities and metropolitan regions develop a traffic analysis zone (TAZ) database to support highway capacity estimates. Often these data include existing land-use estimates as well as population and land-use projections. In the absence of parcel-level data on existing land use, TAZ data can be used to estimate existing development. The TAZ projections are often the best data source for estimating growth projection parameters. Planning-area coverage is another data source that could be developed for forecasting land development. For land-use projection purposes, the study area is subdivided into planning areas, which are delineated to represent development with similar growth characteristics. Planning-area sizes are generally limited to approximately 150 acres to 250 acres and may be done along political or zoning boundaries.

Yet another way is to apply land-use change modeling to forecast the spatial location of new development. The forecast land-use distributions can then be used to assess infrastructure resource requirements and potential environmental impacts. The various approaches and models for land-use change forecasting were summarized in Chapter 3 (GIS Data and Databases), and examples of infrastructure demands based on these models are described in other chapters (e.g., Chapters 5 and 10).

7.3.3 PIPE-NETWORK DESIGN PROCEDURES

Given estimates of water-supply demands for current and future conditions, it is required to design a pipe distribution system to convey potable water from the treatment plant to the connected services. The water-supply distribution system always operates under pressure to maintain positive control

TABLE 7.4
Water-Supply Design Criteria for Pressure

Design Condition	kPa	ft	m	psi
Min. pressure at max. hour	275	92	28	40
Min. pressure at max. day + NFF	140	47	14	20
Max. pressure	700	234	71	102

over water flows and to avoid contamination by leakage into the system. This is in contrast to the irrigation system, which is designed to operate mostly as a controlled drain under gravity flow.

Water-supply distribution system design is accomplished using pipe-network hydraulic models to simulate the performance of the network under various design scenarios, typically for forecast maximum-hour and maximum-day plus fire-flow demands. Assessment of performance is made in view of a variety of design criteria such as listed below and in Table 7.4.

- Adequate pressure shall be maintained throughout the system during the peak demand scenario.
- The water system shall provide potable drinking water to all residents.
- The water system shall provide water for fighting the worst expected fire at all points in the service area.
- The distribution system shall be a looped system of buried pipes.
- All properties shall have a distribution system pipe adjacent to at least one boundary.
- The minimum pipe diameter shall be 200 mm.
- The maximum water velocity in any distribution system pipe shall be 3 m/s.
- Distribution system pipes shall be buried below maximum frost depth.

The pipe-network design process is iterative in character, involving the specification of a pipe-network layout having pipes of certain types and sizes. For reliability, it is desirable that the network be a looped system so that water can be delivered to all services even if a certain pipe is shut off for repairs. Topography can be a primary determinant of system layout to take advantage of gravity distribution that is more reliable and cheaper than pumping. Distribution storage tanks may be placed at strategic locations in the system to provide storage to meet fluctuations in use, provide water for fire-fighting use, and stabilize pressures in the distribution system. Also, pumps may be required to move water to high storage areas.

Before using the model, it is necessary to calibrate the hydraulic model to ensure an accurate representation of the actual distribution system hydraulics. The calibration process consists of setting up the hydraulic model to reflect the actual water demands and operational conditions that existed during a specified time period, running the analysis, and then comparing the calculated flows and pressures with the actual flows and pressures as recorded in the records.

Pipe-network models are based on hydraulic flow and network theory, and use the principles of conservation of mass and energy to represent flows and friction losses throughout a network. Pipe networks create relatively complex problems, particularly if the network consists of a large number of pipes, and solutions of these problems require sophisticated computerized mathematical procedures. The principle of mass conservation, or continuity, simply means that the flow passing through two sections of pipe some distance apart must be equal if the flow is steady. When dealing with junctions of two or more pipes, the continuity principle states that the mass flow into the junction must equal the mass flow out of the junction. Flowing water contains three forms of energy that are of interest in hydraulics: potential energy due to its elevation, pressure, and kinetic energy due to motion. Conservation of energy is based on the Bernoulli equation, whereby the components of total energy are balanced between two sections; this balance includes frictional head losses incurred by flowing water. Jeppson (1982) and Walski (1984) provide reviews of numerical solution procedures for the linear theory, Newton-Raphson, and Hardy-Cross methods. The method used in EPANET to solve the flow continuity and head-loss equations that characterize the hydraulic state of the pipe network at a given point in time is termed a hybrid node-loop approach (Rossman 2000). Solution procedures for the pipe-network equations continue to be advanced through research and development.

7.3.4 GIS-BASED WATER-SUPPLY NETWORK MODELING

There are a number of modeling packages for simulating pipe networks. The most common is the public-domain EPANET model (Rossman 2000). EPANET is a computer program that performs extended-period simulation of hydraulic and water-quality behavior within pressurized pipe networks. A network consists of pipes, nodes (pipe junctions), pumps, valves, and storage tanks or reservoirs. EPANET tracks the flow of water in each pipe, the pressure at each node, the height of water in each tank, and the concentration of a chemical species throughout the network during a simulation period comprising multiple time steps. In addition to chemical species, water age and source tracing can also be simulated. Other commercial pipe-network software includes WaterCAD® (http://www.bentley.com/en-US/Products/Bentley+Water/) and InfoWater®/H2ONET® (http://www.mwhsoft.com/).

GIS provides functions for the development and preparation of accurate spatial information for input into the network design modeling process, including network layout, connectivity, pipe characteristics, pressure gradients, demand patterns, cost analysis, network routing and allocation, and effective color graphic display of results. The GIS provides functionality to accomplish database management operations for both spatial and attribute data, user-friendly dialog interfaces for data manipulation and output display, and model subsystems, including both simulation and optimization. The EPANET and InfoWater (H2ONET) packages provide GIS interfaces and functions for building pipe-network simulation models.

An application of EPANET and InfoWater GIS pipe-network modeling capabilities was conducted by Szana (2006). The process used to create an all-pipes hydraulic model from GIS data for the City of Boulder, Colorado, included (a) processing of initial GIS data sets, (b) importing data into the GIS, (c) formulating the initial hydraulic model based on the GIS pipe data, (d) error checking for network connectivity, (e) assignment of node elevations based on the digital elevation model (DEM), (f) assignment of node demands (current and future), (g) calibration of the model for current conditions, and (h) simulations of alternative scenarios.

Processing of initial GIS data sets involved importing the ArcInfo® data from the city GIS department. The all-pipes model was created from the GIS pipes data set, which contained fields for all information useful to perform hydraulic and system analyses. Erroneous and missing data were filled in during the error-checking and calibration steps. Unique identifiers for each pipe were established to facilitate data transfer between the hydraulic model and the GIS data set in the future. Private lines and hydrant laterals were added for displaying connections from the system mains to the buildings data set. This data set was used to assist with demand allocation from parcel and meter data. Data on valves, fire hydrants, and water main breaks were added; these are important for delineating pressure zones and performing transient and unidirectional flushing analyses. A digital elevation model was imported and georegistered to the other spatial data sets; elevation is a fundamental attribute of the hydraulic model, and all nodes, reservoirs, pumps and tanks must have assigned elevations. A parcels data set was imported to support land-use-demand factors and customer notification for maintenance and emergency conditions. Other GIS data were added, including the street network, parks, and other land features; these were for general map identification purposes and were not used for the hydraulic analyses.

Creating the all-pipes network involved importing the pipes and adding junctions to the pipe endpoints. Next, pressure zones were delineated, and the network was reviewed to ensure connectivity. Finally, elevations were assigned to the new junctions, and the all-pipes network was connected to the pumps, valves, tanks, and reservoirs. Various ancillary procedures are conducted to build an

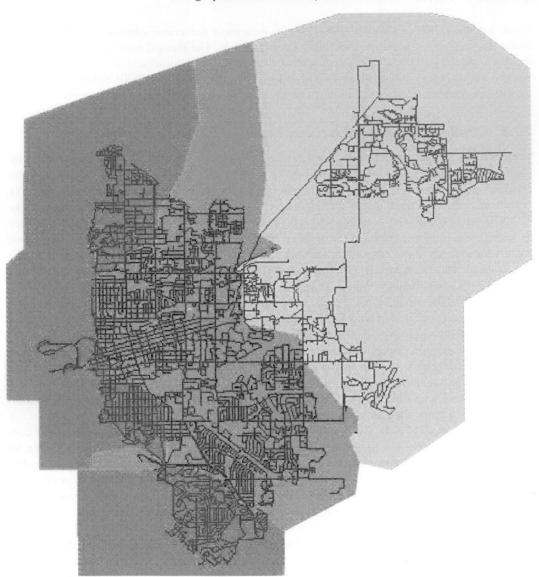

FIGURE 7.4 (See color insert following page 136.)   Pipe network model completed assignments to pressure zones. (From Szana 2006. With permission.)

initial hydraulic model. Errors encountered during the import process included duplicate IDs and nodes, topology not established, and disconnected pipes. The final pipe network was completed and available for nodal demand assignments and simulations (Figure 7.4).

The final requirement after creating the all-pipes network and before attempting to run a hydraulic simulation was to assign demands to the model nodes. Water-use data from the billing system were used to represent the minimum (December 2005) and maximum (July 2005) demands. Demands were allocated to the model using a meter-closest pipe method; this feature locates the pipe that is closest to each meter. Demands were assigned to the nodes connected to the pipes using a distance-weighted calculation where the node closer to the meter will be allocated more demand than the far junction (Figure 7.5). Demands for future growth or rezoning were estimated from the zoning designations for where the system will serve in the future. Land-use-demand factors have been estimated by the city; these were supplemented by land-use-demand factors from meters located within each land-use type.

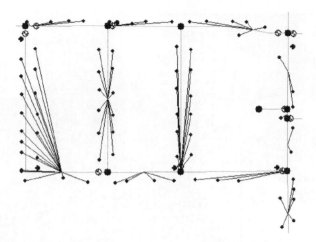

FIGURE 7.5 Meter/pipe association generated during automated demand allocation. (From Szana 2006. With permission.)

Given a candidate pipe-network layout, the GIS can be used to overlay the land-use plans onto the network. Then the nodes (intersections) of the network are assigned to provide water to certain areas of land. The node-area assignments are made for the base (current) year and future design years. Often, a nearest-neighborhood assignment is made using a Thiessen polygon approach. Then, given the land-use mix in each node area, the water demands are computed for each node using the procedures described above. The total average-day demands for residential, industrial-commercial-institutional, and parks are summed separately for each network junction (node). Peaking factors are applied to obtain maximum-day and maximum-hour estimates, as well as characteristic needed fire flows. These nodal demands are then used as input to the pipe-network hydraulic model.

Model calibration was accomplished by adjusting parameters until model outputs matched field data collected with the utility's Supervisory Control and Data Acquisition (SCADA) system. This procedure involved two main processes. First, model parameters and attributes that can be physically tested and measured are adjusted to attain model results that more closely match actual results in the field. Second, pipe roughness, which changes over time and is very difficult to measure, can be optimized to minimize the difference between the model and field results. The initial model calibration was accomplished by trial and error; running model simulations; and then viewing pressures, flow rates, tank levels, and water age. The process is aided by the GIS-based visual displays of pressures, velocities, and differences between monitored and simulated values. Figure 7.6 shows the final hydraulic pressure surface for the calibrated model. Pipe-network model calibration activities typically encounter a number of issues as documented by Szana (2006) and Ray et al. (2007). Some of the issues include input data errors, assignment of water-use records to nodes, different (and nonstandard) coordinate systems, boundaries between pressure zones, discontinuities in the network CAD map, questionable or missing SCADA data, and the general complexity of the system. Calibration problems are resolved by making many adjustments to the model using a mix of advanced data-processing tricks and labor-intensive manipulations.

Establishment of the model calibration provides the foundation for simulations of various scenarios, including maximum day, minimum day, average day, fire flows, system security and vulnerability, transient analysis, and flushing analysis.

7.3.5 GIS-Based Pipeline Routing

GIS finds many applications in the routing of pipelines. Used in conjunction with MCE-based suitability mappings, the GIS helps identify "best" routes that minimize costs, environmental

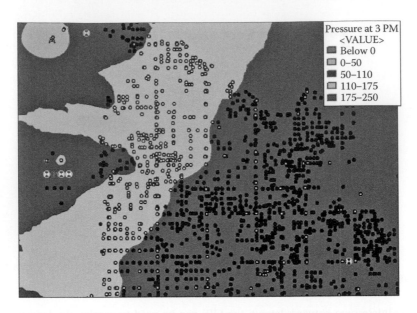

FIGURE 7.6 Thematic map of system pressure during model calibration. (From Szana 2006. With permission.)

impacts, and related objectives. An example of such an application was provided by Luettinger and Clark (2005) for a pipeline routing in the Salt Lake City area. A 1524-mm (≈60 in.)-diameter pipeline will convey finished water approximately 9 km (≈12 mi.) through mostly developed areas of two communities. Construction of a large-diameter transmission pipeline through heavily developed cities creates many engineering, construction, and public relations challenges. A GIS-based route-selection process was used to provide a rational basis for narrowing hundreds of potential alternatives into one final alignment corridor. Figure 7.7 shows a reduced set of routing alternatives. The route-selection process was based on construction costs as well as important noncost issues, including constructability, system compatibility, community disruption, traffic and transportation, utility conflicts, seismic and geologic considerations, and environmental concerns, as well as permit, right-of-way, and land-use issues. These factors were incorporated into the network model by assigning equivalent lengths to the pipes to reflect differences in the factor scores. Network-analysis software was used to model the least-cost path between each of the established routing points, using all of the cost data (as equivalent lengths) compiled for the 440 segments. The GIS allowed large amounts of pipeline cost-related data to be collected, stored, and documented for each alignment alternative. This allowed for a logical selection and ranking of alternatives, resulting in one final alignment corridor that was acceptable to all of the stakeholders in the project.

7.3.6 GIS-BASED WATER NETWORK OPTIMIZATION

Municipal water-distribution systems represent a major portion of the investment in urban infrastructure and a critical component of public works. The goal is to design water-distribution systems to deliver potable water over spatially extensive areas in required quantities and under satisfactory pressures. In addition to these goals, cost-effectiveness and reliability in system design are also important. Traditional methods of design of municipal water-distribution systems are limited because system parameters are often generalized; spatial details such as installation costs are reduced to simplified values expressing average tendencies; and trial-and-error procedures are followed, invoking questions as to whether the optimum design has been achieved. Even with the use of hydraulic-network simulation models, design engineers are still faced with a difficult task.

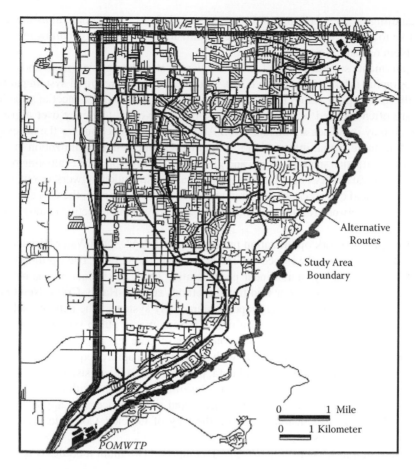

FIGURE 7.7 Pipeline route alternatives for the Point of the Mountain Aqueduct. (From Luettinger and Clark 2005. With permission.)

Optimization methods can be applied to make the designing of water-supply networks easier and to ensure minimum cost and maximized reliability. The complexity of the optimal layout and design problem is underscored by the following factors: (a) the discrete characteristics of the decision variables (i.e., commercially available pipe sizes); (b) complex, discrete cost functions involving material, labor, and geographically based costs related to layout and excavation; (c) the need for multiple demand loading patterns to be considered in the design; and (d) the requirement for a priori knowledge of both pipe flows and pressures for calculation of energy costs in systems requiring pumping (Gessler 1982).

A wide variety of techniques have been proposed over the past two decades for the optimal design of pipe networks. Articles by Walski (1985), Walters (1988), and Goulter (1992) have provided extensive reviews of the most relevant and promising techniques. Alperovits and Shamir (1977) suggested the linear programming gradient (LPG) method, which performs a hierarchical decomposition on the optimization problem. Variations on the LPG method have been developed by Bhave and Lam (1983). Eiger et al. (1994) extended the method to calculate a lower bound on the global optimal solution using generalized duality theory. A duality-gap resolution procedure is applied, but as future work attempts to include pumps and booster sizing, the problem of duality gaps (i.e., inconsistencies between the primal and dual solutions) will likely be magnified. Bhave and Sonak (1992) consider the LPG method to be inefficient compared with other methods.

The desirable simulation and optimization techniques needed for optimal pipe-network design should be incorporated within the framework of a decision-support system (DSS). Effective pipe-network design requires the availability of user-friendly decision-support systems that are reasonably flexible in handling the large amount of design data needed, including constraints the design engineer would like to impose and those that utilize optimization techniques that allow incorporation of as much problem realism as possible.

These requirements suggest the need for an efficient spatial data-management and -analysis tool such as a geographic information system (GIS). Without a GIS, the required system parameters are often generalized to typical values prior to use. Spatial details on pipe connections, such as installation cost, are often reduced to a single value expressing average tendency over a group of connections, which may introduce significant error. A GIS provides functions for the development and preparation of accurate spatial information for input to network design optimization models. It also facilitates postoptimization spatial analysis and graphical output display for evaluating results. Given the spatial and time variability of parameters such as water-distribution network layout, street layout, pipe characteristics and cost, pressure requirements, and demand patterns, the GIS can perform cost analysis, implement network routing and allocation, and provide an effective color-graphic display of results.

A GIS-based pipe-network optimization program, called WADSOP (Water Distribution System Optimization Program), was developed by Taher and Labadie (1996). The optimal design methodology in WADSOP employs a nonlinear programming (NLP) technique as the network solver, which offers certain advantages over traditional methods such as Hardy-Cross, Newton-Raphson, and linear system theory for balancing looped water-supply systems. The network solver is linked to a linear programming (LP) optimization model for automated pump sizing and pipe selection. The two models are applied interactively in a convergent scheme that affords numerous advantages. The models are linked to a GIS in WADSOP with the ability to capture requisite data; build network topology; verify, modify, and update spatial data; perform spatial analysis; and provide both hardcopy reporting and a graphical display of results. Several comparisons with published results from other methodologies show improved performance, rapid convergence, and applicability to large-scale systems. WADSOP was demonstrated on the City of Greeley, Colorado, water-distribution network as a real-world case study.

The optimal design methodology in WADSOP comprises two main modules (Figure 7.8):

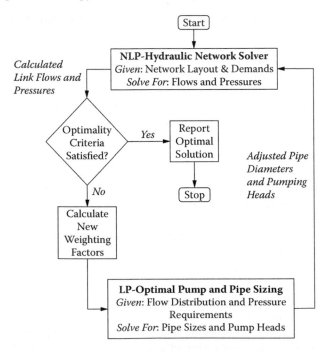

FIGURE 7.8 WADSOP optimization procedure. (From Taher and Labadie 1996. With permission.)

- Hydraulic Network Flow Model: A nonlinear programming (NLP) algorithm is employed as the network solver for hydraulic balancing of pipe flows and nodal pressures for a given system layout, design parameters (i.e., pipe sizes and pump heads), and nodal demands.
- Optimal Pump and Pipe Sizing Model: Linear programming (LP) is utilized to determine pipe sizes, associated lengths, and pumping heads that minimize the cost of the system while satisfying hydraulic criteria imposed by the designer. The LP model determines changes in the design variables (i.e., pipe size and pumping head) based on outputs (i.e., heads and flows) from the network solver.

The phases of the optimal design process are listed as follows:

- Initial pipe layout, flow pattern, and pipe sizes are assumed; the initial layout can be determined from the street network and topography.
- Based on the initial values, the flows and pressures throughout the network are determined using the NLP-based hydraulic network solver.
- Adjusted flows and pressures are input to the LP model, which adjusts pipe sizes and pumping heads such that all minimum pressure requirements are satisfied at minimum cost.
- The new configuration of pipe sizes and pumping heads is provided to the network solver to again calculate adjusted flows and pressures.

Sequential solution of the two models is repeated until convergence to the optimum system configuration is found.

7.3.6.1 WADSOP Decision-Support System

Components of the WADSOP prototype decision-support system for water-distribution system design and analysis include (a) dialog manager, (b) GIS database, and (c) a model base as seen in Figure 7.9.

The dialog manager is an interactive, menu-driven, point-and-click user interface that provides efficient guidance through the application. The dialog manager is responsible for automating input data preparation, executing all associated models, and displaying results. The user is not required to learn a complex macro command language in order to use the package. Several help functions assist the user in proper procedure.

The GIS database provides storage and management capabilities for all spatial and attribute information necessary for the model components and the user. TABLES is a simple data-management retrieval system provided by PC ArcInfo® that creates and maintains feature attribute tables compatible with dBASE file format structures. Attribute data on coverage features are stored in four different database files: a TIC file, a boundary file (BND), a polygon attribute table or point attribute table (PAT), and an Arc attribute table (AAT). These files contain descriptive information about coverages and are accessed by ArcShell® for updating and by ArcView for query, analysis, and reporting.

The model base consists of three submodels: the analysis program (simulator), the design program (optimizer), and GIS spatial analysis functions. The simulation model, as described previously, is a generalized, stand-alone program used for network hydraulic balancing as well as serving as a network solver for the optimization model. The generalized optimization model is employed to size the network pipes and pumps, also described previously. The GIS functions are PC ArcInfo modules offering a wide range of spatial analysis capabilities, including routing, allocation, and cost estimation. The model base (i.e., simulator, optimizer, and GIS functions) is linked to (a) the dialog manager for execution and (b) the database to obtain necessary data for the modeling process and convey results back to the database.

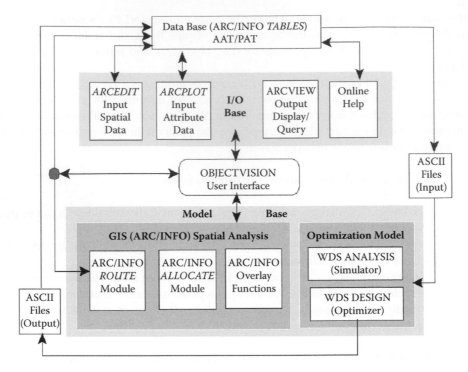

FIGURE 7.9 WADSOP prototype decision-support system for water-distribution system design and analysis includes (a) a dialog manager, (b) a GIS database, and (c) a model base. (From Taher and Labadie 1996. With permission.)

Application of WADSOP to the example network design problems revealed the following:

- WADSOP efficiently handles the design of large systems with multiple loads, pipe sizes, and pipe costs in a single run. Computer time required for the design of the network in Example 1 was 14 s as opposed to the 41 min required by WADISO (Walski et al. 1990). WADISO utilizes pipe groupings and successive optimization procedures to reduce the computational load, whereas such measures are not required for WADSOP. Example 2 also highlighted the efficiency of WADSOP, with computational time savings of approximately 80% over the algorithm of Lansey and Mays (1989).
- The optimization procedure in WADSOP directly incorporates pipe sizes as discrete variables, while several other models require variables to be continuous. Rounding off to the nearest available pipe size can lead to solutions that are nonoptimal or fail to satisfy all requirements.
- The ability of the model to divide links into two pipes of consecutive diameters results in further reduction in network cost. This accounts for the reduced cost of the WADSOP solution to Example 1 over the WADISO solution.

7.4   GIS FOR IRRIGATION

7.4.1   IRRIGATION SYSTEMS PLANNING AND DESIGN

Irrigation systems are designed to provide water applied to the soil for plant growth. In arid regions, agriculture is sustained entirely by irrigation supplies; in humid regions, irrigation supplements deficient rainfall. Irrigation systems are also developed for residential lawns and other vegetation in urban areas. An irrigation system may be comprised of many components, including storage reservoirs, pipes, ditches and canals, and sprinklers. Development of irrigation systems has been central to the evolution of modern civilization, as food production and distribution have so far kept pace with demand. Although more land is continually being placed under irrigation, the rate of food production is slowing, some irrigated lands are being lost to salinity degradation, and there are many environmental impacts associated with irrigated agriculture (e.g., nutrient and pesticide runoff). Also, available water supplies are becoming inadequate to sustain both escalating urbanization and irrigation demands. An outstanding characteristic of irrigated agriculture is the large amount of consumptive use of applied water: up to 85% is lost from source to field. Irrigation crop consumption rates typically range from 40% to 65%; the remaining 35% to 60% of applied crop water percolates to the groundwater, flows off the end of the field, or evaporates as it flows across the field. The groundwater portion will eventually flow down-gradient back to the surface streams (i.e., return flow). Improvements of irrigation efficiencies are therefore assuming greater importance, and GIS tools are helping in various ways to achieve these goals.

There are two basic types of agricultural irrigation systems: flood and sprinkler. Surface flood irrigation consists of releasing water over the surface of the land to flood the fields. Flood irrigation is the oldest form of irrigation, and can be used for any crop. Ideally, the land has low slope and is level enough for the water to distribute evenly over the surface. The more planar the land, the more efficient the flooding will be. Advances with laser-leveled fields aided by GIS mapping have shown quite high efficiencies. Furrow flooding is where small ditches between crop rows are supplied by gated pipes, canals, and ditches with turnouts or siphons. Sprinkler irrigation systems utilize pipes and sprinkler heads to distribute water to the plants. Three common types of sprinkler systems are solid set, continuous move, and periodic move. Sprinkler irrigation systems are usually more efficient than flood irrigation systems, except during windy conditions. Micro irrigation systems utilize rubber tubing and microsprayers to water each plant. Due to the higher cost, micro irrigation is typically used only for high-valued crops such as vineyards or nuts. A conveyance system transfers from a water source to the fields. This can be achieved with canals, ditches, and pipes, or any combination of these. Ditches and canals are open to the air and are more susceptible to seepage and evaporation than pipes. A ditch or canal can be lined or unlined. Lined canals and ditches can have concrete, clay, or impermeable membrane linings on their bottom and sides; these are much more efficient than unlined canals.

Design and operation of irrigation systems require information on available water supply, terrain, soils, and climate (Table 7.1). Soils determine whether the land is arable, i.e., whether there will be sufficient yield to warrant development. To be suitable for irrigation farming, the soil must have a reasonably high infiltration rate and water-holding capacity. The infiltration rate should be low enough to avoid excessive loss of water by percolation below the root zone. The soil must be deep enough to allow root development and permit drainage. It must be free of black alkali (sodium-saturated), other salts not susceptible to removal by leaching, and toxic elements. Land slopes should be such that excessive erosion will not occur. The land may be leveled to enhance irrigation farming, although such leveling may cause problems with the topsoil distribution and thickness. The land should be located in proximity to a water supply without excessive pumping or transmission costs. The general layout and size of the area should be conducive to division into field units that permit effective farming practices. Details on soil suitability, water requirements, facilities, economic analyses, and related information requirements for irrigation are provided by Linsley et al. (1992).

In addition to irrigation project design, data on soils and weather are applied for irrigation operations. Irrigation scheduling is the farmer's decision process relative to "when" to irrigate and "how much" water to apply at each irrigation event (Fortes 2005). It requires knowledge of crop water requirements and yield responses to water, the constraints specific to the irrigation method and respective on-farm delivery systems, the limitations of the water-supply system relative to the delivery schedules applied, and the financial and economic implications of the irrigation practice. Proper irrigation scheduling can reduce irrigation demand and increase productivity. A large number of tools are available to support field irrigation scheduling, from in-field and remote sensors to simulation models. Irrigation scheduling models are particularly useful to support individual farmers and irrigation advisory services (Ortega et al. 2005). The main value of models results from their capabilities to simulate alternative irrigation schedules relative to different levels of allowed crop water stress and to various constraints in water availability (Pereira et al. 2002).

7.4.2 GIS FOR IRRIGATION SYSTEMS DESIGN AND MODELING

GIS tools and integrated models find many applications for irrigation, including financial analyses, determining water requirements, and irrigation scheduling. Knox and Weatherfield (1999) outlined some of these applications. For financial analyses, GIS is used to quantify and map the total financial benefits of irrigation ($/ha) and the financial impacts of partial or total bans on abstraction for irrigation. Irrigation water requirements are determined by mapping the spatial distribution of water requirements based on soil and crop distributions. GIS-based modeling approaches are used to establish irrigation scheduling based on water-balance modeling. Hess (1996) described an irrigation model embedded within the GIS. These data were correlated to digital data sets on soils, agroclimate, land use, and irrigation practice to produce tabular and map outputs of irrigation need (depth) and demand (volume) at national, regional, and catchment levels. The GIS approach allows areas of peak demand to be delineated and quantified by sub-basins. The GIS-based modeling approach is also currently being used to administer irrigation needs for all irrigated crops. Map and tabular output from the GIS model can provide licensing staff with the information necessary to establish reasonable abstraction amounts to compare against requested volumes on both existing and new license applications for spray irrigation.

GIS databases for irrigation include coverages for crops, irrigation methods, and soils. These data are coupled with agroclimatic data to provide information on growing-season and water-use requirements. A representative irrigation database is described (Figure 7.10) for the GISAREG model (Fortes et al. 2005). It includes:

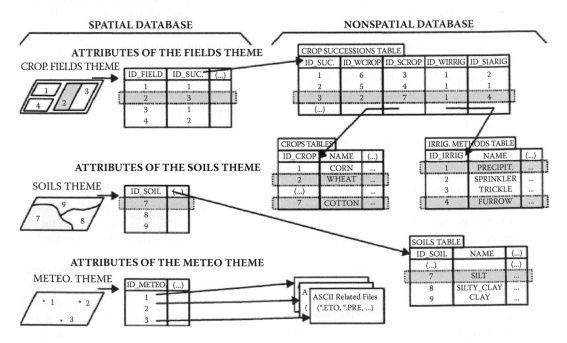

FIGURE 7.10 Schematic representation of the spatial and nonspatial databases and respective interrelations. (From Fortes et al. 2005. With permission.)

- The crop successions table, which defines the annual crops succession (two crops that successively occupy the same field) and includes data on these crops and on the irrigation method used. It includes the codes for identification of the crop succession and respective designation, the winter crop and the respective irrigation method, and the summer crop and irrigation method.
- The crop table, including: the crop identification code and its designation; the crop type code (1 = "bare soil," 2 = "annual crop," and 3 = "annual crop with a soil frozen period"), which identifies the time period to be used to estimate the initial soil moisture; and the crop input data (Table 7.1) required for computing the crop evapotranspiration.
- The irrigation methods table, containing: (a) the code and designation of the irrigation method and (b) the respective fraction of soil surface wetted by irrigation.
- The soils table, including: (a) the soil identification code and designation and (b) the soil data required for performing the water balance of the evaporative layer.

GIS is used to identify the dominant characteristics of each cropped field based on spatial data on crops, soils, and meteorological stations. The crop fields mapped as polygon coverage are associated with a table of field attributes, including the identification code of each crop field, the identification of the annual crop succession, and other relevant information as desired by the user. A polygon theme of soil types is associated with a table of soil attributes. A point theme contains the location of meteorological stations identified by a code. GIS overlay procedures allow the identification of the soil, climate, and cropping characteristics of each cropped field. A Thiessen polygon theme can be generated to define the geographical influence of each meteorological station. Overlaying the meteorological Thiessen polygon theme with the field polygon theme assigns the climate data to each field. Another overlay operation generates the intersection between the soils themes and the cropped fields to assign the dominant soil type to each field. The GISAREG irrigation scheduling model is described in more detail below.

7.4.3 CASE STUDY: EVALUATION OF IRRIGATION AGRICULTURE

GIS is used routinely for irrigation system planning, often involving the application of suitability analysis to evaluate the potential for irrigation agriculture or for irrigation waste management. One example was in southern Australia, where there was a concern with the off-site migration of saline drainage from irrigated lands (Dowling et al. 2000). The strategy was to store drainage disposal water in the irrigated areas using evaporative disposal basins. This approach posed questions regarding the availability of suitable land and the impact of these basins. A GIS-based approach was developed using suitability criteria to minimize the risk of off-site impacts of basin leakage. Not only was it important to ensure that the leakage rate was low, but also that the adverse impacts of any leakage (and associated contamination) would be minimized by not siting basins over good-quality groundwater or by not allowing any contamination plume to move outside the vicinity of the disposal basin.

The criteria were proximity to surface-water features (streams, drains, and irrigation channels) and infrastructure (urban areas and roads), water-table depth and salinity, and soil hydraulic conductivity. In most cases, the parameters were directly measured; however, for hydraulic conductivity surrogates such as various forms of soil classification and for rice, irrigation data were used. Suitability ranges were defined separately for each of the inputs, which were then combined according to relative importance to derive a manageable number of overall suitability classes ranging from optimal to unsuitable. There are a number of factors that influence the extent to which any contaminant plume can move. First, the permeability of the subsoil determines the leakage to the aquifer. Second, the depth to groundwater determines the storage available beneath the basin as opposed to that available laterally. Third, the groundwater gradient and permeability of the aquifer determine the movement away from the basin itself. It was assumed that a low-permeability soil and shallow

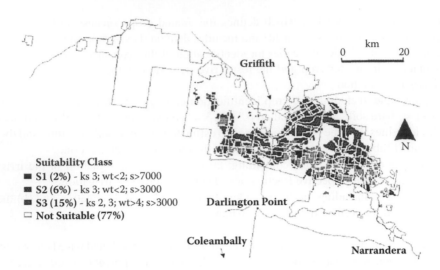

Suitability Class
- ■ **S1 (2%)** - ks 3; wt<2; s>7000
- ■ **S2 (6%)** - ks 3; wt<2; s>3000
- ■ **S3 (15%)** - ks 2, 3; wt>4; s>3000
- ▢ **Not Suitable (77%)**

FIGURE 7.11 Overall suitability of land for disposal basins based on all themes except hydraulic conductivity. S1 is optimal conditions in all criteria, while S2 and S3 have depth to water-table and groundwater salinity criteria relaxed. (From Dowling et al. 2000. With permission.)

saline groundwater (unable to develop gradients) were prerequisites for suitability. Further, current potable groundwater resources were to be protected, buffers to urban settlements were to be maintained to avoid odor and insect problems, and distances to surface-water features were to be maximized to prevent degradation. Figure 7.11 shows the distribution of land suitability for disposal basins.

Another example was demonstrated in the Umatilla River basin of north-central Oregon (Loveland and Johnson 1983). Landsat data and 1:24,000-scale aerial photographs were initially used to map expansion of irrigation from 1973 to 1979 and to identify crops under irrigation in 1979. Using Landsat images obtained late in the irrigation season (late July through August), the land-cover categories were readily mapped on the basis of their image color, size, shape, and location. This resulted in generalized maps displaying the dominant cover types in the basin. The final maps were interpreted with a minimum mapping unit size of 40 acres. The crop data were then used with historical water-requirement figures and digital elevation and hydrographic data to estimate water and power use for the 1979 irrigation season. Soil survey data were manually digitized, and the soil polygons were converted to 63.6-m^2 grid cells and stored in 158 soil-mapping-unit classes. Based on the chemical and physical characteristics of each soil-mapping-unit, each of the 158 classes was assigned to one of five soil irrigability categories (excellent, good, fair, poor, or unsuitable). Land-ownership maps were manually digitized, and the ownership polygons were converted to a grid-cell format. Documents containing the locations of irrigation pumping plants along the Columbia River and its tributaries were used to establish a digital file of the locations of pumping plants.

A composite map of land suitability for irrigation development was derived based on factors of land cover, land ownership, soil irrigability, slope gradient, and potential energy costs. Irrigation energy requirements were based on the equation

$$\text{kWh/acre} = (1.024 \times \text{Diversion} \times \text{TDH})/E_p$$

where 1.024 is a conversion constant, Diversion = acre-feet of water diverted for irrigation based on the types of crops irrigated, and TDH = total dynamic head (i.e., sum of pump lift, friction loss of pipe flow, and system operating pressure). TDH is primarily based on landscape position as derived from the digital terrain data and the pumping-plant source pool elevation. The assumed operating pressure was 240 ft of static lift, and E_p = pump efficiency (assumed 85%).

TABLE 7.5
Umatilla Basin Irrigation Assessment Weighting Scheme

	Model Inputs		
	Land Cover	Soils	Slope (%)
Weights	2	2	2
10	Dryland agric.	Excellent	0–3
9	Dryland agric.	Excellent	0–3
8	Dryland agric.	Excellent	0–3
7	Rangeland	Excellent	4–7
6	Rangeland	Good	4–7
5	Rangeland	Good	4–7
4	Rangeland	Good	8–12
3	Rangeland	Fair	8–12
2	Rangeland	Fair	8–12
1	Rangeland	Poor	8–12
Not considered for irrigation suitability	Existing irrig. Water bodies Urban areas Public lands	Unsurveyed	>13

Source: Adapted from Loveland and Johnson (1983).

TABLE 7.6
Irrigation Suitability, Composite Scores, and Frequency

Category	Composite Scores	Occurrence (%)
Unsuitable	0–31	20
Poor	32–44	21
Fair	45–50	21
Good	51–53	19
Excellent	54–58	19

Source: Adapted from Loveland and Johnson (1983).

A GIS-based additive-overlay model was used to characterize the irrigation development potential of the Umatilla River basin. Table 7.5 lists the factors and weighting scheme used. Energy costs had twice the importance of any other factor; physical factors of land cover, soils, and slope were of equal importance. The analysis was applied only to private lands; wildlife areas, military reservations, state parks, and national forests were excluded. Water bodies, wetlands, urban areas, and forestlands that cannot be irrigated were also excluded. Lands with slopes from 0 to 3% are preferred for irrigation; slopes from 8% to 12% are least preferred; and slopes of 13% and greater were excluded.

The additive-overlay procedure was applied to generate composite suitability scores of irrigation potential for each grid cell of the feasible set. The scores ranged between 0 and 58 and were assigned using study-team judgment to categories of irrigation suitability (Table 7.6). The resultant occurrence frequencies of the categories were evenly divided across the five levels of irrigation suitability.

7.4.4 Irrigation Consumptive-Use Modeling

Consumptive use (CU) is the loss of water from irrigated crops to evapotranspiration (ET). CU is also incurred by other uses such as lawn watering, reservoir storage, and industrial and domestic uses. Accounting for CU is important as a major portion of water-budget analyses. In locations having prior appropriation water-rights policies, such as Colorado and the western United States, the historic CU defines the beneficial use by a water-rights holder. This is the amount of water that can be transferred to other uses through petitions for a change in the water-appropriation process. CU amounts to approximately 50% of water application to an irrigated field; this amount varies, depending on the type of crop and other factors. The percentage of irrigation CU compared with total CU averages around 60% for water uses in Colorado. Historic cropping patterns and areas were determined by field surveys and satellite image processing.

The state of Colorado has developed a GIS-based Consumptive Use Model (StateCU) to provide standardized estimates of both crop and noncrop consumptive use within the state (CDSS 2007). The CU modeling effort is part of the Colorado Decision-Support System (CDSS) development for the waters of the state (http://cdss.state.co.us/). The target is the users' interest in estimating potential or actual consumptive use of agricultural lands. The CU model classifies the uses into two main categories: (a) agricultural consumptive users and (b) other consumptive users. The "other" category includes CU from livestock; stock pond; and municipal, industrial, and reservoir consumption. StateCU includes a fully functional interactive graphical user interface (GUI) that allows all input data to be viewed, modified, and saved.

The purpose of the CDSS consumptive use (CU) model is to estimate the amount of water that leaves a basin as evapotranspiration (ET) due to human activity. The CU model can be used to calculate the historic consumptive use in a basin. The agricultural CU is from irrigated crops and is computed using standardized methods such as the SCS Blaney-Criddle method (Blaney et al. 1952; Jensen et al. 1990). The empirical equations relate ET with mean air temperature and mean percentage daylight hours. The original procedure was developed and shown to be sufficiently accurate for estimating seasonal ET (USDA 1970). However, the procedure was also modified to reasonably estimate short-period consumptive use. The modifications include the use of (a) climatic coefficients that are directly related to the mean air temperature for each of the consecutive short periods that constitute the growing season and (b) coefficients that reflect the influence of the crop-growth stages on consumptive use rates.

StateCU allows several levels of analysis, as follows:

- Crop irrigation water requirement by structure (monthly or daily)
- Water-supply-limited crop consumptive use by structure (monthly)
- Water-supply-limited crop consumptive use by structure and priority (monthly)
- Depletion by structure and priority (monthly)

Estimation of evapotranspiration requires spatial data sets on climatic conditions, soils characteristics, and vegetation patterns. Climate data on temperature, solar energy, wind, and altitude are relevant data pertaining to evaporation demand. Soils characteristics play an important role in water budgeting and hydrological modeling procedures, as they strongly influence the spatial and temporal distributions of soil moisture, water for crops, and return flows of applied irrigation waters. Cropping patterns determine the extent of soil water demand, ET uptake, and delivery. All of these data values vary significantly in space, and the entire process of spatial data management and ET computations is readily handled using GIS tools.

Application of the integrated consumptive-use (CU) model and database system is made for the agricultural lands of western Colorado developed for the Colorado River Decision Support System (Garcia 1995). The CU GIS system permits access and manipulation of the climatic, soils, and crops databases for input to the CU model. The resultant graphical and alphanumeric soils databases are then accessible by an evapotranspiration and soil-moisture accounting simulation model.

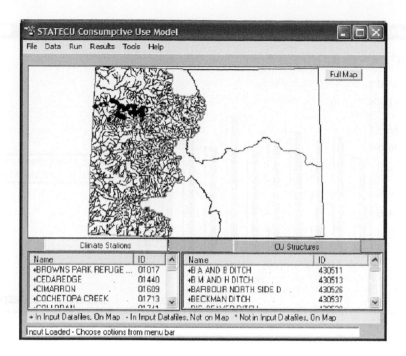

FIGURE 7.12 StateCU main interface window provides user access to input data sets, analysis controls, and output displays. (From CDSS 2007.)

The main StateCU graphical user interface is shown in Figure 7.12. The CU model interface makes use of GIS, specifically the Geographical Resources Analysis Support System (GRASS) (CERL 1993), which is used for spatial analysis and display, and ArcInfo, which is used for spatial coverage development. When a crop consumptive-use input data set is loaded, the map displays the available graphics overlays, which may include climate station locations, CU structure locations, water-district boundaries, and basin hydrography. The user can choose an individual climate station or CU structure directly from the map or from the station listing below the map. The data menu displays the major categories of StateCU input data that can be viewed and edited. These data include model control options, temperature and precipitation, frost dates, crop characteristics, groundwater pumping, direct diversion rights, and various administrative data. Analysis options include those for water supply, water rights, return flows, groundwater, and soil moisture. There are several output options, including a bar chart display summarizing analysis results (Figure 7.13).

The Visual Database Browser (VDB) provides access to map-oriented data products. The VDB supports interactive access to the various spatial data products, color graphics display and browsing of these data, and tabulation of data characteristics. Interactive data review functions include menu selection of data sets, color display, overlay of vectors and sites onto raster images, zooming features, multiple map management, variable resolution control, legend display, map history review, and the graphical selection of site time-series data stored in the Informix database. Spatial data sets accessible for the entire Colorado River basin in Colorado by VDB include elevation; irrigated lands; land use; soils; precipitation; evaporation; river network; basin boundaries; counties, divisions, and districts; road networks; public land survey; cities; river measurement sites; and climate measurement sites, reservoirs, and diversion headgates.

7.4.5 GIS-BASED IRRIGATION SYSTEM SCHEDULING

GIS tools and integrated models find extensive application for scheduling of water applications for an irrigation system. The GISAREG model (Fortes et al. 2005) is representative of a GIS-based

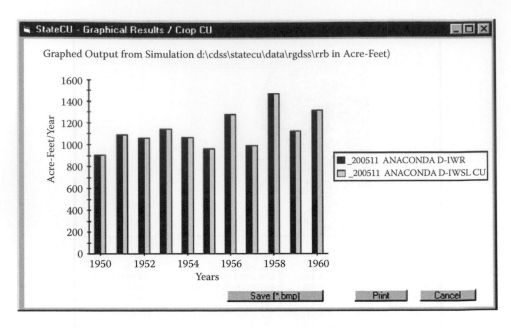

FIGURE 7.13   Colorado StateCU model results for historic consumptive-use accounting.

modeling package directed to improving irrigation scheduling. The model uses the ArcView software and Avenue scripting language to integrate the spatial and attribute databases with legacy irrigation-scheduling models. Spatial and attribute databases on crops, soils, and weather were described above. The model can be applied for different water-management scenarios and produces crop irrigation maps and time-dependent irrigation depths at selected aggregation levels, including the farm scale. GISAREG simulates alternative irrigation schedules relative to different levels of allowed crop water stress as well as various constraints in water availability. The irrigation scheduling alternatives are evaluated from the relative yield loss produced when crop evapotranspiration is below its potential level.

GISAREG inputs are precipitation, reference evapotranspiration, total and readily available soil water; soil water content at planting; and crop factors relative to crop growth stages, crop coefficients, root depths, and water-yield response factors. Various time-step computations are possible, from daily up to monthly, depending on weather data availability. GISAREG model results include annual crop irrigation requirements (mm), readily available water at the beginning and end of the irrigation period (mm), total available water in deep soil layers at the beginning of the irrigation (mm), percolation due to excess of irrigation (mm), precipitation during the irrigation period (mm), nonused precipitation (mm), cumulated actual and maximum evapotranspiration (mm), and monthly irrigation requirements (mm). Results are displayed in map and tabular formats.

GISAREG was applied to assess the crop irrigation requirements at an irrigation project area in the Syr Darya basin, Uzbekistan. The area has an arid climate, with cold winters and hot summers, where only storm rains occur from May to September. The soils are generally heavy and deep, with high water-holding capacity. The main summer crop is cotton, and the winter crop is wheat, generally furrow irrigated. Various simulation scenarios were run to assess performance, including:

- OY, aiming at attaining the optimal yield (i.e., no stress as irrigation threshold)
- RES15, using the same irrigation threshold but imposing a 15-day time interval between successive irrigations
- RES20, as above, but with a time interval of 20 days
- STRESS, adopting a threshold 50% below that for OY

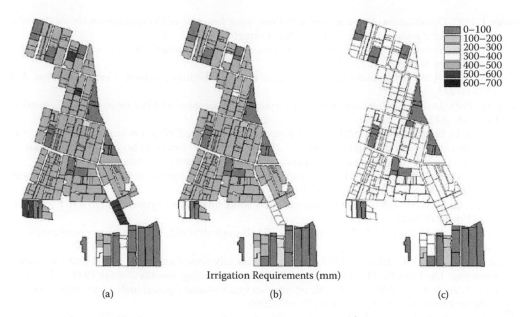

Irrigation Requirements (mm)

(a)   (b)   (c)

FIGURE 7.14   Mapping crop irrigation requirements for the average demand year adopting three irrigation-scheduling scenarios: (a) optimal yield, without water restrictions; (b) RES20, for 20 day's minimum interval between irrigations; and (c) STRESS, adopting a low irrigation threshold. (From Fortes et al. 2005. With permission.)

One scenario was run to assess the interannual variation of crop irrigation requirements. For this purpose, the model was run for all the years of the climatic data set, and the wet, dry, and average demand years were identified. The results for the average demand year are presented in Figure 7.14. Comparing Figures 7.14a and b for the OY and RES20 scenarios, small differences are apparent except for maize, which has a higher demand. In contrast, the STRESS scenario in Figure 7.14c shows evident differences from the other scenarios, indicating large impacts on productivity for most of the field crops.

REFERENCES

Alperovits, E., and U. Shamir. 1977. Design of optimal water distribution systems. *Water Resour. Res.* 13 (6): 885–900.

AWWA (American Water Works Association). 1998. *Distribution system requirements for fire protection (M31)*, 3rd ed. Denver: AWWA. http://www.awwa.org/Bookstore/.

Bhave, P., and V. Sonak. 1992. A critical study of the linear programming gradient method for optimal design of water supply networks. *Water Resour. Res.* 28 (6): 1577–1584.

Blaney, H. F., L. R. Rich, W. D. Criddle, et al. 1952. Consumptive use of water. *Trans. ASCE* 117: 948–967.

CDSS (Colorado Decision Support Systems). 2007. State CU documentation. Water Conservation Board and Division of Water Resources, State of Colorado. http://cdss.state.co.us/.

CERL (U.S. Army Corps of Engineers Construction Engineering Research Lab). 1993. GRASS 4.1 user's reference manual. Champaign, Ill.: CERL. http://www.cecer.army.mil.

Dowling, T., G. Walker, I. Jolly, E. Christen, and E. Murray. 2000. On-farm and community scale salt disposal basins on the riverine plain: Testing a GIS-based suitability approach for regional planning. CSIRO Land and Water technical report 3/00. CSIRO Land and Water Murray-Darling Basin Commission, Australia. March.

Eiger, G., U. Shamir, and A. Ben-Tal. 1994. Optimal design of water distribution networks. *Water Resour. Res.* 30 (9): 2637–2646.

Fortes, P. S., A. E. Platonov, and L. S. Pereira. 2005. GISAREG: A GIS-based irrigation scheduling simulation model to support improved water use. *Agric. Water Manage.* 77: 159–179.

Garcia, L. 1995. Consumptive use model and visual data browser: Colorado River Decision Support System (CRDSS). Denver: Colorado Water Conservation Board.

Gessler, J. 1982. Optimization of pipe networks. In *Proc. Ninth International Symposium on Urban Hydrology: Hydraulics and sediment control*, Univ. of Ky., Lexington, July 27–30.

Goulter, I. 1992. Systems analysis in water distribution system design: From theory to practice. *ASCE J. Water Resour. Plan. Manage.* 118 (3): 238–248.

Hess, T. M. 1996. A microcomputer scheduling program for supplementary irrigation. *Comput. Electron. Agric.* 15: 233–243.

Hubly, D. 1995. Design of water and wastewater systems. Department of Civil Engineering, University of Colorado at Denver.

Jensen, E., R. D. Burman, and R. G. Allen. 1990. *Evapotranspiration and irrigation water requirements*. ASCE manuals and reports on engineering practice, No. 70. New York: American Society of Civil Engineers.

Jeppson, R. W. 1982. *Analysis of flow in pipe networks*. Ann Arbor, Mich.: Ann Arbor Science.

Knox, J. W., and E. K. Weatherfield. 1999. The application of GIS to irrigation water resource management in England and Wales. *Geogr. J.* 165 (1): 90–98.

Lansey, K., and L. Mays. 1989. Optimization model for water distribution system design. *ASCE J. Hydraul.* 115 (10): 1401–1418.

Linsley, R. K., J. B. Franzini, D. L. Freyberg, and G. Tchobanoglous. 1992. *Water resources engineering*. New York: McGraw-Hill.

Loveland, T. R., and G. E. Johnson. 1983. The role of remotely sensed and other spatial data for predictive modeling: The Umatilla, Oregon, example. *Photogrammetric Eng. Remote Sensing* 49 (8): 1183–1192.

Luettinger, J., and T. Clark. 2005. Geographic information system-based pipeline route selection process. *ASCE J. Water Resour. Plann. Manage.* 131 (3): 193–2000.

Ortega, J. F., J. A. de Juan, and J. M. Tarjuelo. 2005. Improving water management: The irrigation advisory service of Castilla la Mancha (Spain). *Agric. Water Manage.* 77: 37–58.

Pereira, L. S., T. Oweis, and A. Zairi. 2002. Irrigation management under water scarcity. *Agric. Water Manage.* 57: 175–206.

Prins, J. G., and M. Bodeaux. 2000. Building land use plans for water and irrigation master plans: A case study. Paper presented at ESRI User's Conference, San Diego, Calif.

Ray, R., T. Walski, T. Gangemi, M. Gennone, and B. Juergens. 2007. Lessons learned in calibrating an IDSE model. In *ASCE World Environmental and Water Resources Congress 2007: Restoring our natural habitat*. New York: American Society of Civil Engineers.

Rossman, L. A. 2000. EPANET 2 users manual. Report EPA/600/R-00/057. Cincinnati: U.S. Environmental Protection Agency, Water Supply and Water Resources Division, National Risk Management Research Laboratory.

Szana, K. 2006. Water distribution systems modeling and optimization in a GIS environment. Master's project report, Department of Civil Engineering, University of Colorado at Denver.

Taher, S., and J. Labadie. 1996. Optimal design of water distribution networks with GIS. *ASCE J. Water Resour. Plann. Manage.* 122 (4): 301–311.

USDA (U.S. Department of Agriculture). 1970. Irrigation water requirements. Tech. release No. 21 (rev.). Washington, D.C.: USDA Soil Conservation Service.

Walski, T. 1984. *Analysis of water distribution systems*. New York: Van Nostrand-Reinhold.

Walski, T. 1985. State-of-the-art in pipe network optimization. In *Proc. of specialty conference on computer applications in water resources*, 559–568. New York: American Society of Civil Engineers.

Walski, T., J. Gessler, and J. Sjostrom. 1990. *Water distribution systems: Simulation and sizing*. Boca Raton, Fla.: Lewis Publishers.

Walters, G. 1988. Optimal design of pipe networks: a review. In *Proc. first international conference on computer methods and water resources*, Vol. 2, 21–31. Southhampton, U.K.: Springer Verlag.

Wu, Z. Y., R. H. Wang, D. Diaz, and T. Walski. 2003. Mining water consumption and GIS-based data for loading water distribution models. *ASCE World Water Congr.* 118: 23.

8 GIS for Wastewater and Stormwater Systems

8.1 WASTEWATER AND STORMWATER SYSTEMS PLANNING AND DESIGN

8.1.1 WASTEWATER AND STORMWATER SYSTEMS COMPONENTS

Wastewater and stormwater systems are major components of urban water resources systems, having the purposes of positive control of contaminated effluent, avoidance of off-site environmental contamination, and flooding. Sewage, or wastewater, is carried by sewer pipes and channels and may originate from domestic and industrial sources, storm runoff, infiltration, and inflow. Domestic and industrial sewage discharges are associated with sanitary facilities of houses and buildings, and from production processes of manufacturing and food processing; these sources are relatively constant but vary with the level of activity (e.g., diurnal variation, batch processing). Storm sewage results from runoff from precipitation events and includes wash-off of litter and wastes from roadways and the land surface (e.g., soil erosion). Infiltration is water that enters sewers from the ground through leaks; it can represent drainage of groundwater and is typically greater during periods of high water levels. Inflow is stormwater that enters sewers through leaks, broken pipes, connected roof drains, and sump pumps during storm runoff events.

It is standard practice to design sanitary and stormwater sewers as separate systems to maintain efficient control of the highly contaminated sanitary wastewater and to avoid complications of mixing sanitary sewage with the large volumes of storm runoff. Some older systems were constructed as combined systems that convey both sanitary and storm sewage; these systems require special consideration, and efforts to separate the two systems are expensve and involve careful cost-benefit and risk assessments. The primary elements of a sanitary wastewater collection system are lateral collection pipes, pumps and force mains, interceptor sewers, detension basins, treatment plants, and effluent disposal facilities (Figure 8.1). Customers are usually connected to the laterals, although some users may be connected to mains. Sanitary sewage collected from each service connection flows by gravity toward the treatment plant. Open-channel flow hydraulics governs, although pumping through force mains may be required to overcome topographic barriers. Stormwater systems differ in that the water source is runoff, and facilities for capture and conveyance are typically larger and include open channels and detention basins.

The pipes in sewer systems are not fully gridded as they are in a water distribution system, but are instead typically laid out as a treelike or branching structure, with flows and pipe sizes increasing in the downstream trunk mains and interceptors. It is common practice to blend these collection systems with the local watershed topography to take advantage of gravity flow. A hierarchy of sewer pipes is usually defined as laterals discharging into mains, mains into trunk mains, and trunk mains into interceptors. Stormwater systems often lead to detention basins, where solids settling and volume control can be achieved. Sanitary wastes are required to be treated to at least the secondary level (85% removal). Treated wastewater can be discharged to a surface water body, such as the river in Figure 8.1, or may be reused for irrigation of agricultural land, parks, golf courses, or other public areas where ingestion of the water is unlikely. Table 8.1 lists components and data for wastewater and stormwater systems; these are GIS and alphanumeric data associated with each data category. Time series data on flows and quality are a notable category.

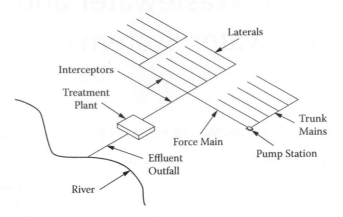

FIGURE 8.1 Wastewater collection system components and interrelations.

TABLE 8.1
Wastewater and Stormwater Systems Data

Terrain
DEM
Slopes
Drainage
Watersheds and sub-basins
Wetlands and ponds

Land use
Land uses (percent impervious)
Population
Parcel boundaries
Zoning
Master plan
TAZ
Forecast land use

Wastewater sources
Residences
Commercial/offices
Industries
Recycling and reuse

Wastewater infrastructure
Collection system
  (laterals, interceptors)
Manholes and junctions
Pumps
Force mains
Treatment plants
Bypasses and overflows
Basement sumps and flooding records
Gages and meters

Water demands
Customer delivery records
Conservation practices

Stormwater infrastructure
Drainage ways
Channels, interceptors, culverts
Collector pipes
Inlets
Manholes and junctions
Diversions
Detention and catch basins
Pumps and lift stations
Outfalls
Impervious areas

Climate
Precipitation—IDF, observations
Temperature
Potential ET
Groundwater levels

Soils
Permeability
Layer depths
Soil textural percentages
Initial soil water content

Administrative boundaries
Political boundaries
Wastewater management districts
Transportation authorities

8.1.2 Wastewater and Stormwater Collection System Design Procedures

The wastewater collection system is designed to operate mostly as a gravity flow system. The sewer pipes, although secure from the environment, generally serve as circular open channels and are seldom designed to flow full. The interconnected sewerage collection system drains wastewater and stormwater from connected services and from some inflow and infiltration downgradient to the wastewater treatment plant. Exceptions occur where topographic and physical conditions require collection in a wet well and pumping through a force main.

A tentative sewerage layout is made by locating lines along streets, rights of way, or utility easements and following the natural slope of the land to the extent possible. This involves delineating the watershed contributing area. The main sewer leaves the service area at the lowest point, with submains and laterals radiating from outlying areas. Ridges within the service area may require the construction of systems with separate discharges or pumping across the high area. In flat terrain, all sewers may be sloped to a common point from which the collected flow is pumped.

The vertical profile of the sewer is limited by the need to provide minimum cover and service to basement sanitary facilities and the desire to minimize excavation. In northern areas, up to 3 m (10 ft) of cover may be required to prevent freezing. In southern areas, minimum cover is dictated by traffic loads and ranges upward from 0.75 m (2.5 ft), depending on the pipe size and type and the anticipated loads. The design is based on hydraulic grade line principles which represent elevation profiles and energy.

Sewage and stormwater flows are computed from the connected service area tributary to a given sewer pipe section. The land-use method and/or connected services for complete buildout are used to determine the average flows and peaking factors. Flows are cumulative in a downstream direction as pipes and flows are joined at junction boxes or manholes. Manning's equation is used to determine the required size of a pipe given the slope. There is a trade-off between slope and pipe size; a smaller pipe can carry the same flow as a larger pipe if the slope is steeper. Thus, in the design of the collection network, the design is begun at the top of the system, and the engineer works downstream, collecting additional flows until the outlet of the system. The procedure is highly interactive, as downstream conditions may require modifications to pipe sizes and slopes in upstream pipes. The results of the sewer design program include sizes, slopes, and depths for the pipes; minimum scour velocities; volumes of excavation required; and the total cost of the resulting sewer network.

8.1.3 GIS Applications for Wastewater and Stormwater Systems

GIS databases, analysis functions, and linked simulation models are extensively used for the design and management of wastewater and stormwater systems. Applications include initial planning and design of these facilities, subsequent operations and maintenance of the facilities, finance and administration (ESRI 2007), and litigation and forensic analyses are becoming common. The important characteristic of wastewater utilities is that network features have defined behaviors, such as connectivity and flow, at the feature level. The GIS can perform basic tracing and network operations that allow data retrievals for design (e.g., network model formulation) and management (e.g., complaint checking). The multitude of applications provides the motivation for the development of an enterprise GIS that can be used across the organization for the various functions.

8.1.3.1 Planning and Design

Planning and engineering tasks that can be accomplished easily using GIS software include automated mapping, watershed modeling, population and demand projections, water flow analyses, and collection system master planning. GISs provide an intelligent database, so that flow analyses become a semi-automated procedure integrated into one system. The GIS can integrate data on the current or planned network, formulate the spatial and attribute data for the various models, and display the model results in map formats. User interaction with the data and models provides an effective way to examine a wide range of alternative designs. Dynamic segmentation can also be used to derive a generalized network that combines hydraulically similar sections into larger strips to make flow-analysis algorithms run more efficiently. GIS software allows representation of a project in three-dimensional form to visualize the impact of facilities on landscape during the design process. These data can then be combined with other computer-aided engineering functions to assist the engineering designer in the planning and scenario testing of alternative designs. Web tools such as Google Earth, YouTube, and Wikipedia are increasingly being applied.

8.1.3.2   Operations and Maintenance

Water and wastewater organizations are often responsible for constructing or overseeing the development of new facilities and replacement of old facilities. GIS aids in tracking information related to projects, permits, construction work order management, inspections, and as-built drawings. Dynamic segmentation and image integration features in GIS allow utilities to store and display images in conjunction with a footage reading and a description of the pipe defect (roots, grease, other problems). Maps, drawings, and pictures can also be associated with valves, manholes, drop inlets, and other features to provide valuable information to the user. GIS spatial selection and display tools allow users to visualize scheduled work, ongoing activities, recurring maintenance problems, and historical information. The topological characteristics of a GIS database can support network tracing and can be used to analyze specific properties or services that may be impacted by such events as stoppages, main breaks, drainage defects, and so forth.

8.1.3.3   Finance and Administration

Finance and administration provides the central administrative oversight to support the planning and operational activities of the organization. Budget forecasting, facility inventory analysis, customer billing, and other key administrative functions can be enhanced through the implementation of an integrated, enterprise-wide GIS solution. Administrative activities can be streamlined for rate justifications, complaint tracking, development review and approval, right-of-way engineering, capital improvement project tracking, and redistricting.

8.2   GIS DATABASE DEVELOPMENT FOR WASTEWATER AND STORMWATER SYSTEMS

8.2.1   GIS Database Development

Wastewater and stormwater systems are characterized as networks typically having a dendritic structure of collectors draining to interceptors and trunk mains in the downstream direction. Data to describe the system and its components (Table 8.1) are collected from various sources. Typically, cities and utilities have gathered geospatial data in various data formats over time. Commonly used formats include CAD design and as-built drawings (as DXF, DWG, and/or shapefiles), geometric networks, land-use maps and master plans, parcel files, spreadsheets, geodatabases, and various databases of alphanumeric attribute data keyed to location. Import of external data into the GIS can be made for customer billing records, digital elevation models, and satellite and aerial imagery of various types. Collation of these data can require extensive efforts for data quality control and maintenance, hardware and software resources development, and interdepartmental cooperation.

Field data collections are often required to determine the current waste- and stormwater system configuration, and GIS tools play an important role for these activities. For example, Landry et al. (2001) and Curley et al. (2006) demonstrated the use of GIS, GPS, and mobile computer technologies for large-area stormwater facility maintenance programs. Data were collected in the field on structures and pipes, noting the location, asset type, condition rating, digital photos, shape and flow direction, and maintenance status. To comply with existing data standards within the county, Landry et al. (2001) used an ArcInfo® coverage as the final data format for the stormwater

FIGURE 8.2 View of stormwater structures and pipes illustrating topological relationships. (From Landry et al. 2001. With permission.)

inventory; all stormwater pipes would be stored as arcs, and each arc would be connected to exactly two nodes, one at each end (Figure 8.2). By utilizing the new protocol, which coupled GPS data collection and GIS postprocessing, the county surveyed roughly one-third of its municipality in less than a year, a marked improvement over the six years it took to do the other two-thirds while using the old method, which included manually plotting and digitizing the data. The stormwater information collection system by Curely et al. (2006) was implemented using ArcPad® Application Builder running on portable GPS-integrated handheld computers.

Procedures for wastewater demand estimations rely on land-use data of various types. GIS applications are used to convert and validate sewer data, process population, land use, and topographic data and export the data to model format. Data sets and processing typically include:

- Extensive land base information such as digital orthophotos, address-coded streets, political boundaries, parcel boundaries, and topographic contours.
- Sewer system facilities data covering the pipe collection system, including pipe diameters, invert and ground elevations, and materials. This is the database used for producing sewer maps, and it is also used for the sewer maintenance management system. The attributes for modeled pipes need to be verified from as-built drawings and field surveys.
- Population, employment, and land-use information (existing and projected) by geographic units as small as a parcel and city block. These data are maintained and updated by the local or regional sewer authority. These data are the basis for computing wastewater flows in all parts of the sewer system.
- Major industrial dischargers' estimated wastewater flows and other discharge characteristics. These data are typically maintained by the industrial permitting agency. This database can be imported into the GIS using geocoding using the addresses and address-matching tools.
- Flow monitoring sites, as digitized and maintained by the public works flow monitoring group and modeling team.

Some good practices for creating GIS data for later use in a model include (Govindan et al. 2006): (a) "snapping" pipe ends to other element types; (b) standardized element labeling conventions; (c) customer service lines in separate features classes from system pipes; and (d) wet wells, pumps, and other system components as separate feature classes. Conversely, the following are some of the possible errors in data that need attention (and may be driven by the capabilities of the numerical solver):

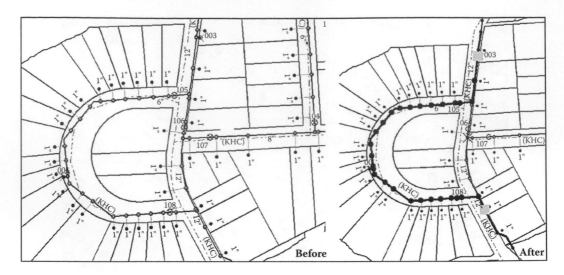

FIGURE 8.3 GIS tools are instrumental in defining network connectivity and flow tracing as well as junction isolation operations. (From Oppman and Przybyla 2003. With permission.)

missing attributes, features not properly connected, features digitized backwards, feature types with no model counterparts, identifiers incompatible with the model, and GISs containing hydraulically insignificant or "short" pipe segments. GIS tools provide the functionality for performing network connectivity tracing and junction isolations (Figure 8.3), which aid in creating an error-free database.

Parcel, zoning, and traffic coverages are primary data sources to be used in the development of the land-use plan for sewer master plan development. The parcel coverage links each parcel to an activity table that defines the current type of development and the number of residential dwelling units. Each building would have an associated unit sanitary sewage generation factor. Stormwater generation and associated water pollution would be associated with the impervious surface area of the parcel and contributing street runoff. The zoning coverage identifies the zoning of existing parcels. A general plan coverage identifies the proposed zoning of undeveloped areas. These two coverages may not be compatible because the land-use categories are not always consistent. Land-use zones for the general plan are in gross areas that include street rights-of-way, while the zoning coverage uses net land use where rights-of-way and other areas have been zoned separately. There may be defined sewer plan land-use categories as part of the master plan. Traffic analysis zone (TAZ) coverages are commonly used to identify planning areas having similar growth characteristics. A lookup table can be developed to define the anticipated land-use growth rate for each TAZ planning area and to define the expected year when development in currently undeveloped areas will commence.

8.2.2 WASTEWATER AND STORMWATER GEODATABASES

There are many reasons for consolidation of GIS and related attribute data sets into a single system to manage and store spatial and tabular infrastructure data and support engineering and construction, operations and maintenance, and financial and administrative applications. Although it may involve considerable effort both technically and administratively, the benefits of an enterprise GIS for wastewater utilities are realized over time by the numerous users addressing the multitude of utility design and management functions across departments and jurisdictions. Typical requirements of municipal utilities are to:

- Update GIS databases with as-built data
- Produce standard and custom map products
- Integrate computer-aided design (CAD) drawings into the GIS environment

- Integrate with other enterprise systems, such as work management systems (WMSs), document management systems (DMSs), infrastructure management systems (IMSs), materials management systems (MMSs), and customer information systems (CISs)
- Analyze installed network for capacity planning and capital improvement projects
- Manage operations activities, such as leaks, repairs, and inspections

Water infrastructure data models have been developed by ESRI (http://support.esri.com/data-models), and these provide a guide for the development of customized data models for a particular application. The Water Utilities Data Model (Grise et al. 2001) contains a ready-to-use data model that can be configured and customized for use at water utilities. A keystone of this water networks data model is capturing the behavior of real-world water objects such as valves and lines; object databases combine data and application behavior modeling. As a result, the model not only includes an essential set of water object classes and properties, it also includes rules and relationships that define object behaviors.

Thematic groups in the Water/Sewer/Stormwater data model include those for Lines, Equipment, and Facility (Grise et al. 2001). Figure 8.4 shows the objects in the Line thematic group. Complex edges are the base class for all network lines. The WaterLine class was created as a general top-level class for any type of network line. In other words, all classes beneath the WaterLine class will contain the properties of this feature class. The inherent behavior of complex edges is very different than the traditional ArcGIS® topology model. The ArcGIS® system automatically maintains the relationships between complex edges, any attached devices, and other edges, so you can choose how to physically segment your network. For instance, you may choose to physically segment sewer pipes between manholes, since, among other things, it is important to capture InvertElevation data on mains where they connect with manholes. This can only be captured for the starts and ends of gravity mains, so sewer/stormwater networks should be segmented at manholes. The Equipment thematic group contains a large number of feature classes that do not have a spatial representation. Equipment objects are modeled separately from their location in the ground so as to maintain

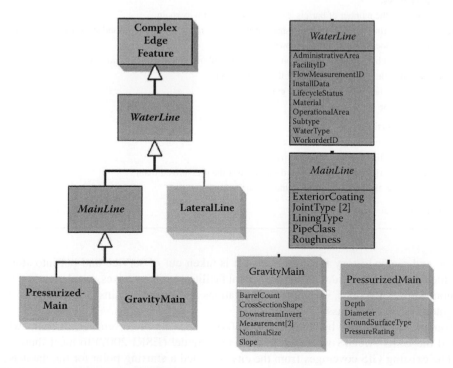

FIGURE 8.4 Line objects in ArcGIS® Water. (From Grise et al. 2001. With permission.)

TABLE 8.2
Attributes of Line Objects in ArcGIS® Water

WaterLine abstract class attributes:

AdministrativeArea: string; a general-purpose string that is used to store information such as the name of the municipality/owners

FacilityID: string; the line facility identifier

FlowMeasurementID: string; the flow measurement identifier of the line

InstallDate: date; the date the line was installed

LifecycleStatus: string; the status of the line

Material: string; the material of the line

OperationalArea: string; a general-purpose attribute that records information such as the basin or pressure zone of the line

Subtype: integer; a general attribute that is used to store subtypes

WaterType: string; the type of water found in the line; these types are CombinedWastewater, PotableWater, RawWater, ReclaimedWater, SaltWater, Sewage, StormRunoff, and WastewaterEffluent (Note that some of these types apply to a water distribution system. They exist, however, because this class is shared with the water data model.)

WorkorderID: string; the identifier of the work order that is associated with the installation of this line

LateralLine class attributes:

LocationDescription: string; the description of the lateral line connection location

Size: integer; the size of the lateral line

MainLine class attributes:

ExteriorCoating: string; the exterior pipe coating of the main line

JointType[2]: string; the joint type of each end of the main line; the joint types (implemented as coded value domains) are CM, FL, MECH, PO, RCCB, SOL, and WELD

LiningType: string; the lining type (interior coating) of the main line

PipeClass: string; the class rating of the main line

Roughness: double; the roughness coefficient of the main line

GravityMain properties:

BarrelCount: integer; the number of barrels associated with the gravity main

CrossSectionShape: string; the cross-section shape of the gravity main; for the sewer and stormwater model, the shapes are Circular, Horseshoe, Oblong, and Unknown

DownstreamInvert: double; the invert elevation (interior bottom) of the downstream end of the main

Measurement[2]: double; the measurement of the gravity main

NominalSize: double; the nominal size of the gravity main

Slope: double; the slope of the gravity main

PressurizedMain properties:

Depth: integer; the depth of the pressurized main

Diameter: integer; the diameter of the pressurized main

GroundSurfaceType: string; the type of ground surface over the pressurized main

PressureRating: string; the pressure rating of the pressurized main

Source: From Grise et al. (2001).

---

inspection and repair history for the object if it is taken out of service and put into storage. The Facility thematic group includes all network point facilities such as valves, manholes, and fittings. It also contains other junctions, including pump stations and storage basins. Table 8.2 lists attributes of the WaterLine abstract class.

An example was provided by Gough et al. (2004) for the City of Torrence, California, where they simplified the ESRI ArcGIS water/sewer/storm data model (ESRI 2007) to meet their particular needs. The existing GIS coverages from the city provided a starting point for the database definition. Additional feature classes and associated attributes were identified using the existing sewer

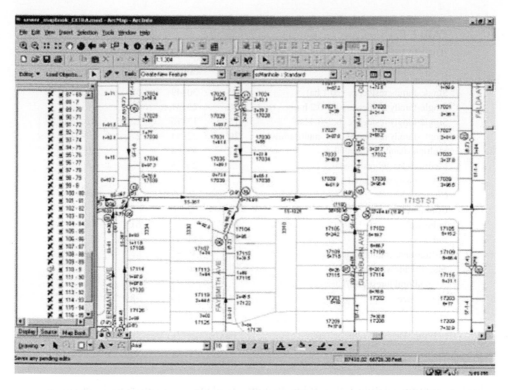

FIGURE 8.5 Section of Torrence, California, sewer system with annotations. (From Gough et al. 2004. With permission.)

atlas maps from the city. The list was refined with some additional changes appropriate for the city's future needs through a pilot study, and additional attributes and domain values as new ones are discovered in the production area of the sewer system GIS updates. Figure 8.5 shows a screen shot of a section of the Torrence sewer system as retrieved from the database. Since the installation of the updated sewer data, the city has seen some benefits from the adoption of the geodatabase and attention to detail on the design. Also, the incorporation of domain values that were encountered on the maps has facilitated more accurate attribute data entry by city staff.

8.2.3 Impervious Surface Mapping

Percent impervious area is a primary land attribute for stormwater modeling. These data can be developed from analysis of high-resolution aerial photography or satellite imagery. Impervious surfaces are mainly constructed surfaces—rooftops, sidewalks, roads, and parking lots—covered by impermeable materials such as asphalt, concrete, brick, and stone. These materials seal surfaces, repel water, and prevent precipitation and melt-water from infiltrating soils and generally increase pollutant loading such as heat and traffic-related debris. Development of a scientific basis for the relationship between land use and impervious surface has been the subject of considerable study beginning in the 1970s. In earlier studies, aerial photos were used. Fankhauser (1999) described a method to estimate impervious area from color infrared aerial photographs and orthophotos; estimated impervious surface based on color photo analysis was within 10% of hand field measurements of impervious surfaces. Lee and Heaney (2003) used five levels of measuring and GIS-based analysis to measure Directly Connected Impervious Area (DCIA) for a small residential neighborhood. The GIS-only method overestimated DCIA by 1.5 times the value obtained through field surveys, while field methods required more labor resources.

TABLE 8.3
Impervious Percentages for Various Land-Cover Classes

Land-Cover Class	Mean [a]	Range [a]	Mean [b]	Range [b]
Single-family residence (<0.1 ha)	39	30–49	35	19–48
Single-family residence (0.1–0.2 ha)	26	22–31	26	18–38
Single-family residence (0.2–0.4 ha)	15	13–16	22	17–28
Single-family residence (incl. multifamily)	30	22–44	27	19–37
Multiple-family residential	57	53–64	57	50–71
Medium-density residential	...	20–25		
Low-density residential	...	5–10		
Commercial	88	66–98	92	88–95
Commercial	81	52–90		
Industrial	60	...	61	48–69
Industrial	40	11–57		
Institutional/commercial	...	35–55		
Major roads with median	...	50–70		
Agricultural/forest/golf/idle	...	2–7		
Open space	5	1–14	8	2–15

[a] Adapted from Brabec et al. (2002) and Dougherty et al. (2004).
[b] From Wang and Johnson (2007).

As high-resolution satellite imagery has become available, and supervised classification procedures have been adopted and are the norm today (e.g., Slonecker et al. 2001; Trauth et al. 2004). Land-cover and elevation models have been derived from satellite imagery and LIDAR (LIght Detection And Ranging) data. Lu and Weng (2006) demonstrated an impervious surface classification method that incorporates surface temperature and albedo, derived from Landsat 7 ETM+ satellite images; 30-m resolution thermal infrared data were extracted from the ETM+ reflective band. Surface temperatures were derived from the infrared data. Overall, the thermal and albedo method results in 83.8% accuracy, with greater accuracy in areas with more than 30% impervious surfaces. Table 8.3 summarizes typical values of percent impervious area for various land-cover classes. There are some caveats to be acknowledged for these data, including (a) there is considerable variability within some land classes and development density, and (b) the data do not account for the drainage areas that are directly connected by storm sewers to the streams (i.e., the "effective" impervious area).

A recent study by Wang and Johnson (2007) was directed to assessing the validity of the Table 8.3 values of impervious percentages for 20 land-use classes at the Anderson Classification Levels II and III (LU Code). Three methods were applied to a study area in Ft. Collins, Colorado, (a) digitizing from 0.15-m (0.5 ft)-high resolution aerial photography, and (b) supervised classification of the 0.15-m imagery. For the manual method, selected sample sites having singular land use were identified from zoning and parcel maps, and digitized to outline impervious areas as polygons. Although the manual digitizing method required extensive time and effort, the sample site data obtained were helpful for the supervised classification. The 1-m natural color NAIP (National Agriculture Imagery Program) imagery was helpful as a check for building-shadow areas. Statistical summaries of these data corresponded well with the literature values (Table 8.3). Classification of satellite imagery by itself did not support Level III detail.

Impervious-surface classification using a feature-extraction approach was applied to map impervious surfaces for the Croton Watershed (approximately 400 mi^2) using digital aerial imagery at 0.3-m ground resolution (Washburn et al. 2003). The feature-extraction approach recognizes that the linear shape of human-made features (e.g., roads and buildings) is as important as the radiometric signature. The imagery source, collected during the fall of 2001 (minimal vegetative cover or leaf-off

FIGURE 8.6 (See color insert following page 136.) Impervious surface areas (shown in yellow) extracted by image processing. (From Washburn et al. 2003. With permission.)

conditions), was delivered radiometrically balanced and orthorectified. The study team mosaicked the three-band color-infrared orthoimagery for data sets that are radiometrically consistent across daily collections. The resulting mosaics were the basis for the impervious-feature identification performed using Feature Analyst® and ArcGIS. The first step for Feature Analyst extraction was to specify samples of the desired features. The software uses these samples to automatically extract all similar regions by "learning" the samples' spectral and neighborhood characteristics. For optimum performance, the selected features represented the generalized morphology of the feature class; these included double-lane roadways with and without the double yellow dividing line, highways, driveways, and houses with a variety of roof configurations and materials. The accuracy assessment for the entire Croton Watershed region showed an overall accuracy of 97.5%. Research has shown that impervious-surface studies to date report accuracies as high as 85% to 90% for the overall classification in a watershed; these studies were for impervious or urban classes within a comprehensive land-cover classification. The feature-extraction methodology attains higher accuracies because the process concentrates on finding only the impervious surfaces rather than on a general classification of the entire image. Figure 8.6 illustrates the impervious areas identified.

8.3 GIS-BASED WASTEWATER COLLECTION SYSTEM DESIGN AND MANAGEMENT APPLICATIONS

8.3.1 GIS-BASED ESTIMATION OF SANITARY WASTEWATER DEMANDS

The basic structure for estimating sanitary wastewater flow demands is similar to the structure described for estimating water demands (Chapter 7); however, there are some significant differences. For example, there are basic differences in the structure of the sanitary collection system (i.e., dendritic as opposed to gridded); this makes the flows cumulative as one progresses downward through the system. A principal difference between the estimation of water demand and wastewater flow is the definition of the service area. In a water distribution system, the service area for a subject node is often defined as the area around the node that is not closer to another node in the system. In wastewater collection systems, the service area is defined as the total area producing wastewater flow that must be carried by the subject line. Another difference is in the estimation of peaking factors. In the wastewater flow estimation process, the selection of the maximum-hour and minimum-hour peaking factor is usually based on the total population served or the average-day (AD) flow rate expected. Also, irrigation demand is not a factor in estimation of wastewater flows because irrigation water is not expected to enter the wastewater system. Finally, the sanitary system estimation process involves the addition of infiltration and inflow (II).

Variations of the population and land-use methods are used to estimate dry-weather wastewater flows for medium-sized service areas that include residential, commercial, and industrial developments and small communities. Table 8.4 presents some typical sanitary unit flow factors based on water use; these should be checked using data specific for the area obtained from water-consumption records and/or monitoring. The human-consumption method could also be used to estimate wastewater flow, less the irrigation demand. GIS and related databases support the assignment of flows from land-use-based flows, point-source flows such as industrial flows, and wet-weather infiltration or storm-related inflow and infiltration in addition to calibrating the unit flow generation factors and infiltration and inflow parameters. Peaking factors for wastewater are commonly based on guidance provided by regulatory agencies (e.g., ASCE 1970) or by flow monitoring.

Determination of the flows that enter the system involves several steps, as summarized by Govindan et al. (2006). The first step in determining loading would be to first understand the current year dry-weather flows and then add the complexities of wet weather and future conditions. Loading data can be obtained manually from customer or flowmeter data or automatically using software import tools. Automated model-loading tools can take loading information from a variety of GIS-based sources, such as customer meter data, system flowmeter, or polygons with known population or land use, and assign those flows to elements. GIS tools provide automated model-loading tools oriented to the types of data available to describe dry-weather flows. For example, some of the loading options for the SewerGEMS® model LoadBuilder® modules (Bentley 2005) include:

1. Water-consumption data can be imported by geocoding customer water-consumption data to the nearest manhole in the sewer network. Consumption data can be tagged to the nearest pipe and then specifying how the demand will be split among the bounding manholes. The aggregated consumption data for service polygons (buffers) along meter routes or drainage basins can then be assigned to the associated manhole.
2. Flow-monitoring data can be assigned by distributing the flow from each monitor equally across all elements in a specified area (polygon), usually a drainage basin or sub-basin. The flows can be proportionally distributed to elements based on the actual service areas of the elements involved. Alternatively, the flows could be distributed based on the population in the service area.
3. Land-use parcel and census data can be imported based on user-defined requirements developed for each type of land use or per capita. The loads can be allocated either equally or proportionally to the elements located within the polygons contained in land-use and census maps.

Infiltration/inflow (I/I) is the term used to describe groundwater infiltration and stormwater inflow into wastewater collection systems through pipe or manhole defects. The I/I enters the collection system through cracked clay pipe and other forms of deteriorated pipe, tree root damage, leaky manhole seals, cracked manhole lid casings, and brick manholes. Other ways for I/I to enter a collection system are attributed to improper or illegal plumbing practices such as building footing tile drains, sump pumps, building downspouts, or storm sewer discharges that are connected to

TABLE 8.4
Sanitary Flow Estimation Based on Water Use

Zoning		Densities		Water Demand Per Capita		Wastewater[a]	
		(units/ha)	(capita/unit)	(L/[capita×d])	(L/[d×ha])	Consumption (%)	Use (L/[d×ha])
R1	Single-family residences, low density	3	2.5	680	5,100	75	3,825
R2	Single-family residences, medium density	6	3	500	9,000	80	7,200
R4	Single-family residences, high density	12	3.5	360	15,120	90	13,608
M1	Multifamily residences, low density	20	3	270	16,200	95	15,390
M2	Multifamily residences, medium density	35	3.5	250	30,625	95	29,094
C1	Commercial				5,000	95	4,750
I1	Light industrial				7,300	90	6,570
I2	Dense industrial				9,400	75	7,050

Source: Hubly 1995.

[a] Wastewater is computed as water demand less consumptive use.

the sanitary sewer system. Field data must be collected to analyze a wastewater collection system's problems and determine whether it would be cost effective to reduce the I/I. The data consist of manhole and pipe inspection information, collection system inventory, and system hydraulic flow information. GIS database and analysis functions in conjunction with hydraulic modeling software provide an efficient means for conducting such analyses (e.g., Cannistra and Purves 1992; Christensen et al. 1999).

8.3.2 GIS-BASED HYDROLOGIC AND HYDRAULIC MODELING

Hydraulic models, GIS, and data-management tools are routinely used to assist in the planning and design of sewer systems. Often a GIS is used as a simple preprocessor of spatial data in the modeling of urban stormwater. Here, the GIS may simply store geographic information in a database, or it may be used to calculate model-input parameters from stored geographic data. As a postprocessor, GIS may be used to map water surface elevations and concentrations, or to perform spatial statistical operations on model output. As an example, Meyer et al. (1993) linked a physically based urban stormwater runoff model to a low-cost, PC-based raster GIS package to facilitate preparation, examination, and analysis of spatially distributed model inputs and parameters. Shamsi (1998) generated SWMM (Storm Water Management Model) input data from a GIS. Xu et al. (2001) describe a mixed land-use distributed hydrologic model that uses GIS as a pre- and postprocessor of model information. For that application, the model output of time series of simulated flows could be depicted dynamically through an ArcView interface.

Application of GIS in urban stormwater systems has lagged behind that for more-natural watershed systems because of the need for large and detailed spatial, facility, and temporal databases (Heaney et al. 1999). Integration of GIS with urban hydrologic and hydraulic models was identified as an emerging trend by Heaney et al. (2000). As more urban jurisdictions create GIS coverages, the integration of modeling software and GIS software has become more prevalent. There are now a number of sewer and stormwater modeling software packages with GIS interfaces (Table 8.5).

Some packages build on the public-domain Stormwater Management Model (SWMM) software (Rossman 2007). SWMM is a dynamic rainfall-runoff simulation model used for single-event or long-term (continuous) simulation of runoff quantity and quality from primarily urban areas. The runoff component of SWMM operates on a collection of subcatchment areas that receive precipitation and generate runoff and pollutant loads (Figure 8.7). The routing portion of SWMM transports this runoff through a system of pipes, channels, storage/treatment devices, pumps, and regulators. SWMM tracks the quantity and quality of runoff generated within each subcatchment as well as the flow rate, flow depth, and quality of water in each pipe and channel during a simulation period comprising multiple time steps. Since its inception, SWMM has been used in thousands of sewer and stormwater studies throughout the world. Typical applications include:

- Design and sizing of drainage system components for flood control
- Sizing of detention facilities and their appurtenances for flood control and water quality protection
- Floodplain mapping of natural channel systems
- Designing control strategies for minimizing combined sewer overflows
- Evaluating the impact of inflow and infiltration on sanitary sewer overflows
- Generating nonpoint-source pollutant loadings for waste load allocation studies
- Evaluating the effectiveness of best management practices (BMPs) for reducing wet-weather pollutant loadings

SWMM conceptualizes a drainage system as a series of water and material flows between several major environmental compartments. These compartments and the SWMM objects they contain include:

TABLE 8.5
Urban Wastewater and Stormwater Modeling Packages with GIS

Product/Model	Description	Company	Web Site
SWMM	Dry- and wet-weather flow simulator; Windows interface	U.S. EPA	http://www.epa.gov/ednnrmrl/models/swmm/
H2OMAP Sewer, InfoSWMM	Sanitary and storm sewer collection systems; ArcGIS, AutoCAD	MWH Soft	www.mwhsoft.com
HYDRA	Sanitary and storm sewer system design; exchange ArcGIS & AutoCAD	Pizer, Inc.	http://www.pizer.com
Infoworks, InfoNet	Drainage and sewerage master planning & asset management; OpenMI	Wallingford Software Limited	http://www.wallingford-software.co.uk/
MIKE URBAN, MIKE SWMM	Stormwater and sanitary systems using SWMM5; ArcGIS	DHI Water & Environment	http://www.dhi.us/software.html
PCSWMM.NET	Windows–GIS interface to SWMM5; ArcGIS, others	Computational Hydraulics Int.	http://www.computationalhydraulics.com/
PondPack v. 8	Analyzes watershed networks & sizing detention ponds; uses NRCS unit hydrograph Tc calculation formulas	Haestad Methods, Inc. (now Bentley Systems)	http://www.haestad.com/software/
SewerGEMS, SewerCAD, StormCAD, Bentley Wastewater	Sanitary and stormwater system design & asset management; AutoCAD, ArcGIS, and MicroStation	Bentley Systems	http://www.bentley.com http://www.haestad.com
StormNET	H&H models for storm and sanitary systems; ArcGIS, AutoCAD, MapInfo	BOSS International	http://www.bossintl.com
XPswmm, XPstorm	Link-node models for storm and wastewater systems; GIS exchange	XP Software	http://www.xpsoftware.com/

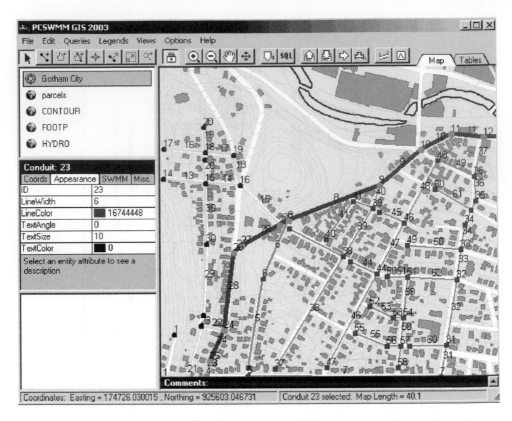

FIGURE 8.7   PCSWMM interface window. (*Source:* http://www.computationalhydraulics.com/.)

- Atmosphere compartment, from which precipitation falls and pollutants are deposited onto the land-surface compartment. SWMM uses Rain Gauge objects to represent rainfall inputs to the system.
- Land Surface compartment, which is represented through one or more Subcatchment objects. It receives precipitation from the Atmospheric compartment in the form of rain or snow; it sends outflow in the form of infiltration to the Groundwater compartment and also as surface runoff and pollutant loadings to the Transport compartment.
- Groundwater compartment receives infiltration from the Land Surface compartment and transfers a portion of this inflow to the Transport compartment. This compartment is modeled using Aquifer objects.
- Transport compartment contains a network of conveyance elements (channels, pipes, pumps, and regulators) and storage/treatment units that transport water to outfalls or to treatment facilities. Inflows to this compartment can come from surface runoff, groundwater interflow, sanitary dry-weather flow, or from user-defined hydrographs. The components of the Transport compartment are modeled with Node and Link objects.

Subcatchments can be divided into pervious and impervious subareas. Surface runoff can infiltrate into the upper soil zone of the pervious subarea, but not through the impervious subarea. Impervious areas are themselves divided into two subareas: one that contains depression storage and another that does not. Runoff flow from one subarea in a subcatchment can be routed to the other subarea, or both subareas can drain to the subcatchment outlet. Infiltration of rainfall from the pervious area of a subcatchment into the unsaturated upper soil zone can be described using three different models: (a) Horton infiltration, (b) Green-Ampt infiltration, and (c) SCS Curve Number

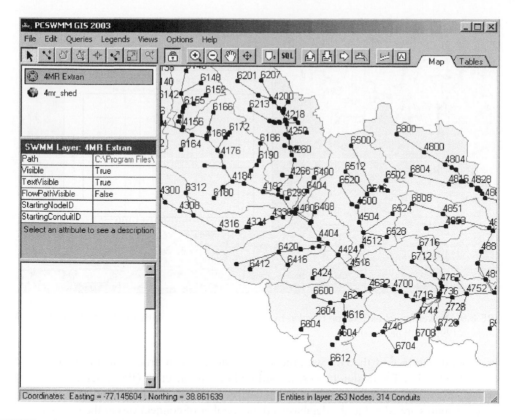

FIGURE 8.8 Urban watersheds linked by the drainage network. (*Source:* http://www.computationalhydraulics.com/.)

infiltration. Other principal input parameters for subcatchments include assigned rain gauge, outlet node or subcatchment, assigned land uses, tributary surface area, imperviousness, slope, characteristic width of overland flow, Manning's *n* for overland flow on both pervious and impervious areas, depression storage in both pervious and impervious areas, and percent of impervious area with no depression storage. Figure 8.8 illustrates urban watersheds linked by the drainage network.

Junctions are drainage system nodes where links join together. Physically, they can represent the confluence of natural surface channels, manholes in a sewer system, or pipe connection fittings. External inflows can enter the system at junctions. Excess water at a junction can become partially pressurized while connecting conduits are surcharged, and this excess can either be lost from the system or be allowed to pond atop the junction and, subsequently, drain back into the junction. The principal input parameters for a junction are invert elevation, height to ground surface, ponded surface area when flooded (optional), and external inflow data (optional). There are also outfall nodes, flow divider nodes, storage nodes, various types of conduits, pumps, orifices, and weirs.

8.3.3 GIS-Based Wastewater and Stormwater System Modeling

GIS supports a series of spatial-analysis functions that are ideally suited for design of sewer systems (Greene et al. 1999). These spatial-analysis functions are used to preprocess the data for use by the sewer design program. They include: (a) identification of the desired locations for manholes, (b) creation of buffers around prohibited areas, (c) creation of a triangular irregular network (TIN) for topographic information, (d) determination of surface elevations for manholes and vertices of buffers, (e) creation of a preliminary sewer network as a TIN using manhole locations and their elevations, (f) identification of potential lines for the sewer network through overlay operations that remove preliminary sewer lines that cross prohibited areas, and (g) a graphical display of results.

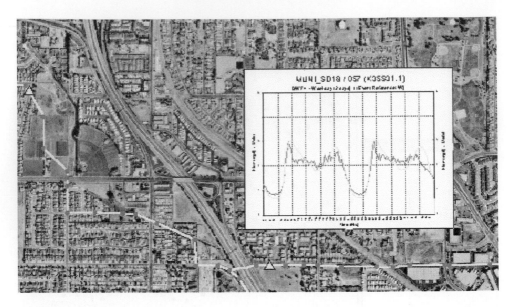

FIGURE 8.9 Sewer model calibration is aided by integrating flow data with a GIS interface. (From Wang and Baldwin 1999. With permission.)

The general interface of the GIS with the sewer system design process involves an interactive sequence of activities. The design process as outlined by Greene et al. (1999) begins with the user interactively creating a layer in the GIS with the desired locations for the manholes. For each manhole, two attributes are added to the database: (a) the number of connections to the manhole and (b) whether the manhole is in an outfall location. An attribute value of 0 or 1 is used to indicate if the location is unsuitable (0) or suitable (1) for an outfall. In selecting the locations for the manholes, the user must meet the standards set by the ASCE and the Water Pollution Control Federation (ASCE 1970). Additional requirements may include the spacing and maximum depths of manholes, minimum depth of cover for sewer lines, and the maximum allowable infiltration. The user must have knowledge of these regulations and how they are satisfied.

Examples are illustrated by Wang and Baldwin (1999), as seen in Figure 8.9, and Greene et al. (1999), who compiled modeling and data-management tools that could be accessed from customized menus, buttons, and tools, all within the standard GIS windows. The first stage of the development process involved designing the database and graphic file structure. The attribute data (e.g., pipe size, length, inverts, and material type) were stored in separate database file (DBF) tables and joined to corresponding graphic files using a key field. This approach maximized the use of the graphic files by joining different forms of attribute data to the same graphic file, in effect "normalizing" the tabular and graphic data. The following details the main hydraulic modeling process tools used for constructing and calibrating the city's sewer model.

- Data Validation: Performs extensive data validation checks based on specified rules such as missing lengths, sizes, reverse slopes, connectivity errors, etc. The results are displayed on the plan view with error descriptions "joined" to the attribute data.
- Network Trace Manager: Used to verify connectivity errors, select profile views, define model networks, and identify subarea basins.
- Profile Viewer: Creates a theme displaying a profile of a selected pipe section. Node and link data can be accessed from the profile view.
- Model Export: Extracts a selected portion of the city sewer system for reformatting and export to the city's modeling software package (HydroWorks®). The routine has the capability of exporting to various model formats.

- Model Results Import: Imports selected results from the modeling program for graphical viewing and analysis (e.g., highlighting of problem areas on the sewer map).
- Hydrograph Viewer: Imports selected flow-monitoring data and links those data to the monitor symbols on the GIS map, for the purpose of displaying flow and depth graphs along with modeling-results graphs during the model-calibration process.

The data requirements of the models are an additional challenge. Models require data in specified formats. The GIS interface capabilities claimed by modeling software suppliers are limited to importing and exporting data in a predefined format and data structure. Consequently, data must be translated from a customer-specific format to the input format required by the model. To utilize modeling results in GIS, the interface must also convert model output to a format that can be linked back to the GIS.

8.4 GIS-BASED DECISION-SUPPORT SYSTEMS FOR WASTEWATER AND STORMWATER SYSTEMS

Integration of GIS with databases and simulation models of wastewater and stormwater systems provides the foundation for decision-support systems (DSSs). An integrated suite of decision-support system tools may be seen as a natural extension of simulation models (e.g., SWMM), GIS (e.g., ArcGIS, IDRISI®, MicroStation®), relational databases (e.g., dBASE®, Oracle®, Access®), and evaluation tools (e.g., optimization software). The direction of DSS development indicates that the best value for the significant time and investment in use of GIS in urban stormwater modeling and management is when it is integrated into a DSS. Also, the trend is toward enterprise GIS implementation where the data and models are used across the organization and are integrated into maintenance and operations, as well as planning and design. For example, a number of commercial vendors explain how a GIS can be shared among many varied users, e.g., gas utilities, water utilities, stormwater, etc., thus maximizing the benefits derived from data collection and management (Table 8.6).

TABLE 8.6
Municipal Asset-Management Software with GIS

Product/Model	Description	Company	Web Site
ArcFM	Managing and modeling facility data for water/wastewater, electric, gas	Telvent Miner & Miner (TM&M)	www.miner.com
CityWorks	Manage capital assets and infrastructure; ArcGIS	Azteca Systems, Inc.	www.azteca.com
GeoMedia® Public Works Manager	Manage a complex wastewater project; GeoMedia	Intergraph	http://www.intergraph.com/
Hansen	Enterprise asset management; integrated map viewer; ArcGIS	Hansen Information Technologies	www.hansen.com
Municipal Infrastructure Management System (MIMS)	Track assets for roads, water, storm, and sanitary sewer networks; AutoCAD, Microstation	Sierra Systems	www.albertamims.org
Oracle Water Applications	Monitor and manage inventory, equipment and personnel; Oracle Spatial, ArcGIS	Oracle Corporation	http://www.oracle.com/industries/utilities/water.html
Real-Time Infrastructure Valuation Analysis (RIVA)	Web-based asset inventory and management; supports ArcGIS and Intergraph	Bentley Systems	www.rivaonline.com

Examples of wastewater and stormwater decision-support systems were summarized by Sample et al. (2001), who also demonstrated a neighborhood-scale stormwater DSS to identify the optimal mix of best management practices (BMPs). Xue et al. (1996) and Xue and Bechtel (1997) described the development of a model designed to evaluate the effectiveness of BMPs. This model, called the best management practices assessment model (BMPAM), was linked with ArcView to create an integrated management tool to evaluate the pollutant load reduction potential of a hypothetical wet pond. Kim et al. (1998) used ArcView with an economic evaluation model and a hydraulic simulator to evaluate storm sewer design alternatives. The hydraulic simulator was used to generate initial design alternatives, which in turn were evaluated with an economic model. The GIS was used to store spatial information, generate model input, and present alternative solutions. The complete package of GIS, economic evaluation model, and hydraulic simulator was termed a *planning support system* (Kim et al. 1998).

GIS has achieved significant use as part of municipal infrastructure asset-management systems. A municipal asset-management system is used to store and manage asset data, and to support operational and strategic decision-making processes. Municipal infrastructure asset-management systems are used for land and property management, facilities and infrastructure management, and utility management. Halfawy et al. (2005) present a review of these systems. Danylo and Lemer (1998) envisioned the role of an asset-management system as "an integrator, a system that can interact with and interpret the output coming from many dissimilar systems." The general purposes of asset-management software are to address typical asset-management issues such as: what do you own? what is it worth? what is the condition? what is the remaining service life? how much should you invest to ensure sustainability? and what needs to be done and when?

Asset-management software can be classified into two broad categories: general-purpose software and asset-specific software. General-purpose systems typically offer generic functionality that needs to be customized and adapted for specific data and work processes related to specific classes of assets. Asset-specific software solutions provide a set of built-in data models and processes to support the management of a specific class of municipal assets (e.g., facilities, sewers, roads, bridges, etc.). The use of GIS is found in both categories, as they support a range of data import/export options with relational database-management systems. Examples of asset-specific software are the ESRI data models provided as UMLs for various municipal sectors (see www.esri.com). Asset-specific software is also used, for example, for condition assessment and rating of sewers based on closed-circuit television (CCTV) inspection.

There are a number of municipal asset-management software vendors (Table 8.6). Examples of the capabilities follow. Figure 8.10 shows the MIMS MapViewer and forms for managing water, sanitary, stormwater, and road networks.

An example of the use of GIS for a comprehensive stormwater asset-management system was provided by Landry et al. (2001). Their work was noted earlier concerning the field inventory of stormwater infrastructure. Prior to the stormwater infrastructure inventory, the county committed to using the Hansen Information Systems asset-management system for all public works infrastructure (i.e., roads, signs, bridges, stormwater, etc.). While this system accommodates many of the service requests and financial-reporting-related issues, it did not provide a GIS-based graphical interface to view and query inventory data. To complement the existing Hansen system, the team developed an ArcView GIS application to allow users (including non-GIS trained users) to access inventory data via a geographic interface. An ArcView interface was customized using the Avenue programming language so that users could perform a variety of specialized tasks. Separate menu choices provided users the ability to run precoded queries, select assets using a custom query form, print reports of selected assets, generate preformatted printable maps, and view photographs and attribute information by selecting individual assets (see Figure 8.11). The customized ArcView interface extended the county's system to accommodate the following user needs:

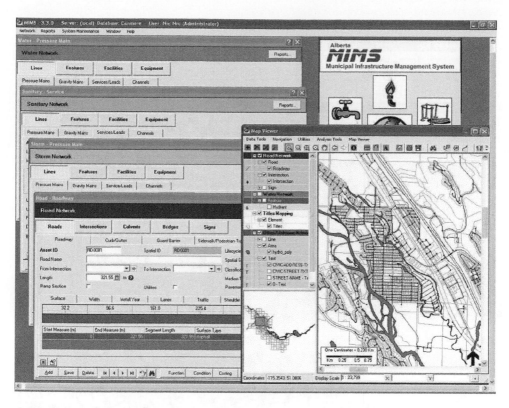

FIGURE 8.10 MIMS MapViewer and forms for managing water, sanitary, stormwater, and roads networks. (Courtesy of Sierra Systems. With permission.)

- Users would need to view stormwater infrastructure data together with reference data sets such as roads, parcels, natural water features, aerial photography, etc.
- Stormwater structures would need to be symbolized to resemble the 1:200-scale maps that users had been accustomed to using.
- Query tools would be needed to quickly zoom into a particular street address or intersection, parcel folio number, section-township-range, or, if known, to a specific stormwater asset.
- Additional query tools would be necessary to select and then present infrastructure assets based on reference data layers, such as section-township-range, political districts, maintenance service areas, etc.
- It would be necessary to create simple but well-formatted maps upon request to distribute to maintenance staff or citizens.
- Photographs taken of each stormwater asset would need to be viewed within the ArcView interface.
- Asset attributes would need to be readily accessible, including the unique identification number of each asset that would be used when issuing work orders for maintenance-related activities.
- Users would need to view the location of historic information such as past flood events in relation to stormwater infrastructure.

The use of GIS tools to formulate a DSS facilitates the tracking of inventory, the budgeting process, and the comprehensive planning related to a complex sanitary sewer system. Clear (2002) demonstrated this functionality to support decisions on a comprehensive planning level instead of a design level. To accomplish the objective of creating a DSS tool for these purposes, a number

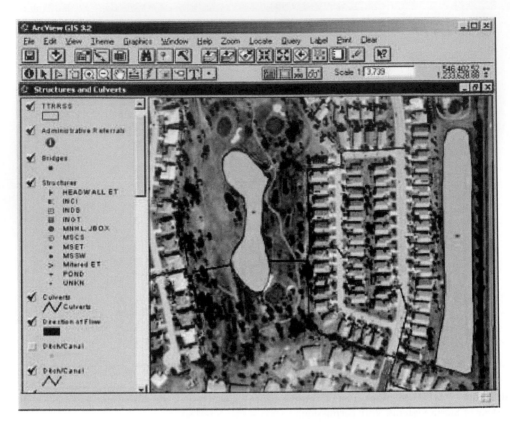

FIGURE 8.11 Screen shot of customized ArcView interface showing stormwater assets. (From Landry et al. 2001. With permission.)

of elements were collected, including pipe information, lift-station information, graphic features, political boundaries, residential boundaries, television-inspected sites, capital-improvement projects, and septic-system locations. Once these data were collected, they were inserted into a GIS through a number of steps, including the formation of the inventory themes, the formation of the necessary scripts utilized to maintain the DSS, and the formation of a shapefile depicting televised pipes (including an example on how to hot-link a video to these pipes). Budgeting applications included maintenance management and priority setting as well as delinquent account and assessment tabulations. A comprehensive planning application addressed where sewer systems should extend in order to capture failing septic systems (Figure 8.12). County assignment of the urban growth area (UGA) could impact the sewer district revenues by constraining where sewer services could be extended. A figure displaying the areas where the most revenue would be lost on a subbasin region was created as a result of this study. The darker regions in Figure 8.12 display the areas where a greater revenue loss would be seen.

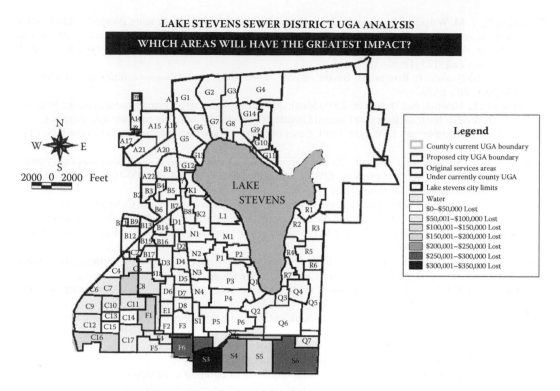

FIGURE 8.12   Sewer district mapping showing which areas would have the greatest impact on revenues. (From Clear 2002. With permission.)

REFERENCES

ASCE. 1970. *Design and construction of sanitary and storm sewers: Manual of engineering practice*, No. 37. New York: American Society of Civil Engineers.

Bentley Systems Inc. 2005. *SewerGEMS users' manual*. Waterbury, Conn.: Haestad Press.

Brabec, E., S. Schulte, and P. L. Richards. 2002. Impervious surfaces and water quality: A review of current literature and its implications for watershed planning. *J. Plann. Lit.* 16 (4): 499–514.

Cannistra, J., and A. Purves. 1992. Stormwater management: Lynchburg uses GIS to disconnect rainleaders. *Geo Info Systems* 2 (7): 52–59.

Christensen, J., M. Carter, P. Hsiung, and K. Bonner. 1999. Infiltration/inflow analysis using GIS. In *Proc. ESRI International User Conference*. Paper 704. San Diego, CA. http://training.esri.com/campus/library/index.cfm.

Clear. S. 2002. Lake Stevens Sewer District: Comprehensive planning decision support system based on GIS. Master's report, Dept. Civil Engineering, Univ. Colorado–Denver.

Curley, P., C. Anderson, and R. S. Clarke. 2006. A mobile stormwater facility information collection system for Fulton County. In *Proc. Twenty-Sixth Annual ESRI User Conference*. San Diego, Calif.

Danylo, N., and A. Lemer. 1998. Asset management for the public works manager: Challenges and strategies. Findings of the APWA Task Force on Asset Management. Kansas City, Mo.: American Public Works Association.

Dougherty, M., R. L. Dymond, S. J. Goetz, C. A. Jantz, and N. Goulet. 2004. Evaluation of impervious surface estimates in a rapidly urbanizing watershed. *Photogrammetric Eng. Remote Sensing* 70 (11): 1275–1284.

ESRI (Environmental Systems Research Institute). 2007. GIS technology for water, wastewater, and storm water utilities. http://www.esri.com/water.

Fankhauser, R. 1999. Automatic determination of imperviousness in urban areas from digital orthophotos. *Water Sci. Technol.* 39 (9): 81–86.

Gough, J., T. Tu, L. Palakur, and H. Vendra. 2004. Creating a sewer geodatabase model for the City of Torrance. In *Proc. ESRI International User Conference*. Paper 2048. San Diego, CA. http://training.esri.com/campus/library/confproc/.

Govindan, S., T. M. Walski, R. Mankowski, J. Cook, and M. Sharkey. 2006. Leveraging geospatial data to solve storm sewer issues. Paper presented at *Map India 2006*. New Delhi, India. http://www.bentley.com/haestad.

Greene, R., N. Agbenowoshi, and G. F. Loganathan. 1999. GIS based approach to sewer system design. *J. Surveying Eng.* 125 (1): 36–57.

Grise, S., E. Idolyantes, E. Brinton, B. Booth, and M. Zeiler. 2001. *ArcGIS water utilities data model.* San Diego, Calif.: ESRI.

Halfawy, M., L. Newton, and D. Vanier. 2005. Municipal infrastructure asset management systems: State-of-the-art review. National Research Council Canada. NRCC-48339. http://irc.nrc-cnrc.gc.ca/ircpubs.

Heaney, J. P., D. Sample, and L. Wright. 1999. Costs of urban stormwater systems. Report to the U.S. EPA, Edison, NJ.

Heaney, J. P, D. Sample, and L. Wright. 2000. Geographical information systems, decision support systems, and urban stormwater management. EPA# EPA/600/R-00/027, NTIS# PB2000-104077. Edison, N.J.: U.S. Environmental Protection Agency.

Hubly, D. 1995. Design of Water and Wastewater Systems. Dept. of Civil Eng., Univ Colorado-Denver.

Kim, H.-B., K.-M. Kim, and J.-C. Lee. 1998. Sewer alternative generation using GIS and simulation models in a planning support system. In *Proc. ESRI 1998 Int. User Conf.* San Diego, Calif. http://www.esri.com/library/userconf/proc98/PROCEED/.

Landry, S., N. M. Oliver, and K. Campbell. 2001. Stormwater: From data collection to interface development. In *Proc. Twenty-First Annual ESRI User Conference*. San Diego, Calif.

Lee, J. G., and J. P. Heaney. 2003. Estimation of urban imperviousness and its impacts on storm water systems. *J. Water Res. Plann. Manage. ASCE* 129 (5): 419–426.

Lu, D. S., and Q. H. Weng. 2006. Use of impervious surface in urban land-use classification. *Remote Sensing Environ.* 102 (1–2): 146–160.

Meyer, S. P., T. H. Salem, and J. W. Labadie. 1993. Geographic information systems in urban stormwater management. *J. Water Resour. Plann. Manage. ASCE* 119: 206.

Oppman, S., and J. Przybyla. 2003. A GIS-centric approach to building a water/sewer/storm geodatabase. *ArcNews Online* Summer. http://www.esri.com/news/arcnews/summer03articles/oakland-county.html.

Rossman, L. A. 2007. Storm water management model user's manual, version 5.0. Report EPA/600/R-05/040. Cincinnati, Ohio: National Risk Management Research Laboratory, Office of Research and Development, U.S. Environmental Protection Agency.

Sample, D. J., J. P. Heaney, L. T. Wright, and R. Koustas. 2001. Geographic information systems, decision support systems, and urban stormwater management. *J. Water Resour. Plann. Manage. ASCE* 127: 155.

Shamsi, U. M. 1998. ArcView applications in SWMM modeling. In *Advances in modeling the management of stormwater impacts*, vol. 6, ed. W. James, 219–233. Guelph, Ont.: Computational Hydraulics International.

Slonecker, T. E., D. B. Jennings, and D. Garofalo. 2001. Remote sensing of impervious surfaces: A review. *Remote Sensing Reviews* 20: 227–255.

Trauth, K. M., A. C. Correa, H. Wang, and J. Adhityawarma. 2004. Predicting environmental impacts associated with population distribution and land use using remote sensing and GIS: A stormwater runoff example. In *Proc. ASCE World Water Congress*. Salt Lake City, Utah.

Wang, G., and A. Baldwin. 1999. Integrating sewer modeling and GIS technologies. In *Proc. Annual ESRI Users Conference*. San Diego, Calif. http://gis.esri.com/library/userconf/proc99/proceed/papers/pap221/p221.htm.

Wang, H., and L. Johnson. 2007. Imagery for impervious surface estimation, Fort Collins, Colo. Dept. Civil Engineering, University of Colorado at Denver.

Washburn, G., J. Barber, and K. Opitz. 2003. New York City Department of Environmental Protection uses GIS to locate impervious surfaces. *ESRI ArcNews* Winter. http://www.esri.com/news/arcnews/winter0304articles/nyc-dep-uses.html.

Xu, Z. X., K. Ito, G. A. Schultz, and J. Y. Li. 2001. Integrated hydrologic modeling and GIS in water resources management. *J. Computing Civ. Eng.* 15 (3): 217–223.

Xue, R. Z., and T. J. Bechtel. 1997. Integration of stormwater runoff and pollutant model with BMP assessment model using ArcView GIS. In *Proc. 1997 ESRI User Conf.* San Diego, Calif. http://www.esri.com/library/userconf/proc97/PROC97/TO700/PAP656/P656.htm.

Xue, R. Z., T. J. Bechtel, and Z. Chen. 1996. Developing a user-friendly tool for BMP assessment model using a GIS. In *Proc. AWRA Annu. Symp. GIS and Water Resources*, eds. C. A. Hallam, J. M. Salisbury, K. J. Lanfear, and W. A. Battaglin, 285–294. Herndon, Va.: American Water Resources Association.

9 GIS for Floodplain Management

9.1 INTRODUCTION

Floodplains are low areas subject to flooding, and these lands are regulated in most communities to minimize threats to loss of lives and damages. Floodplains occupy about 5% of the land in the United States (Linsley et al. 1992), and because they are nearly level, they are attractive sites for railroads, highways, and cities. Over the years, reservoirs, levees, and improved channels have been constructed to provide some level of flood protection. However, encroachment on lands subject to flooding has outpaced protective measures, and infrequent severe floods have overwhelmed flood-control works in some locations. All of these factors have made floodplain management a high priority in many communities.

GIS concepts and tools are extensively applied for floodplain mapping. The objectives of this chapter are therefore to (a) review floodplain-management concepts and information needs, including floodplain data development, hydraulic analyses, and zone definitions, and (b) consider the role of GIS in floodplain mapping and management. GIS procedures are described for floodplain mapping of flood zones, land use, habitat, and hydrography as well as application of GIS for determination of management-related information, such as flood damage estimation and building-permit reviews. First, a short review of the floodplain-management problem is presented along with the regulatory background and approaches. Floodplain mapping requirements as established by the Federal Emergency Management Agency (FEMA) are summarized, and the use of GIS concepts and tools for this purpose is described. Floodplain data development procedures and databases are then presented. GIS applications for integrated hydraulic modeling for floodplain delineation are reviewed. The chapter concludes with descriptions of selected examples of floodplain impact assessments, including tabulation of potential and actual flood damages.

9.2 FLOODPLAIN MANAGEMENT

Some 20,000 communities in the United States are subject to substantial risk of flooding, and worldwide the increase in population and property values in flood-threatened lands requires continued vigilance in floodplain management. Efforts to reduce the risks associated with excess water in streams have led to the implementation of a variety of land-use regulatory policies and flood-damage-mitigation measures. Commonly accepted approaches for reducing flood damages include structural and nonstructural measures directed toward controlling and avoiding floodwaters. *Floodplain management* is a general term that includes various strategies and actions taken by the various governmental and interest groups involved. Table 9.1 lists four primary strategies and various tools for floodplain management.

A primary motivation for floodplain management in the United States was the passage of the National Flood Insurance Act of 1968, which established the National Flood Insurance Program (NFIP). The NFIP is administered by the Federal Emergency Management Agency (FEMA). The 1968 act subsidized flood insurance within communities that were willing to adopt floodplain-management programs to mitigate future flood losses. This legislation also required the identification of all floodplain areas within the United States and the establishment of flood-risk zones within those areas. The risk data to identify floodplain areas, as required by the 1968 act, have been acquired through flood insurance studies (FIS), which involve hydrologic and hydraulic studies of flood risks.

TABLE 9.1
Floodplain-Management Strategies and Tools

Modifying susceptibility to flood losses
    Regulations
    Development and redevelopment policies
    Disaster preparedness
    Flood forecasting, warning, and emergency plans
    Floodproofing and elevation

Modifying flooding
    Dams and reservoirs
    Dikes, levees, and floodwalls
    Channel alterations
    High-flow diversions
    Stormwater management
    Shoreline protection

Modifying the impacts of flooding
    Land-treatment measures
    Information and education
    Flood insurance
    Tax adjustments
    Flood emergency measures
    Disaster assistance
    Postflood recovery

Restoring and preserving the natural and cultural resources of floodplains
    Regulations
    Development and redevelopment policies
    Information and education
    Tax adjustments
    Administrative measures

Source: Adapted from FIFMTF (1992).

Using the results of an FIS, FEMA prepares a flood insurance rate map (FIRM) that depicts the spatial extent of special flood hazard areas (SFHA) and other thematic features related to flood-risk assessment. An SFHA is an area subject to inundation by a flood having a 1% or greater probability of being equaled or exceeded during any given year. This 100-year flood (or base flood) is the national standard on which the floodplain management and insurance requirements of the NFIP are based. Flood insurance premiums are based on a number of conditions and physical factors, including the formally accepted flood insurance rate map (FIRM).

The most widespread floodplain regulations are the minimum requirements of the NFIP, which must be enacted and enforced by communities participating in the program. The minimum regulations vary, depending upon risk studies and mappings that have been done in the community. These include, for example, (a) permitting for all proposed new development and (b) enforcing risk zone, base flood elevation, and floodway requirements after the flood insurance map for the area becomes effective. Many of the measures involve consideration of whether a particular property is in a designated floodplain or not; thus the level of mapping detail is at the so-called cadastre (i.e., property boundary) level. Applications include selection of sites for relocation, prioritizing eligibility for home-buyout programs, and identifying repeatedly damaged properties in SFHAs.

Some communities are more assertive in their management of floodplains, recognizing that avoidance of a 100-year flood level does not completely eliminate the threat of flooding. These communities direct their efforts toward property acquisition, in the form of easements or fee interest, for the purpose of preserving floodplain areas. Residential structures located in the high-hazard flood area are demolished, and the land is used to create a "greenway corridor" along the stream to support various multiple objectives, including pedestrian and bicycle paths, riparian habitat enhancement, and water-quality improvements. The dedication of conservation easements, whereby the landowner receives a tax credit, is another possibility that is becoming increasingly popular.

At present, FISs have been completed for all floodplains, and FIRMs have been published for virtually all communities in the nation having flood risks—approximately 20,000 communities in the United States. These studies have resulted in the publication of over 80,000 individual FIRM panels. In addition to initial FISs, FEMA is responsible for maintaining the FIRMs as communities grow, as new or better scientific and technical data concerning flood risks become available, and as older FISs become outdated by the construction of flood-control projects or the urbanization of rural watersheds. Developments of GIS technology led FEMA to adopt these procedures for NFIP mapping and engineering. Currently, there is a major effort to convert the FIRMs to a digital format. Thus, FIRMs will become DFIRMs—digital flood insurance rate maps. DFIRMs allow for automatic updates, automatic recalculations, and, most importantly, they give property owners the ability to access information on the Internet to determine whether their property is located in a floodplain.

There are other motivations for floodplain management besides the requirements of flood insurance and damage-reduction programs. Floodplains are part of the riparian zone of streams and thus are important ecosystem features. Riparian zones comprise some of the richest ecosystem diversity of the landscape, and protection of these areas from development is an important consideration for the maintenance of ecosystem functions. Periodic flood flows are critical to maintaining this diversity through disturbance of vegetation and sediment.

9.3   FLOODPLAIN MAPPING REQUIREMENTS

Given the regulatory authority of FEMA, it is their standards that establish primary mapping requirements for floodplains. These requirements are detailed in a sequence of documents under the heading "Guidelines and Specifications for Flood Hazard Mapping Partners" (FEMA 2005). FEMA defines technical requirements, product specifications for flood hazard maps, related National Flood Insurance Program (NFIP) products, and associated coordination and documentation activities. Table 9.2 lists the data required for floodplain mapping.

9.4   GIS FOR FLOODPLAIN MAPPING

9.4.1   FLOODPLAIN DATA DEVELOPMENT

Data required for floodplain mapping and management purposes are composed of three general categories (Table 9.2): (a) floodplain and watershed topography data to support hydrologic and hydraulic modeling, (b) physical data on built facilities for drainage control and buildings, and (c) administrative data on jurisdiction boundaries. These data are developed from a variety of sources using various GIS procedures and technologies and are ultimately collected into a comprehensive data set supportive of project needs and longer-term multiple-purpose management goals. GIS tools for floodplain map updates have become standard practice. Floodplain maps and information prepared using traditional methods during the 1970s and 1980s are the basis for most current regulatory programs. Programs for map modernization are progressing (e.g., http://www.fema.gov/), and GIS standards are promulgated for these (see Section 9.4.2).

Floodplain terrain data differ from the coarser watershed data in their accuracy and level of precision. Development of watershed topographic and stream system data has been described in detail in Chapter 5 (GIS for Surface Water Hydrology). Vertical and horizontal accuracy of floodplain

TABLE 9.2
Floodplain Mapping Data Requirements

Watershed features
  Hydrography lines and areas
  Stream networks, junctions, and diversions
  Sub-basins
  Reservoirs
  River and rain gauges

Hydrologic modeling features
  Hydrologic models used
  Regression equations
  Intensity duration–frequency relations
  Precipitation event types and patterns

Survey and elevation features
  Contours
  ERM (elevation reference mark) points
  Permanent benchmark points
  PLSS (public land survey system) line and areas
  Spot elevations
  Cross-section lines

Flood features
  BFE (base flood elevation) lines
  Flood insurance risk areas and lines
  General structures (bridges, dams, weirs, etc.)
  Levees
  CBRS (coastal barrier resources system) areas
  Coastal transect lines
  Channel cross-section lines

Hydraulic modeling features
  Hydraulic models used
  Hydraulic structures
  Channel overflow segments

Property features
  Buildings
  Docks
  Parcels
  Recorded easements

Reference information
  DOQ index
  Firm panel areas
  Horizontal reference lines
  Horizontal reference points
  Quad areas
  Study-area boundaries

Political and transportation features
  Political jurisdiction areas
  Political lines
  Transportation (roads, train tracks, and airports)

Source: Adapted from FEMA (2005).

channel cross-sections and overland flow areas is required to be highly precise in order to support delineation of flood zones and depths for specific buildings. Whereas watersheds typically can be dealt with at a scale of 1:24,000 (e.g., USGS topographic maps), floodplain features are typically mapped at 1:400 or larger. Digital elevation models (DEMs) with a sufficient resolution in stream channels for hydraulic modeling are not widely available and must be acquired through land surveys and/or aerial remote sensing. The choice of data structure can be important for accurate mapping at these scales. A TIN (triangular irregular network) data structure can precisely represent linear (banks, channel thalweg [hydraulic centerline], ridges) and point features (hills and sinks) using a dense network of points where the land surface is complex.

GIS tools provide a means for making maximum use of channel cross-section data obtained by land surveying. Tate et al. (2002) demonstrated the creation of a floodplain DEM by merging surveyed data for the stream channel with comparatively lower-resolution DEM data for the floodplain. As input, the approach requires a completed HEC-RAS (Corps of Engineers Hydrologic Engineering Center–River Analysis System) simulation data set, a digital elevation model (DEM), and a GIS representation of the stream thalweg. The process begins with the export of the channel data from HEC-RAS to GIS, followed by data conversion from hydraulic model coordinates to geographic coordinates. The stream thalweg may be defined from NHD Reach files, DEM-based delineation, heads-up digitizing from digital orthophotography, or land surveys; the NHD Reach files were found to be inadequate, as was the DEM delineation method based on coarse-resolution data (Tate et al. 2002). Various interpolation and conflation

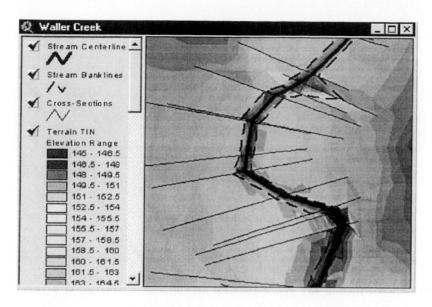

FIGURE 9.1 Terrain triangular irregular network (TIN) with important input features. (From Tate et al. 2002. With permission.)

operations were applied to resolve a "best fit" floodplain and channel TIN from the thalweg, cross-sections, and DEM data sets (Figure 9.1).

Complete land-use and building footprints are increasingly being mapped at high resolution by aerial photogrammetric and LIDAR (LIght Detection And Ranging) techniques. LIDAR technology allows for the collection of very high-resolution elevation data from an airborne platform. Postprocessing filters and editing of the raw LIDAR data yield an elevation data set where vegetation, buildings, and other human-made features are effectively removed from the terrain model. The data obtained represent both the reflective surface, or "first return," elevation data, which incorporate vegetation canopy and human-made features as well as the "bare Earth" surface-elevation data. The final elevation data are represented at 1–5-m horizontal grid postings and have a vertical accuracy of ±15–25 cm. Field surveys may be required to establish horizontal control links to benchmarks, to establish building occupancy status, and to provide requisite details of hydraulic structures.

An emphasis on high-accuracy elevation data was recently reiterated by a study of the National Research Council (NRC 2007). The report concluded that there is sufficient two-dimensional (2-D) "base map imagery" available from digital orthophotos to meet FEMA's flood map modernization goals. These data include the National Digital Orthophoto (http://www.ndop.gov) and National Agriculture Imagery (http://165.221.201.14/NAIP.html) programs. However, the three-dimensional (3-D) "base elevation data" that are needed to determine whether a building should have flood insurance are not adequate. FEMA needs land-surface elevation data that are about ten times more accurate than data currently available for most of the nation. The report recommended that new, high-accuracy digital elevation data be collected nationwide using LIDAR technology. The new data should be input into the National Elevation Dataset that the U.S. Geological Survey (USGS) maintains for use in support of flood map modernization and other applications.

A study by the USGS for the Nisqually River near Puget Sound, Washington (Jones et al. 1998), demonstrated the effectiveness of modern elevation data and GIS for updating flood maps. Existing floodplain maps were shown to have a number of shortcomings, including the fact that they (a) are based on out-of-date flood-probability estimates, (b) are hand-drawn and difficult to manage, (c) have limited vertical accuracy, and (d) are expensive and time consuming to update. The GIS approach was shown to be (a) relatively inexpensive (10%–20% of traditional methods), (b) equally accurate and more detailed, (c) able to provide depth-of-flood details, (d) able to identify areas of

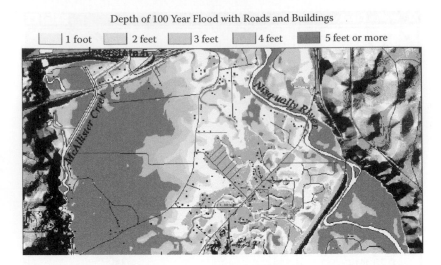

FIGURE 9.2   Flood depth map with locations of roads and buildings. (From Jones et al. 1998.)

uncertainty, and (e) digital, so analyses could be extended to other themes (e.g., roads and build-ings) in support of risk assessments. GIS was used to create and manipulate digital elevation models representing the land surface and the flood surface. Determining the inundated area is a simple calculation: the flood surface elevation model is subtracted from the land surface elevation model at each location, resulting in negative values wherever the flood elevation is greater than the land elevation. A by-product of this calculation is flood depth, which is important for damage and insur-ance assessments when intersected with building floor elevations (Figure 9.2).

GIS as a data manager, manipulator, and analyzer for both pre- and postprocessing of LIDAR data was demonstrated by Heinzer et al. (2000) for a floodplain mapping project in Utah. They found that LIDAR data sets can become very large, typically millions of points per flight swath. This became problematic when using ArcInfo® on current operating system capabilities, and they found that it was better to use non-GIS data-filtering techniques to extract the areas of interest and subsequently use the GIS to operate on smaller data sets. The building- and vegetation-removal algo-rithms essentially eliminated those points that were suspect. It was found that these procedures are about 90% effective, and further editing was required using orthorectified photography. The result of this was a data set containing "no LIDAR points" where the vegetation and buildings existed, and interpolation across these areas was performed. The ESRI Spatial Analyst® extension facilitated the creation of a TIN data structure, which was then converted to a GRID5 at the desired sampled cell size (2 m in their case). After the processing, there were two GRID structures: one consisting of "raw" data containing buildings and vegetation, and the other consisting of bare-Earth data with the buildings and vegetation mostly removed. They used two methods to replace the buildings on the surface: (a) orthorectified photography to create a polygon coverage of the structures—this was converted to a GRID, where a Map Algebra CON statement was used to extrude the buildings back on the bare-Earth elevation model, (b) and the raw LIDAR GRID and the "bare surface" GRID to create a difference GRID by subtracting the two. Because buildings are higher and have solid returns, Map Algebra statements were written against the difference GRID to extrude the buildings back onto the bare-Earth surface-elevation model. Additional cleanup was then performed using the orthophotography. This method was the easiest to implement.

Although medium-resolution satellite imagery, such as the nominal 30-m data from Thematic Mapper (TM), is inadequate for hydraulic modeling purposes, it does provide valuable data for monitoring flood events. An example of this type of application is described in Chapter 11 (GIS for Water Resources Monitoring and Forecasting).

9.4.2 FLOODPLAIN GEODATABASE

Given the extensive and disparate data sources required for floodplain mapping, there is a strong motivation to integrate the data into a comprehensive geodatabase. Doing so would provide advantages that a geodatabase provides in terms of standardization, removal of redundancy, concurrency control, and transferability, to name a few.

As noted above, FEMA has developed various data collection and reporting standards for floodplain studies as part of its nationwide program of map modernization. The two principal documents relevant to the design of a geodatabase for FEMA flood hazard mapping are (a) Appendix L of the Guidelines and Specifications for Flood Hazard Mapping Partners: Guidance for Preparing Draft Digital Data and DFIRM Database (FEMA 2003, Appendix L) and (b) Appendix N of the Guidelines and Specifications for Flood Hazard Mapping Partners: Data Capture Standards (DCS) (FEMA 2005, Appendix N). These standards specify the current GIS databases used for archiving flood hazard models and results. Figure 9.3 presents the DCS relationship diagram for hydraulics.

FEMA DFIRM and DCS databases were reviewed by Walker and Maidment (2006), and they provided guidance on merging those databases into a more coherent Flood Study Geodatabase (FSG) based on the Arc Hydro data model (Maidment 2002) described in Chapter 4. Their study involved (a) merging the DFIRM database with the DCS database and removing overlapping tables,

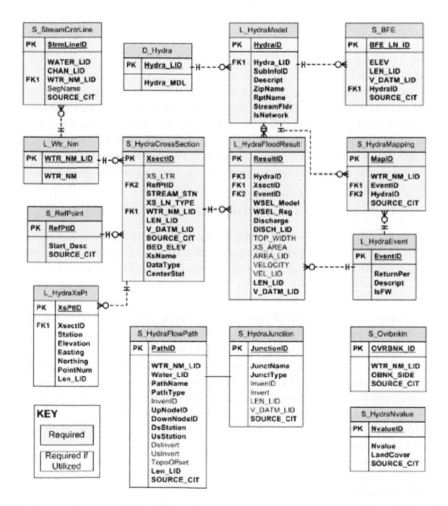

FIGURE 9.3 Data-capture standards relationship diagram for hydraulics. (From FEMA 2005.)

FIGURE 9.4 Flood study geodatabase: Feature data sets and classes (From Walker and Maidment 2006. With permission.)

(b) converting the DFIRM/DCS relational database system into a geodatabase (i.e., the FSG), (c) developing a correlation between the FSG and the Arc Hydro data model, (d) developing an XML version of the FSG schema, and (e) relating the flood hazard data to the National Hydrography Dataset (NHD). Feature data sets and classes of their integrated FSG are shown in Figure 9.4. They concluded their study by recommending that the FEMA DFIRM and DCS databases could be merged into a consolidated geodatabase, thus making the floodplain study data submittals more standardized and compatible with water resources databases nationwide (i.e., Arc Hydro and the NHD).

9.5 FLOODPLAIN HYDRAULIC MODELING WITH GIS

The technical core of floodplain studies is the hydrologic and hydraulic modeling activities that lead to the delineation of the floodplain boundary. GIS has become central to the conduct of such modeling studies, providing the means for integration of the various data involved, coordinating the various models, and providing high-resolution maps required for supporting flood-management strategies.

There are a number of floodplain hydrologic and hydraulic (H&H) modeling packages that have been developed by the public and private sectors (Table 9.3). The software packages vary in their capabilities for representing floodplain hydraulics and the level of GIS integration. In the one-dimensional (1-D) approach, water flow is assumed to occur in one dominant spatial dimension aligned with the centerline of the main river channel. The geometry of the problem is represented in the model by channel and floodplain cross-sections perpendicular to the channel centerline. Predictions of changes in water level and flow along the river are obtained from numerical solutions of the St. Venant equations for volume and momentum conservation. Two-dimensional approaches solve for water level and depth-averaged velocities in two spatial dimensions using finite-difference, finite-element, or finite-volume computational grid approaches (Pender and Neelz 2007). The 2-D models are appropriate for situations where there is opportunity for floodwaters to spill out of primary channels and flow overland, such as alluvial fans.

As with other applications of GIS and modeling, there can be varying degrees of integration of GIS database and analysis tools with the models. These range from using GIS as a pre- and

TABLE 9.3
Floodplain Hydraulics Software with GIS

Product/Model	Description	Company	Web Site
HEC-RAS, HEC-GeoRAS	1-D hydraulics for channel networks; GeoRAS with ArcGIS	USACE, HEC	http://www.hec.usace.army.mil/ software/hec-ras/
FLO-2D	2-D flood routing model; GIS shape file interface	FLO-2D Software, Inc.	http://www.flo-2d.com/
GIS Stream Pro, PrePro2002	Develop HEC-RAS cross-sections based on a terrain model	Dodson & Associates, Inc.	http://www.dodson-hydro.com/
Infoworks RS (ISIS)	1-D hydraulics for open channels, floodplains; InfoView GIS	Wallingford Software Ltd.	http://www.wallingfordsoftware. com/products/infoworks_rs/
LISFLOOD-FP	2-D floodplain hydrodynamic model	Geog. Sciences, University of Bristol, U.K.	http://www.ggy.bris.ac.uk/ research/hydrology/models/ lisflood
MIKE 11, MIKE 21, MIKE FLOOD	Coupled 1-D and 2-D floodplain hydrodynamic models	DHI Water & Environment	http://www.dhigroup.com/
PCSWMM.NET	Windows GIS interface to SWMM5, ArcGIS, others	Computational Hydraulics Int.	http://www. computationalhydraulics.com/
RiverCAD	Display HEC-RAS output in AutoCAD	BOSS International	http://www.bossintl.com/
TELEMAC-2D, 3-D	2-D hydrodynamic finite-element model	SOGREAH Telemac	http://www.telemacsystem.com/
TUFLOW	Coupled 1-D and 2-D floodplain hydrodynamic model	BMT WBM	http://www.tuflow.com/

postprocessor to full integration through interface development, e.g., HEC-GeoRAS (Shamsi 2002). With full integration, modeling modules are developed in or are called from a GIS. All tasks of creating model input, editing data, running the model, and displaying output results are available in GIS (e.g., EPA BASINS). Alternatively, there may be a model-based interface where GIS modules are developed in or are called from a computer model (e.g., PCSWMM GIS). Because development and customization tools within most GIS packages provide relatively simple programming capability, the first approach provides limited modeling power. Because it is difficult to program all the GIS functions in a floodplain model, the second approach provides limited GIS capability.

9.5.1 HEC-RAS AND HEC-GEORAS

The widely distributed and used water resources computer programs of the Corps of Engineers Hydrologic Engineering Center (HEC) include those for analyzing watersheds and floodplain hydraulics. HEC-RAS (River Analysis System) (HEC 2006) is designed to perform 1-D hydraulic calculations for a full network of natural and constructed channels. The HEC-RAS system contains four 1-D river analysis components for: (a) steady flow water surface profile computations; (b) unsteady flow simulation; (c) movable boundary sediment transport computations; and (d) water quality analysis. A key element is that all four components use a common geometric data representation and common geometric and hydraulic computation routines. The effects of various obstructions such as bridges, culverts, weirs, and structures in the floodplain may be considered in the computations. The steady-flow system is designed for application in floodplain-management and flood insurance studies to evaluate floodway encroachments.

HEC-GeoRAS is a set of procedures, tools, and utilities for processing geospatial data in ArcGIS® using a graphical user interface (GUI) (HEC 2006). The interface allows the preparation

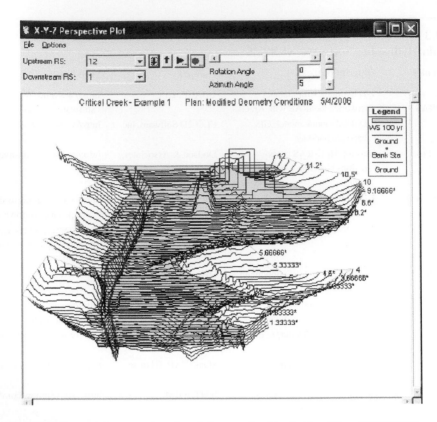

FIGURE 9.5   HEC Geo-RAS perspective plot of river reach with a bridge. (From HEC 2006.)

of geometric data for import into HEC-RAS and processes simulation results exported from HEC-RAS. To create the import file, the user must have an existing digital elevation model (DEM) of the river system in the ArcInfo TIN format. The user creates a series of line themes pertinent to developing geometric data for HEC-RAS. The themes created are the Stream Centerline, Flow Path Centerlines (optional), Main Channel Banks (optional), and Cross Section Cut Lines, referred to as the RAS themes. Additional RAS themes may be created to extract additional geometric data for import in HEC-RAS. These themes include Land Use, Levee Alignment, Ineffective Flow Areas, and Storage Areas. Water surface profile data and velocity data exported from HEC-RAS simulations may be processed by HEC-GeoRAS for GIS analysis for floodplain mapping, flood damage computations, ecosystem restoration, and flood warning response and preparedness. Figure 9.5 illustrates the type of display obtainable from HEC's GeoRAS. Ackerman et al. (2000) provide a demonstration of the application of GeoRAS with details on the ArcGIS interface.

The HEC-RAS and its GeoRAS graphical interface are widely used for floodplain modeling and mapping studies. GIS tools and procedures provide an enhanced basis for preparation of the extensive detailed data on channel cross-sections required for the hydraulic simulation. These methods were demonstrated by Kraus (1999) and Tate et al. (2002). Dodson and Li (1999) tabulated map-creation savings using GIS methods. The general procedure is to create a DEM for the channel and the adjacent floodplain areas. The channel DEM is typically derived from a triangulated irregular network (TIN), developed from a combination of field survey data and topographic map data, and a GIS representation of the stream centerline. The data consisted of floodplain cross-sections and reach lengths, which are located to coincide with field-surveyed cross-sections of the channels being studied. The floodplain data developed in the GIS are imported into HEC-RAS, where they are combined with the field-surveyed channel data in order to construct full floodplain cross-sections that reflect accurate channel and overbank data for the HEC-RAS model. Some conversions to standardized coordinate are required to merge the various data. The HEC-RAS program computes water surface elevations along the channels, and the results are transferred back to the GIS, where the floodplain limits are automatically and accurately mapped.

9.5.2 TWO-DIMENSIONAL FLOODPLAIN MODELING

Some floodplain topographic conditions do not lend themselves to hydraulic modeling using the 1-D approach. These situations occur where overbank flows are distributed across the land surface, such as might occur with an alluvial fan topographic structure. For these and other situations, 2-D flow hydraulic models of the finite-difference or finite-element type are increasingly being used.

An example of 2-D hydraulic modeling was demonstrated for the South Boulder Creek floodplain study (HDR 2007). The MIKE FLOOD® model (DHI 2004) used in the study combines the traditional channelized 1-D flow analysis model with a more physically based 2-D model that analyzes distributed flow patterns away from the channel and across the floodplain. Due to the complexity of flow paths, the city updated the older 2-ft mapping with detailed 1-ft contour-interval maps. These topographic data were developed using LIDAR technology, which provided a DEM having 1-m grid spacing and a vertical resolution of 15 cm. This allowed representation of the ground and structures in greater detail. Figure 9.6a shows the definition of 1-D channel and canal segments on a high-resolution digital orthophoto. Figure 9.6b shows a segment of the LIDAR DEM used as the land base for the 2-D model.

MIKE FLOOD integrates the 1-D model MIKE 11 and the 2-D model MIKE 21 into a single, dynamically coupled modeling system. The model is capable of routing the entire flood hydrograph downstream through the floodplain and therefore captures the impacts of (a) obstructions, both natural and human-made, in the channel and across the floodplain; (b) floodplain storage; and (c) rising and falling flow and volume characteristics of the flood hydrograph. Channel cross-sections were selected along the 1-D channel reaches to represent the typical geometry of the channel and to depict physical features that may impact the movement of flood flows down the channel. In all, approximately 350 cross-sections were used to define the main stem of South Boulder Creek; 950 additional cross-sections were used to represent the secondary channels that had significant flood-carrying capacity. Each channel cross-section is defined by two important elements: elevation and roughness. All bridges and culverts in excess of 36 in. in diameter were included in the floodplain model as 1-D elements.

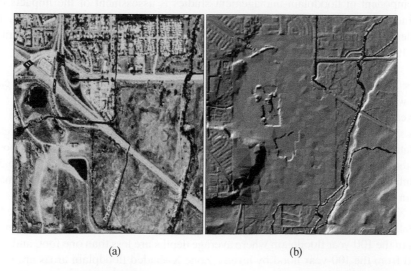

(a)   (b)

FIGURE 9.6 Channel and floodplain hydraulic features were defined by (a) digital orthophotos and (b) LIDAR DEM for South Boulder Creek floodplain study. (Courtesy of City of Boulder, Colo.)

The 2-D model represents flooding using a grid of pixels representing varying ground conditions (DHI 2004). These pixels each define a 4 × 4-m square with a horizontal location and a specific elevation. Mathematical computations of the finite-difference type then determine flow rate, depth, and flow direction across each pixel surface and between adjoining pixels, ensuring that the hydraulic principles of energy and momentum conservation and continuity are maintained. One-dimensional channels interface to the overland flow segments. Once in the 2-D grid, floodwater can travel away from the channel, parallel to the channel, or back into the channel. The overland flow direction is determined by hydraulic equations for each pixel that account for changes in ground elevation and slope, water surface elevations in surrounding pixels, and the flow characteristics within the pixel. The 2-D model ultimately computes water surface elevations, flow velocities and direction, and fluxes (mass entering and leaving a cell) for each and every "wetted" pixel in the grid for each and every time step during the applied flood hydrograph. An issue with any floodplain model is calibration; this applies to the 2-D models as well. Given the number of degrees of freedom present in such calibration, whereby separate friction parameters can be assigned at each computational node and at each time step, a reasonable correspondence between observed reach outflow data and a calibrated flood routing model (of any dimensionality or spatial resolution) can be obtained. But having adequate data on actual flooding conditions is often problematic. The Boulder Creek study attempted calibration using air photos taken of a significant flood event. Figure 9.7 shows the proposed floodplain with zone designations for the South Boulder Creek study as derived from the 2-D hydraulic modeling for the 100- and 500-year frequency design events.

An interesting aspect of the 2-D hydraulic modeling was reconciling the level of detail of hydraulic information and inundation depths that are provided by the 2-D model. There has been little experience with using the 2-D modeling approach for the purpose of developing a flood insurance rate map. The standard procedures used by FEMA to represent the water surface elevations and profiles are very rudimentary compared with the information provided by the 2-D model. The South Boulder Creek engineers performed an analysis that indicated that the water surface elevations determined by interpolating between conventional profiles and used for insurance rating purposes may be considerably different than the actual water surface elevation determined by the 2-D model.

9.5.3 Floodplain Impact Assessment with GIS

A major component of floodplain-management studies is assessment of the impacts of flooding. FEMA defines flood zones for insurance purposes according to varying levels of risk. These flood zones are identified on a community's FIRM, which is derived from the hydrology/hydraulic studies for the 1% (100 year-) and 0.2% (500 year-) frequency floods. The flood zones include:

- Zones X-unshaded: Areas of low flood risk determined to be outside the 500-year floodplain with less than a 0.2% chance of flooding each year.
- 500-year floodplain: Areas of the floodplain subject to the 500-year flood that have a 0.2% chance of being equaled or exceeded in any given year. Some communities choose to regulate areas within the 500-year floodplain based on the understanding that floods larger than the 100-year event can and will occur. In some instances, regulation of the 500-year floodplain is applied to critical facilities to ensure that such facilities are not damaged and remain operational during and after a flood event.
- Zone X-shaded: Areas of moderate flood risk that are determined to be within the 500-year floodplain, where there is a 0.2% chance of being equaled or exceeded in any given year, areas in the 100-year floodplain where average depths are less than one foot, and areas protected from the 100-year flood by levees. Zone X-shaded floodplain areas are not subject to FEMA required floodplain regulation or mandatory flood insurance purchase requirements. A local community may require floodplain regulation of Zone X-shaded areas, if

Proposed Floodplain

FIGURE 9.7 (See color insert following page 136.) Proposed South Boulder Creek 100-year floodplain zone designation map (provisional, not fully approved). (Courtesy City of Boulder, Colo.; http://www.south-bouldercreek.com/.)

only to inform property owners and residents of an increased flood hazard. It is noteworthy that FEMA is now requiring that all berms and levees be certified. Otherwise, they must be removed as a land form for the purposes of defining the flood hazard areas.

- Zone AE: Areas of special flood hazard (high risk) inside the 100-year floodplain where water surface elevations have been determined. A 100-year flood event is a flood that has a 1% chance of being equaled or exceeded in any given year. As a condition of participating in the NFIP, local communities must adopt floodplain regulations to manage and restrict construction inside areas of special flood hazard (Zone AE).

- 100-year floodplain: Areas subject to a flood that has a 1% chance of being equaled or exceeded in any given year. Some communities choose to regulate all areas of the 100-year floodplain, recognizing that in many cases flooding of shallow flood depths (that may be considered Zone X under FEMA standards) can still cause significant damage to basements and structures built close to grade. A floodplain development permit is required for any development or construction in the 100-year floodplain or Zone AE. Where construction is permitted, it must conform to flood-protection standards that require, at a minimum, the lowest floor of any residential building to be at least two feet above the base flood elevation. Nonresidential buildings must be elevated to the same flood-protection elevation or may be flood-proofed such that, below the flood-protection elevation, the structure is water tight with walls substantially impermeable to the passage of water.

- Conveyance (floodway) zone: Those portions of the floodplain required for passage (or conveyance) of the 100-year flood, based on an encroachment of the floodplain from the edges of inundation to a point where the 100-year flood profile (or water surface elevations) will be raised by no more than 6 in. or 1 ft, depending on the jurisdiction. In the conveyance zone or floodway, any development, encroachment, obstruction, or use that would result in any increase in the base flood elevation is prohibited. The most common floodway is defined by a rise of 1 ft; however, a jurisdiction may use other criteria (e.g., City of Boulder uses 6 in.).

- High-hazard zone: Those portions of the 100-year floodplain where an unacceptably high hazard to human safety exists. For some communities, this is defined as those areas where the product number of velocity (measured in feet per second) times flow depth (measured in feet) equals or exceeds 4, or where flow depths equal or exceed 4 ft. In the high-hazard zone, the construction, expansion, or enlargement of any structure intended for human occupancy, or the establishment of a new parking lot, is prohibited. Additionally, any change in use of an existing structure intended for human occupancy from nonresidential to residential is prohibited.

An added complication arises in defining the floodway when using a 2-D model because the model accounts for storage effects. By constraining the floodway in attempting to define the floodway, the benefits of storage are lost and will have negative impacts in the form of increases in the water surface elevation downstream.

GIS techniques are also used to provide a high-resolution basis for assessment of risk for individual properties and buildings. Standardized procedures for flood-damage analysis (FDA) are provided by the U.S. Army Corps of Engineers Hydrologic Engineering Center (HEC 1998). The HEC-FDA software is developed to calculate expected annual flood damages associated with existing conditions and proposed alternatives of flood control projects. Integration of GIS tools into the FDA software has provided enhanced functionality for flood-damage assessments.

Flood-damage analysis methods in the past have required tedious inventories of all of the structures in a floodplain to determine the water surface elevation–damage curves for each land use in a stream reach. GIS is now used to develop the elevation–damage curves in the flood impact zones. Using a digital elevation model (DEM) and land-parcel data, the GIS can determine the parcel and building elevations from the DEM. The GIS can also be used to overlay the land use, parcel, structure value, and water surface data on the DEM and to compile the elevation-damage information. The resultant elevation-damage curves are then input back into the GIS to develop a graphical representation of the damages as the water surface elevations increase.

FIGURE 9.8 South Boulder Creek floodplain delineation provides detail on properties and water depths in the regulatory zones. (*Source:* http://gisweb.ci.boulder.co.us/website/pds/sbc_flood/viewer.htm.)

An example was described by Tri (2002) for an urban drainage situation in Louisville, Kentucky, having a large number of structures in the study area (68,000) and a large amount of data generated from the urban hydraulics software (SWMM), which modeled a combined sewer area with 4800 sewer manholes. GIS data were used to develop input for the FDA model, including assessed property value, year of assessment, first-floor elevation (derived from TIN points), style of structure, property use classification, parcel identifier, and address. The value of each structure was obtained from data provided by the county Property Valuation Administration (PVA) office. The PVA maintains a Real Estate Master File (REMF) database of each parcel in the county, with its assessed value as well as certain structural characteristics (floor area, number of stories, type of construction, etc.). The first-floor elevation of each structure was calculated by adding the ground elevation to the foundation height (distance from the ground to the first floor).

Similar procedures were applied for the South Boulder Creek floodplain study to identify individual properties (Figure 9.8). Using GIS overlay techniques, a list of all properties that are completely or partially located in the floodplain are identified and provided on-line in a table keyed to parcel ID, address, building class, jurisdiction, and floodplain classification.

9.5.4 NEW ORLEANS FLOOD DAMAGE ASSESSMENTS

A major effort was directed to assessment of flood damages incurred due to the Hurricane Katrina event in 2005, and GIS played a major role in the study. An Interagency Performance Evaluation Task Force Team (IPET) was convened to conduct a performance evaluation of the New Orleans and Southeast Louisiana Hurricane Protection System during Hurricane Katrina (IPET 2007). Approximately 80% of New Orleans was flooded, in many areas with depth of flooding exceeding 15 ft. The majority, approximately two-thirds overall in areas such as Orleans East Bank and

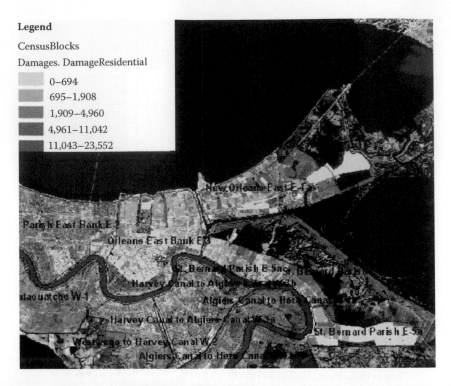

Legend

CensusBlocks

Damages. DamageResidential

- 0–694
- 695–1,908
- 1,909–4,960
- 4,961–11,042
- 11,043–23,552

FIGURE 9.9 (See color insert following page 136.) Distribution of Hurricane Katrina–generated residential direct property damages by census block; damages in thousands of dollars. (From IPET 2007.)

St. Bernard, of the flooding and half of the economic losses can be attributed to water flowing through breaches in floodwalls and levees. Loss of life in the Gulf region was staggering, with almost 1600 fatalities accounted for and another 400 missing and presumed dead. The New Orleans metropolitan area suffered over 727 storm-related deaths as of a February 2006 analysis, and more than 70% of the fatalities were people over age 70. The poor, elderly, and disabled, the groups least likely to be able to evacuate without assistance, were disproportionately impacted. Direct property losses exceeded $20 billion, and 78% of those losses were in residential areas. There was an additional loss of over $7 billion in public structures and utilities.

The consequences of flooding were characterized by the IPET for both property and loss of life. GIS data comprising DEMs, property records, census data, and land use were compiled to support the assessments. The estimates of property damages are based on the type of property, the estimated value of the property, and estimates of the degree of damage/loss as a function of depth of flooding. The fundamental distribution and character of property values were derived from available government databases that provided information to census-block resolution (Figure 9.9). Loss-of-life–flood-depth relationships developed for use in the risk assessment are based on the LifeSim model (Aboelata and Bowles 2005), which makes specific assumptions concerning evacuation, the ability of individuals by age to seek safety, and how these factors vary with different depths of flooding. Uncertainty was computed for both the property-damage–depth-of-flooding relationships and the loss-of-life–depth-of-flooding relationships. There is significantly more uncertainty in the loss-of-life relationships than in the property-loss relationships.

A significant factor in New Orleans's susceptibility to flooding is subsidence of the land surface, due in large part to historic dewatering of swamp and marsh soils, and because of the lack of new sediment from reaching low-lying areas on the floodplain because of levee confinement. Site-specific, cumulative changes in surface elevation in Orleans Parish have been identified in a study by URS (2006). Comparison of historic 5-m horizontal-grid elevation data, relative to a

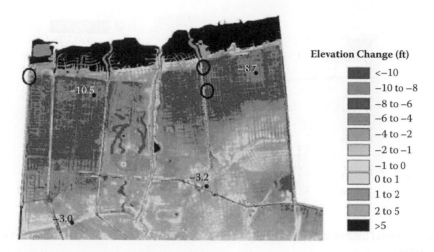

Elevation Change (ft)

- ■ <−10
- ▨ −10 to −8
- ■ −8 to −6
- ▨ −6 to −4
- ▨ −4 to −2
- ▨ −2 to −1
- ▨ −1 to 0
- ▨ 0 to 1
- ■ 1 to 2
- ▨ 2 to 5
- ■ >5

FIGURE 9.10 (See color insert following page 136.) Changes in ground surface elevation due to land subsidence between 1895 and 2001 (URS 2006). Comparison is based on 1895 historic topographic map and 1999–2001 digital elevation models (DEMs). (From IPET 2007.)

constant datum between 1895 and 2001, indicates total subsidence and/or "settlement" in some parts of the city ranges from 8 to 10 ft (Figure 9.10). Positive land gain corresponds to early 1900s dredge filling to create the New Orleans lakefront. The highest values of elevation decline during the past 100 years are in areas underlain by thick marsh and swamp deposits.

Linkage of GIS with hydraulic models can provide valuable information for hazards response planning. For example, hydraulic and companion evacuation modeling for the New Orleans levee failure was demonstrated by Nelson et al. (2008). GIS was linked with a 2-D hydrodynamic model for determining the best evacuation routes from the New Orleans area in the event of levee breaches similar to what actually occurred as a result of Hurricane Katrina. Available DEM data provided elevations to the nearest meter, as required for accurate 2-D hydrodynamic flood simulation in the relatively flat New Orleans area. Land-use maps were utilized for providing roughness values for the hydrodynamic model. For selected discrete time periods, results from the model were imported into ArcGIS and converted into raster files. The dynamic, spatially distributed inundation predictions were used in ArcGIS to model the optimal directions of evacuation by mapping highways and arterials throughout the New Orleans area. Based on available population-density maps, optimal directions for all New Orleans residents were found that minimize travel time to the nearest of seven predefined "outlet" locations using the cost-weighted-distance spatial-analysis tool available in ArcGIS (Figure 9.11). The number of persons whose property would become inundated as a function of time after initiation of levee breach was also calculated using ArcGIS. Results from this study could be used for developing hazard maps for various levee-breach scenarios, as well as possibly utilized in a real-time flood warning system.

9.5.5 FLOODPLAIN HABITAT MODELING WITH GIS

Many river systems include natural and human-made backwater areas that provide habitats for a diverse community of aquatic and avian species, including several listed as endangered by the U.S. Fish and Wildlife Service. Concern has been raised over ecological impacts of river operations on these species and their habitats in these backwater areas. A methodology was presented by Rieker and Labadie (2006) that utilized a GIS in association with a numerical hydraulic model to assess these impacts. The GIS provided geostatistical estimates of water surface elevations within the backwaters during passage of a hydrograph created by reservoir releases, and then quantified and provided animated visualization of the changes in habitats for many species dwelling in these areas.

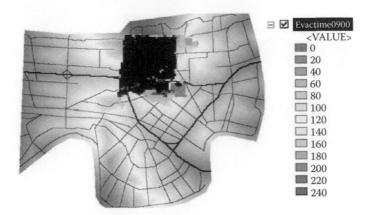

FIGURE 9.11   Travel time (min) to "outlet" layer if leaving 9 h after levee breach. (From Nelson et al. 2008. With permission.)

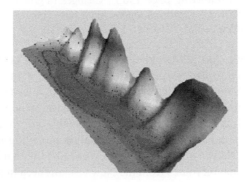

FIGURE 9.12   Combined set of data points and corresponding geostatistically generated surface. (From Rieker and Labadie 2006. With permission.)

This GIS tool was applied to a portion of the Lower Colorado River (LCR) in Arizona/California, which includes several dams and diversion structures controlling flow for a variety of important purposes.

Proper analysis and visualization of changes occurring in the backwaters due to varying water levels require high-resolution bathymetric data. For the LCR study (Rieker and Labadie 2006), three sources of bathymetric data were used in association with geostatistical procedures to obtain the high-resolution depth and inundation area data required for the habitat assessment. These included bathymetric survey data, infrared photos, and USGS 30-m DEM data. Ordinary kriging was employed to improve the estimates of points within most of the bathymetric surfaces used in the study. In the backwater data sets, anisotropy usually becomes apparent where the backwaters are oblong in shape. In these data sets, points along the major axis of the backwater (i.e., along the length) are more likely to be similar to each other. Points along the minor axis of the backwater (i.e., along the width, or the cross-sections) are more likely to differ and usually display trends similar to a fourth-order polynomial-type trend model. Anisotropy was taken into account by utilizing an elliptical neighborhood of known points for the kriging estimation, rather than the entire data set. To produce the bathymetric surfaces for each backwater used in the study, the bathymetric survey data were combined with the water-edge data to produce a data set of points representing known elevation values in the backwater. This data set was then processed using the Geostatistical Analyst® extension to the ArcGIS software to produce a statistically estimated bathymetric surface. USGS DEM values were used to provide an estimate of the terrain surrounding the backwater and fill data gaps within the bathymetric surveys. The ordinary kriging method of analysis produced bathymetric surfaces that were acceptable for use in the habitat analysis and visualization effort. Figure 9.12 shows a set of combined data points and their corresponding geostatistical surface for a representative backwater area.

REFERENCES

Aboelata, M. A., and D. S. Bowles. 2005. LIFESim: A model for estimating dam failure life loss. Institute for Dam Safety Risk Management, Utah State University, Logan, Utah. http://uwrl.usu.edu/people/faculty/DSB/lifesim.pdf.

Ackerman, C. T., T. A. Evans, and G. W. Brunner. 2000. HEC-GeoRAS: Linking GIS to hydraulic analysis using ARC/INFO and HEC-RAS. In *Hydrologic and hydraulic modeling support with geographic information systems*, ed. D. R. Maidment and D. Djokic. Redland, Calif.: ESRI.

DHI (Danish Hydraulic Institute). 2004. MIKE FLOOD, 1D-2D modelling user manual. Agern Allé 5-DK-2970, Hørsholm, Denmark. http://www.dhigroup.com/Software/WaterResources/MIKEFLOOD.aspx.

Dodson, R. D., and X. Li. 1999. The accuracy and efficiency of GIS-based floodplain determinations. In *Proc. ESRI User's Conference*. San Diego, Calif.

FEMA (Federal Emergency Management Agency). 2003. Guidelines and specifications for flood hazard mapping partners: Appendix L. http://www.fema.gov/pdf/fhm/frm_gsana.pdf.

FEMA (Federal Emergency Management Agency). 2005. Guidelines and specifications for flood hazard mapping partners: Appendix N: Data capture standards and guidelines. http://www.fema.gov/pdf/fhm/frm_gsana.pdf.

FIFMTF (Federal Interagency Floodplain Management Task Force). 1992. Floodplain management in the United States: An assessment report. Volume I summary. FEMA publication FIA 17. http://www.fema.gov/hazard/flood/pubs/lib17.shtm.

HDR Engineering, Inc. (HDR). 2007. South Boulder Creek hydraulics report. Prepared for City of Boulder, Colo.

HEC (Hydrologic Engineering Center). 1998. HEC-FDA: Flood damage reduction analysis. User's manual, Version 1.0., CPD-72. U.S. Army Corps of Engineers, Davis, Calif. http://www.hec.usace.army.mil/software/.

HEC (Hydrologic Engineering Center). 2000. HEC-GeoRAS: An extension for support of HEC-RAS using ArcView. User's manual, Version 3.0. U.S. Army Corps of Engineers, Davis, Calif.

HEC (Hydrologic Engineering Center). 2006. HEC-RAS: River analysis system. User's manual, Version 4.0 Beta, CPD-68. U.S. Army Corps of Engineers, Davis, Calif. http://www.hec.usace.army.mil/software/hec-ras/hecras-document.html.

Heinzer, T., M. Sebhat, B. Feinberg, and D. Kerper. 2000. The use of GIS to manage LIDAR elevation data and facilitate integration with the MIKE21 2-D hydraulic model in a flood inundation decision support system. In *Proc. ESRI International Users Conference*. San Diego, Calif.

IPET (Interagency Performance Evaluation Task Force). 2007. Final report of the performance evaluation of the New Orleans and Southeast Louisiana Hurricane Protection System. Volume I: Executive summary and overview. https://ipet.wes.army.mil/.

Jones, J. L., T. L. Haluska, A. K. Williamson, and M. L. Erwin. 1998. Updating flood maps efficiently: Building on existing hydraulic information and modern elevation data with a GIS. U.S. Geological Survey open-file report 98-200. http://wwwdwatcm.wr.usgs.gov/reports/floodgis/.

Kraus, R. A. 1999. Flood plain determination using ArcView and HEC-RAS. Paper 808 presented at ESRI 1999 User Conference. San Diego, Calif.

Linsley, R. K., J. B. Franzini, D. L. Freyberg, and G. Tchobanoglous. 1992. *Water resources engineering*. New York: McGraw-Hill.

Maidment, D. R., ed. 2002. *Arc Hydro: GIS for water resources*. Redlands, Calif.: ESRI.

Nelson, T., S. Uematsu, and J. W. Labadie. 2008. Application of GIS to New Orleans flood inundation and evacuation route analysis. In *Proc. 2008 AWRA Spring Specialty Conference*. San Mateo, Calif.

NRC (National Research Council). 2007. Elevation data for floodplain mapping. Committee on Floodplain Mapping Technologies. National Academies Press. http://books.nap.edu/catalog/11829.html.

Pender, G., and S. Neelz. 2007. Use of computer models of flood inundation to facilitate communication in flood risk management. *Environ. Hazards* 7: 106–114.

Rieker, J. D., and J. W. Labadie. 2006. GIS visualization and analysis of river operations impacts on endangered species habitat. *ASCE J. Water Resour. Plann. Manage.* 132 (3): 153–163.

Shamsi, S. 2002. GIS applications in floodplain management. Paper presented at ESRI International Users Conference. San Diego, Calif.

Tate, E. C., D. R. Maidment, F. Olivera, and D. J. Anderson. 2002. Creating a terrain model for floodplain mapping. *ASCE J. Hydrologic Eng.* 7 (2): 100–108.

Tri, T. G. 2002. Using ArcGIS to link urban stormwater software (SWMM) and economic flood damage analysis software (HEC-FDA). In *Proc. 22nd Annual ESRI International User Conference*. San Diego, Calif.

URS. 2006. A century of subsidence: Change in New Orleans DEMs relative to MGL: 1895 to 1999/2002. Poster prepared for Federal Emergency Management Agency. Baton Rouge, La.: URS Inc.

Walker, W. S., and D. R. Maidment. 2006. Geodatabase design for FEMA flood hazard studies. CRWR online report 06-10. Center for Research in Water Resources, The University of Texas at Austin.

10 GIS for Water Quality

10.1 WATER-QUALITY MONITORING AND MODELING

10.1.1 INTRODUCTION

Maintenance of water quality is sought to avoid impacts of pollution on water users. A primary category of impacts are threats to life through transmission of infectious diseases such as cholera. Other impacts include reduced benefits for recreation, diminished health of aquatic ecosystems, and increased treatment costs for other water users. Urban settlements were highly polluted during the Middle Ages, causing epidemics that wiped out entire populations of people and reduced the lifespan of those who survived. Recognition of the linkage between human waste, water supply, and public health did not occur until the 19th century. John Snow's classic epidemiological study of cholera deaths in London in 1849 identified the Broad Street pump as the source of contamination; removal of the pump handle eliminated the contamination source, and the epidemic subsided. Interestingly, Snow developed a map identifying the residences of the people who contracted the disease as a primary tool to pinpoint the cause of the epidemic (Figure 10.1). This was a clear case of the utility of a GIS-like investigation tool in environmental monitoring. Since that time, efforts to identify, control, and mitigate water pollution have advanced, although somewhat slowly and with varying degrees of commitment by the governing societies.

Major concerns about water quality in industrialized societies, especially drinking water, increased during the period 1880 to 1920. Attention focused on World Wars I and II, and it was not until the 1960s that environmental awareness in the United States led to the passage of the 1972 Water Pollution Control Act Amendments (the Clean Water Act). The 1972 WPCA focused on regulating "point source" discharges of wastewaters. Point sources are those where the wastewater flows in conduits from municipalities and industries, and are therefore more amenable to control. Since that time, scientists and engineers have come to realize that "nonpoint" or diffuse sources of water pollution also play a major role in degradation of water quality. Nonpoint sources originate from urban runoff, erosion from agricultural (cultivated land, irrigated land, and rangeland) and deforested lands and construction sites, groundwater (leaking septic tanks), deforestation, and even air deposition (acid rain). The diffuse-pollution problem has been addressed in the Clean Water Act through the total maximum daily load (TMDL) concept, which is the focus of modern watershed water-quality management planning efforts. Nonpoint-source water pollution is also integrally tied to the hydrology of the land, and characterization of the various sources involves monitoring and modeling the hydrologic characteristics of the contributing watersheds. Thus the data and models described in Chapter 5 (GIS for Surface-Water Hydrology) and Chapter 8 (GIS for Wastewater and Stormwater Systems) are particularly relevant here as well.

GIS performs a central role in support of efforts to monitor water-quality changes within a water body such as a river or bay, to calculate pollutant concentrations and loads to a surface-water body, and modeling water quality of aquatic systems. GIS has also become a valuable set of tools for the important field of epidemiology. The objectives of this chapter are to review water-quality concepts and pollution sources (point and nonpoint); to summarize water-quality monitoring and modeling procedures; and to describe the role of GIS in water-quality monitoring, modeling, and management.

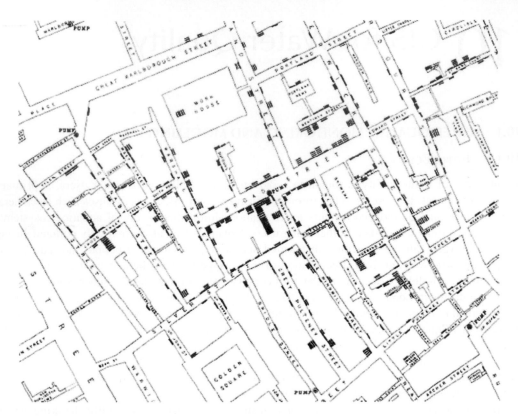

FIGURE 10.1   Map that John Snow used in describing the Broad Street pump cholera outbreak of 1854. The map portrays the locations of cholera deaths (using the bars) in relation to the Broad Street pump. (From Frerichs 2001. With permission.)

10.1.2   Water Quality and Pollution

The quality of water in rivers and other water bodies receiving runoff and discharges is the result of a complex mix of natural and human-influenced processes. Under natural conditions, water draining from a forest, for example, may be considered pristine. However, it still contains dissolved solids, decaying vegetation, and soil particles, to name a few. The term *pollution* refers to changes caused by humans and their actions that result in water-quality conditions that negatively impact the integrity of the water for beneficial purposes, including natural ecosystem integrity. Determining the extent of pollution is difficult, given the wide range of constituent measures that characterize water quality (e.g., dissolved and suspended solids, organics, bacteria, toxics, and metals).

Water-quality management practices are guided by established criteria and standards, typically expressed as constituent concentrations, or narrative statements describing water-quality levels that support particular uses. Criteria may be based on chronic toxicity of various substances to human and aquatic life. Water-quality criteria and standards are either effluent or stream. Effluent standards are typically based on how much of a constituent can be discharged from municipal and industrial sources. Stream standards are related to the beneficial or protected uses of the receiving waters. Effluent standards may be determined by computing the acceptable level of wastes that can be accommodated by a receiving water given the stream standards. This implies that a receiving stream has some waste-assimilative capacity and can accept some degree of degradation before exceeding the standards. In the United States, the Clean Water Act assigns responsibility for issuing scientifically based water and sediment quality criteria and effluent limitations to the Environmental Protection Agency.

TABLE 10.1
Point and Nonpoint Sources of Pollution

Point Sources	Nonpoint Sources
Municipal and industrial wastewater effluents	Return from irrigated agriculture and orchards
Runoff and leachate from solid-waste disposal sites	Runoff from crops, pasture, and rangelands
Runoff and drainage from animal feedlots	Runoff from logging operations, including logging roads and
Runoff from industrial sites	all-terrain vehicles
Storm sewer outfalls from urban centers	Urban runoff from small communities and unsewered settlements
Combined sewer overflows and treatment plant	Drainage from failing septic tank systems
bypasses	Wet and dry atmospheric deposition over water bodies
Mine drainage and runoff (also oil fields)	(e.g., acid rain)
Discharges from storage tanks, chemical waste	Flow from abandoned mines and mining roads
piles, and ships	Runoff and snowmelt from roads outside urban areas
Runoff from construction sites	Wetland drainage
Airport snowmelt and runoff from deicing	Mass outdoor recreation and gatherings
operations	Military training, maneuvers, shooting ranges

Source: Adapted from Novotny (2003).

10.1.3 POLLUTION SOURCES

Water-quality management practice has evolved toward determination of the waste-assimilative capacity of receiving waters. A total mass daily loading (TMDL) assessment is made taking account of all sources of a pollutant, from both point and nonpoint sources, and the waste assimilative capacity of the receiving water body (EPA 1991). Point sources have identifiable single- or multi-point locations where discharges to receiving waters occur. A presumption is that these sources are amenable to control, given the character of the conveyance. Nonpoint sources are diffuse and include everything else that is not a point source. Given the diffuse character of these sources, they are very difficult to control with engineered facilities (e.g., collection and treatment). Point and non-point sources are listed in Table 10.1.

Water-quality assessments require a broad range of environmental and administrative data. Major categories of data include: (a) pollutant source data, (b) watershed characteristics, (c) environmental and ecological data for receiving waters, and (d) administrative data. Table 10.2 lists typical data types of interest. Pollutant-source data include the location and magnitude of point-source discharges.

10.2 GIS FOR WATER-QUALITY MONITORING AND DATABASE DEVELOPMENT

10.2.1 REMOTE SENSING FOR WATER-QUALITY MONITORING

Satellite and airborne remote-sensing techniques have achieved some success with mapping of water quality for monitoring purposes (Ritchie et al. 2003). Suspended sediments, algae, dissolved organic matter, oils, aquatic vascular plants, and thermal releases change the energy spectra of reflected solar and/or emitting thermal radiation from surface waters, and this radiation can be measured using remote-sensing techniques. Most chemicals and pathogens do not directly affect or change the spectral or thermal properties of surface waters, so they can only be inferred indirectly from measurements of other water-quality parameters affected by these chemicals. With empirical approaches, statistical relationships are determined between measured spectral/thermal properties and measured water-quality parameters. For example, suspended sediments increase the radiance

TABLE 10.2
Data Required for Water-Quality Management

Watershed Characteristics	Pollutant Sources	Environmental and Ecological Data	Administrative Data
Topography and land slopes	Locations of point sources and associated discharges and loads	Climate and weather data	Jurisdiction and administrative boundaries
Watershed and sub-basin boundaries		Water-quality monitoring stations: locations and time-series records	
Stream or storm drainage network			Water-quality standards
River segments	Land use–land cover (LU/LC)	Water-quality data (temperature, TDS, SS, BOD, DO, N, P, …)	NPDES permits
Hydraulic infrastructure	Roads and streets		Compliance records
Flow velocities (by season)	Construction sites		Land ownership
Imperviousness of each sub-basin, including directly connected portion	Solid-waste disposal sites	Ecological data and assessments	
Streets, curbs and gutters, and gutter inlets	Urban impervious areas	Drinking-water supplies	
Land use and vegetative cover, including cropping patterns		Beaches and swimming sites	
Soils and characteristics (composition, texture, permeability, erodibility)			
Geologic data			
Surface storage (depression)			
Digital imagery			

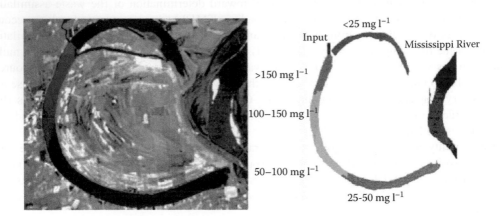

FIGURE 10.2 (See color insert following page 136.)   Landsat TM image of Lake Chicot, Arkansas (left), and a derived image (right) showing categories of suspended sediments mapped in Lake Chicot based on the radiance in the TM image. (From Ritchie et al. 2003. With permission.)

emergent from surface waters (Figure 10.2) in the visible and near-infrared proportion of the electromagnetic spectrum (Ritchie et al. 1976; Schiebe et al. 1992). Thermal plumes in river and coastal waters can be accurately estimated by remote-sensing techniques (Stefan and Schiebe 1970; Gibbons et al. 1989). Maps of suspended chlorophyll were shown to have spatial patterns consistent with field data and circulation of the Neuse River, and these patterns indicated potential point and nonpoint sources (Karaska et al. 2004). Mapping of absolute temperatures by remote sensing provides spatial and temporal patterns of thermal releases that are useful for managing thermal releases. Aircraft-mounted thermal sensors are especially useful in studies of thermal plumes because of the ability to control the timing of data collection. Although limited by the spectral and spatial resolutions of current sensors, the wider availability of new satellites having higher-resolution sensors is expected to improve water-quality monitoring from space platforms.

10.2.2 GIS FOR LAND-USE AND IMPERVIOUS-SURFACE MAPPING

Land-use characteristics pertaining to water quality are often determined using land-use classification maps developed from various sources, including aerial photography and remotely sensed imagery. The amount of urban runoff and its impacts on stream conditions and water quality have been shown to be strongly correlated with the percent area of impervious surfaces within a watershed (Schueler 1994; Arnold and Gibbons 1996; Clausen et al. 2003). Imperviousness influences hydrology, stream habitat, chemical water quality, and biological water quality (Schueler 1994; Arnold and Gibbons 1996). This strong relationship implies that impervious surfaces can serve as an important indicator of water quality, not only because imperviousness has been consistently shown to affect stream hydrology and water quality, but because it can also be readily measured at a variety of scales (i.e., from the parcel level to the watershed and regional levels) (Schueler 1994). GIS-based impervious cover tabulations have been used to evaluate the performance of the Delaware Water Resources Protection Area (WRPA) ordinance (Kauffman et al. 2006). As a result, new development has been restricted to less than 20% impervious cover to protect water supplies. Determination of impervious areas is described in Chapter 8 (GIS for Wastewater and Stormwater Systems), which summarizes typical values of percent impervious area for various land-cover classes.

Remote-sensing and image-processing concepts and tools are extensively used for mapping land use and impervious surfaces. For example, Dougherty et al. (2004) compared impervious-surface (IS) estimates from a recently developed satellite-imagery/land-cover approach with a more traditional aerial-photography–land-use approach. Both approaches were evaluated against a high-quality validation set consisting of planimetric data merged with manually delineated areas of soil disturbance. Results showed that photo-interpreted IS estimates of land class are higher than satellite-derived IS estimates by 100% or more, even in land uses conservatively assigned high IS values. Satellite-derived IS estimates by land class correlated well with planimetric reference data ($r^2 = 0.95$) and with published ranges for similar sites in the region. Results of that site-specific study support the use of automated, satellite-derived IS estimates for planning and management within rapidly urbanizing watersheds where a GIS system is in place, but where time-sensitive, high-quality planimetric data are unavailable.

The National Land Cover Database (NLCD) has been developed by the Multi-Resolution Land Characteristics Consortium (MRLC) of federal agencies (http://www.mrlc.gov/). The data set is described in more detail in Chapter 3 (GIS Data and Databases). In 1993 they purchased Landsat 5 imagery for the conterminous United States and developed the initial NLCD. In 1999 they purchased three dates of Landsat 7 imagery for the entire United States and developed an updated version, called NLCD 2001. NLCD 2001 products include 21 classes of land cover, percent tree canopy, and percent urban imperviousness at 30-m cell resolution. The land-cover maps are of interest, as they support nonpoint water-pollution assessments; the percent-urban-imperviousness maps are of particular interest in this regard. Imperviousness and tree canopy were classified using a commercial regression tree (RT) software called Cubist (Yang et al. 2002). Training data were generally derived from 1-m-resolution digital orthoimagery quarter-quadrangles (DOQQs), classified categorically into canopy/noncanopy or impervious/nonimpervious for each 1-m pixel, and subsequently resampled to 30-m grid proportions. Early products were classified with one or two DOQQs of 6–8 km^2 per Landsat path/row, but in later products, accuracy and prediction quality were improved by distributing three to four smaller chips of 1–4 km^2 in size throughout the Landsat scene. Increasing the sampling frequency not only improved the reliability of training distributions to capture the total range of zonal estimates, but also improved classification efficiency and reduced costs. The completed training images were then extrapolated across mapping zones, using RT models, to derive continuous canopy and imperviousness estimates. Figure 10.3 shows an example of the interactive Web-based mapping utility for the MRLC.

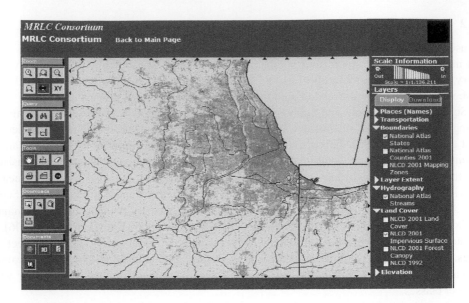

FIGURE 10.3 MRLC Web-based map viewer interface allows users to select an area of interest and to download the data. Data shown are impervious cover for the Chicago area. (*Source:* http://gisdata.usgs.net/ website/MRLC/viewer.php.)

10.2.3 GIS FOR DATA COLLATION AND PROBLEM IDENTIFICATION

GISs provide powerful functions for data capture and collation. It is quite common that water-quality studies obtain data from multiple sources, and these data need to be collated and converted into common formats of the water-quality geodatabase. Water-quality database development activities begin with field data collection, collation, quality control, and database posting; these activities apply for a specific water-quality assessment project. Where the use of archival and repetitive and historic water-quality data is involved, it is required that the data be archived to a formal database that can then be used for exploratory and regulatory purposes by the involved agencies.

The role of GIS for data-capture and collation operations for a specific project was illustrated for an assessment of sediment mercury concentrations in Lavaca Bay, Texas (Benaman and Mathews 2001). The study was conducted by various teams, each collecting differing types of data. Data were required to be in GIS-ready or dbf format, with spatial coordinates attached to the data. Once imported into the ESRI software, ArcView, the spatial locations of sampling points were joined with the related attribute tables, and displays of the mercury concentrations were generated. Mapping of bottom-sediment conditions provided a useful guide for locating where sample cores could be taken; hard-rock bottom sites were avoided (Figure 10.4).

GIS spatial statistical functions provide a means of characterizing the extent and patterns of contamination. The Lavaca Bay study used GIS spatial interpolation of the sediment core samples to examine the sediment mercury concentrations throughout the bay (Figure 10.5).

Collation of data across a large area using GIS provides analysts and managers the capability to examine spatial trends. A study in the U.K. on the 24,000-km Humber Watershed, mapped for key inorganic chemical constituents, used GIS and an extensive water-quality monitoring database to reveal major factors affecting the general characteristics of regional water quality (Oguchi et al. 2000). Sewage inputs from industrial and domestic sources accounted for the high concentration of many constituents in urban areas. Some constituents also exhibited localized high concentrations related to coal mine drainage, soil pollution caused by past ore mining, bedrock geology, agricultural use of fertilizers, and the ingression of seawater into the estuary. Figure 10.6 illustrates the GIS products obtained.

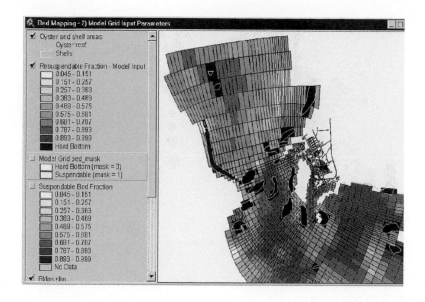

FIGURE 10.4 Locations of mercury-sampling stations in Lavaca Bay, Texas, were guided by mapping bottom-sediment conditions. (From Benaman and Mathews 2001. With permission.)

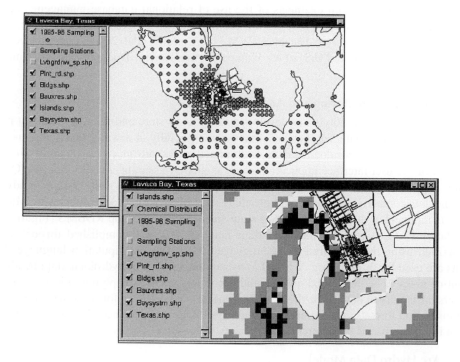

FIGURE 10.5 Sediment mercury concentrations interpolated from point samples. (From Benaman and Mathews 2001. With permission.)

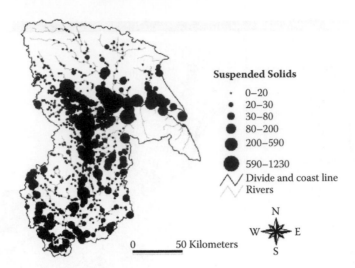

FIGURE 10.6 Map of suspended-solids concentrations (mg/L) for the Humber catchment, U.K. (From Oguchi et al. 2000. With permission.)

10.2.4 GIS for Water-Quality Databases

Given the large amounts of data involved with water-quality monitoring and management applications, there is strong incentive to adopt a database approach. Hydrologic and water-quality data are often stored in a spreadsheet program. However, users quickly realize that spreadsheets are not intended for storage of large data sets, and a proliferation of files, redundant data, and lack of security cause problems for users. Data must be stored, checked for errors, manipulated, retrieved for analysis, and shared. Several examples of the use of relational database-management system (RDBMS) technology illustrate approaches for dealing with these problems. These include the Watershed Monitoring and Analysis Database (WMAD) (Carleton et al. 2005); the Arc Hydro data model (Maidment 2002); and the U.S. EPA's Watershed Assessment, Tracking and Environmental Results (WATERS) program.

10.2.4.1 Watershed Monitoring and Analysis Database

The WMAD (Carleton et al. 2005) was developed as a tool to store and manage data that pertain to a stream's hydrologic characteristics and functions, especially at a watershed scale, including flow, water quality, and meteorological data. Both remedial and advanced tasks can be simplified with the help of the user-interface application, such as quality-assurance/quality-control (QA/QC) calculations, application of correction and conversion factors, retrieval of desired data for advanced analysis, and data comparisons among multiple study sites. WMAD was developed using Microsoft Access®, a component of the Microsoft Office suite and a popular desktop RDBMS (Figure 10.7).

Once data are stored within a database, their manipulation is accomplished through queries that are written using the structured query language (SQL). Data-manipulation language (DML) statements are used for this purpose. WMAD assists in the analysis of hydrologic data by allowing many routine tasks to be performed quickly and accurately, such as applying uniform correction and conversion factors to data through the use of queries. This allows for the retrieval of data in different forms and formats, depending on what analysis is to be performed. For example, the SQL expression in Figure 10.8 is used to retrieve water-quality results.

10.2.4.2 Arc Hydro Data Model

The Arc Hydro data model (Maidment 2002) has been used as a guide for integration of the large variety of spatial and attribute data into a relational geodatabase. The database provides expanded

WMAD Data Structure

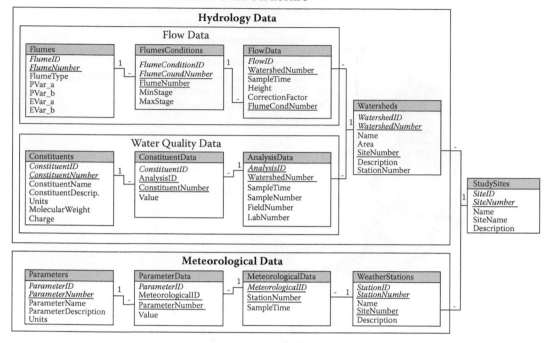

FIGURE 10.7 Overview of WMAD database design with table relationships. Fields in italics are autonumbered (contain a unique value within each table), and underlined fields relate records to other tables. (From Carleton et al. 2005. With permission.)

```
TRANSFORM Avg(CD.Value * C.ConversionFactor) AS ConvertedValue
SELECT AD.SampleTime
FROM AnalysisData AS AD INNER JOIN (Constituents AS C INNER JOIN
ConstituentData AS CD ON C.ConstituentNumber = CD.ConstituentNumber) ON
AD.AnalysisID = CD.AnalysisID
WHERE ((AD.SampleTime Between #1/1/1999# And #12/31/1999#)
AND (AD.WatershedNumber = 1))
GROUP BY AD.SampleTime
ORDER BY AD.Sample Time, C.ConstituentName
PIVOT C.ConstituentName
```

FIGURE 10.8 SQL queries from an RDBMS can be structured to retrieve data in preferred formats and to perform error-checking and unit-conversion calculations. (From Carleton et al. 2005. With permission.)

functionality for data retrievals and processing, as well as cross-institutional coordination. McKinney and Patino-Gomez (2006) demonstrated this concept for the Rio Bravo/Grande River basin between the United States and Mexico; their database was called the Water Quality Data Model (WQDM). For the WQMD, the Arc Hydro data model was used to represent the river network and to define attributes, relations, and connectivity between hydrologic features. A general description of the Arc Hydro data model is presented in Chapter 3. Regional integrated geodatabase projects characteristically involve large amounts of data. Automated flow and monitoring stations produce data at daily, hourly, and quarter-hour intervals. Activities involved with building the regional database include:

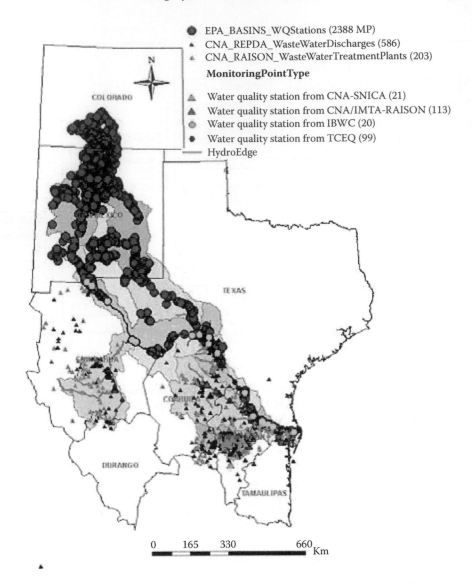

FIGURE 10.9 (See color insert following page 136.)   Monitoring points included in the WQDM geodatabase. (From McKinney and Patino-Gomez 2006. With permission.)

(a) identifying and collecting the data necessary to build an Arc Hydro data model, (b) assembling the watersheds into a regionalized data mode, (c) populating time-series tables with water-quality and stream flow data, and (d) providing the geodatabase to various internal and external agency users. Figure 10.9 shows a data retrieval for water-quality monitoring stations.

10.2.4.3   EPA Watershed Assessment, Tracking, and Environmental Results

The EPA Watershed Assessment, Tracking, and Environmental Results (WATERS) is an integrated information system for the nation's surface waters (http://www.epa.gov/waters/). The EPA Office of Water manages numerous programs in support of the agency's water-quality efforts, and many of these programs collect and store water-quality-related data in separate databases. These databases are managed by the individual water programs, and this separation often inhibits the integrated application of the data they contain. Under WATERS, the databases are connected to the

National Hydrography Dataset (NHD). By linking to the NHD, one water-program database can reach another by virtue of location, and information can be shared across programs.

The WATERS architecture consists of four major components:

- National Hydrography Dataset (NHD): A comprehensive set of digital geospatial data that contains information about surface-water features. The NHD includes an addressing system for linking EPA Water Program data to the underlying NHD surface-water drainage network to facilitate its geographic integration, analysis, and display.
- Reach Address Database (RAD): Stores the reach address for each water-program feature that has been linked to the NHD. These reach addresses record the geographic extent of water-program features in both tabular and spatial formats. The reach addresses link to a static copy of the NHD obtained from the USGS, which also resides in the RAD.
- Water-program databases: The EPA Office of Water has numerous programs that collect and store water-related information in separate databases. By linking the features stored in these water-program databases to the NHD, the collective information held within the databases can be shared across programs to better facilitate water-quality management.
- WATERS tools: Tools and applications that generate and use the reach addresses in the RAD to query, analyze, and display information across water programs. The process of assigning NHD reach addresses to water-program features is often referred to as reach indexing, and the tools that support this assignment as reach indexing tools (RIT).

WATERS databases primarily include (or are projected to include) the following:

- *Water Quality Standards Database* (WQSDB) contains information on the uses that have been designated for water bodies. Examples of such uses are drinking-water supply, recreation, and fish protection. As part of a state's water-quality standards, these designated uses provide a regulatory goal for the water body and define the level of protection assigned to it. WQS also include the scientific criteria to support a given use.
- *STORET* (STOrage and RETrieval) is a repository for water-quality, biological, and physical data and is used by state environmental agencies, the EPA and other federal agencies, universities, private citizens, and many others. The Legacy Data Center (LDC) contains historical water-quality data dating back to the early part of the 20th century and collected up to the end of 1998.
- *National Assessment Database* (NAD) contains information on the attainment of water-quality standards. Assessed waters are classified as either Fully Supporting, Threatened, or Not Supporting their designated uses. This information is reported in the National Water Quality Inventory Report to Congress under Section 305(b) of the Clean Water Act.
- *Total Maximum Daily Load (TMDL) Tracking System* contains information on waters that are not supporting their designated uses. These waters are listed by the state as impaired under Section 303(d) of the Clean Water Act. The statuses of TMDLs are also tracked. TMDLs are pollution-control measures that reduce the discharge of pollutants into impaired waters.
- Other databases include those for the National Pollutant Discharge Elimination System (NPDES) Permits; Safe Drinking Water; Fish Consumption Advisories; Section 319 Grants Reporting and Tracking System (GRTS) for nonpoint sources; Nutrient Criteria Database; Beaches Environmental Assessment, Closure & Health (BEACH) Watch; Vessel Sewage Discharge; and the Clean Watersheds Needs Survey (CWNS).

WATERS uses the NHD addressing system to link EPA Water Program features, such as water-monitoring sites, to the underlying surface-water features (streams, lakes, swamps, etc.). This linkage enables geographic integration, analysis, and display of data stored in different EPA Water

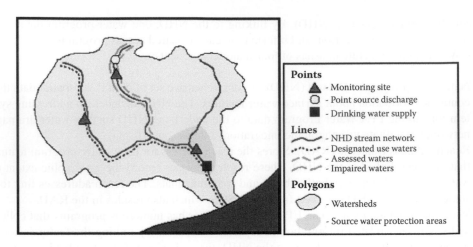

FIGURE 10.10 EPA Water Program data can be displayed and analyzed geographically. Individual Water Program features are represented as points, lines, and polygons. (*Source:* http://www.epa.gov/waters/.)

Program databases using computer-based tools, such as a geographic information system (GIS). The NHD addressing system employs a unique, standard identifier, known as the *reach code*, for each segment of water across the country (Figure 10.10).

The NHD has its own addressing system based upon reaches. A reach is the portion of a stream between two points of confluence. A confluence is the location where two or more streams flow together. In the NHD addressing system, a reach code identifies each reach in the same manner that a street name identifies each street. Reach codes are composed of the eight-digit watershed code followed by a six-digit arbitrarily assigned sequence number. The NHD addressing system utilizes GIS dynamic segmentation functionality, where the reach addresses for water-program features are implemented as events, with NHD reaches serving as the routes and the positions (%) along the reach serving as the measures. Position along a reach is determined relative to the reach extent. The downstream end of a reach is labeled as 0% and the upstream end as 100%; if a water-program feature lies halfway along a reach, its position is 50%. To link EPA Water Program data to the NHD, the EPA as well as individual states and tribes record and store the reach address of each water-program feature in an EPA database called the Reach Address Database (RAD). Within the RAD, an individual water-program feature and its associated reach address are recorded in a row in a table similar to the one below. Once the reach addresses have been stored in the RAD, the water-program databases are logically connected through the NHD, enabling the sharing of information across water programs.

The EPA WATERS program provides an interactive Web-based GIS interface to the various databases called EnviroMapper (EM) (http://map8.epa.gov/enviromapper/). The EM has a variety of GIS tools and features to create a custom map of an area of interest. The EM also allows a user to display multiple feature layers; identify and label features; and zoom, pan, and show latitude and longitude coordinates for a specific location. The map features in EM are organized into feature layers, such as water-program features (water-quality standards, assessed waters, and impaired waters) and facilities (Superfund sites, dischargers to water, hazardous waste handlers), which are drawn one on top of another to produce a map. Figure 10.11 shows a typical map display from EnviroMapper (EPA 2004).

10.3 GIS FOR WATER-QUALITY MODELING

10.3.1 Point- and Nonpoint-Source Water-Quality Modeling with GIS

Models to predict water-quality impacts of point-source discharges began in the 1920s with the development of the Streeter-Phelps model, which represented the fate of organic materials such as biochemical oxygen demand (BOD) on dissolved oxygen (DO) concentrations in rivers. Built on

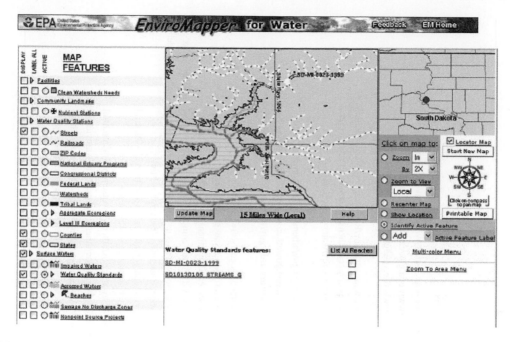

FIGURE 10.11 EnviroMapper is an interactive Web-based GIS and provides functions for retrieval of water-quality standards, pollution discharge sources, and regulatory status for a selected water body. (*Source:* http://map8.epa.gov/enviromapper/.)

concepts of mass balance, chemical and biological rate kinetics, and the coupled BOD/DO equations, the early models have evolved to represent multiple constituents, time variability, complex hydrodynamics, and aquatic ecosystem interactions (see, for example, Thomann and Mueller 1987; Chapra 1997). Point-source water-quality models typically deal with water in channels and estuaries, where pollutants enter at specific locations, primarily from wastewater systems.

Nonpoint models must account for other aspects of the hydrologic cycle (e.g., precipitation and overland flow); these may be thought of as loading models that track the movement of pollutants from the land to the water courses. GISs have found particular application for modeling nonpoint-source water pollution. Data involved include land use and land cover, watersheds and topography, drainage ways, and soils, to name a few. The emphasis is on land runoff; therefore, many of these models incorporate hydrologic routines such as described in Chapter 5 (GIS for Surface-Water Hydrology) and Chapter 8 (GIS for Wastewater and Stormwater Systems). There are a number of nonpoint-source water-quality models having GIS interfaces; many have been developed to assess potential impacts of agricultural activities (Parsons et al. 2004); others build on the urban stormwater models described in Chapter 8. Some popular water-quality models that have been integrated with GISs are listed in Table 10.3.

Water-quality models may be designed to be steady state or time varying. Steady-state models apply when loadings and quality dynamics are relatively time-invariant; this is often a reasonable assumption for point-source models representing low-flow conditions. Introduction of storm runoff loadings from urban and agricultural areas requires time-variable modeling approaches. Models may be deterministic or stochastic. Deterministic models assume fixed values of pollutant loadings, flows, and quality dynamics. Stochastic models incorporate the randomness of physical, chemical, and biological processes, thereby generating probability estimates of water-quality outcomes. A river receiving pollutant discharges is often represented as one-dimensional, and complete mixing in the vertical and lateral directions is assumed. A primary data structure for river models is the network where reaches having similar characteristics can be established as computational elements. The basic equations of mass balance on flows and pollutant mass are applied at the edge

TABLE 10.3
Water-Quality Models for Point and Nonpoint Water-Pollution Assessments

Model Abbreviation	Model Name	Point/Nonpoint	Source	Web Link
AGNPS	Agricultural nonpoint source	Nonpoint	USDA-ARS	http://www.ars.usda.gov/Research/docs.htm?docid=5199
AGWA	Automated geospatial watershed assessment	Nonpoint	USDA-ARS	http://www.epa.gov/esd/land-sci/agwa/
ANSWERS-2000	Aerial nonpoint-source watershed environmental response simulation	Nonpoint	Virginia Tech	http://www.bse.vt.edu/ANSWERS/index.php
APEX	Agricultural policy/environmental eXtender	Nonpoint	USDA-ARS	http://www.ars.usda.gov/Research/docs.html?docid=9792
Aquatox	Simulation model for aquatic ecosystems	Hybrid	US EPA	http://www.epa.gov/athens/wwqtsc/html/aquatox.html
BASINS	Better assessment science integrating point and nonpoint sources	Nonpoint	US EPA	http://www.epa.gov/waterscience/basins/
EFDC	Environmental fluid dynamics code	Hydrodynamics	US EPA	http://www.epa.gov/athens/wwqtsc/html/efdc.html
EPD-RIV1	1-D riverine hydrodynamic & WQ model	Point	US EPA	http://www.epa.gov/athens/wwqtsc/html/epd-riv1.html
GLEAMS/CREAMS	Groundwater loading effects of agricultural management systems	Nonpoint	USDA-ARS	http://www.ars.usda.gov/Research/docs.htm?docid=9797
HSPF	Hydrological simulation program (Fortran)	Nonpoint	USGS	http://water.usgs.gov/software/HSPF/
KINEROS2	Kinematic erosion simulator	Nonpoint	USDA-ARS	http://www.tucson.ars.ag.gov/kineros/
LSPC	Loading simulation program in C++	Nonpoint	US EPA	http://www.epa.gov/athens/wwqtsc/html/lspc.html
MIKE SHE	Simulator of land phase of hydrologic cycle	Hybrid	DHI Group	http://www.dhigroup.com/Software/WaterResources/MIKESHE.aspx
NLEAP	Nitrate leaching predictor	Nonpoint	USDA-ARS	http://arsagsoftware.ars.usda.gov/asru/index.asp?page=models&model=nle
PRMS	Precipitation-runoff modeling system, with MMS	Nonpoint	USGS	http://water.usgs.gov/software/prms.html
QUAL2E, QUAL2K	River and stream water-quality model	Point	US EPA	http://www.epa.gov/athens/wwqtsc/html/qual2k.html
SWAT	Soil and water assessment tool	Nonpoint	USDA-ARS	http://www.ars.usda.gov/Research/
SWMM	Stormwater management model	Hybrid	US EPA	http://www.epa.gov/athens/wwqtsc/html/swmm.html
WAM	Watershed assessment model	Nonpoint	US EPA	http://www.epa.gov/athens/wwqtsc/html/wamview.html
WARMF	Watershed analysis risk-management framework	Point	US EPA	http://www.epa.gov/athens/wwqtsc/html/warmf.html
WASP7	Water-quality analysis simulation program	Point	US EPA	http://www.epa.gov/athens/wwqtsc/html/wasp.html
WCS	Watershed characterization system	Hybrid	US EPA	http://www.epa.gov/athens/wwqtsc/html/wcs.html

of each reach, and the biochemical reactions are computed within the reach. Mass-balance relations are applied at the junctions between river branches. For estuaries and some large rivers, two-dimensional modeling is required, and vertical mixing is assumed. Some situations (e.g., ocean discharges) would require a three-dimensional modeling approach.

Point, nonpoint, and receiving-water hydrodynamic models are often applied in an integrated manner for development of comprehensive water-quality management plans, e.g., for total mass-discharge loading (TMDL) plans. Watershed models play an important role in linking sources of pollutants to receiving water bodies as nonpoint-source loads. Watershed models are driven by precipitation, land use, impervious areas, slope, soil types, and drainage area. GIS programs like BASINS provide the data that are needed for watershed models to predict both water and pollutant runoff from a watershed. Hydrodynamic and water-quality models integrate inputs from point and nonpoint sources to determine impacts on water quality in receiving water bodies. These models allow determination of assimilative capacities of the water body, determine the level of best management practices, or predict the time required for a system to recover after being altered.

10.3.2 Point-Source Water-Quality Modeling with GIS

The most popular point-source water-quality model is the QUAL2E/QUAL2K model provided by the U.S. EPA (http://www.epa.gov/athens/wwqtsc/html/qual2k.html; Chapra et al. 2007). QUAL2K is the latest release and provides the following functionality:

- One-dimensional: The channel is well-mixed vertically and laterally.
- Branching: The system can consist of a main-stem river with branched tributaries.
- Steady-state hydraulics: Nonuniform, steady flow is simulated.
- Diel heat budget: The heat budget and temperature are simulated as a function of meteorology on a diel time scale.
- Diel water-quality kinetics: All water-quality variables are simulated on a diel time scale.
- Heat and mass inputs: Point and nonpoint loads and withdrawals are simulated.
- QUAL2K also divides the system into reaches and elements. However, in contrast to QUAL2E, the element size for QUAL2K can vary from reach to reach.
- Multiple loadings and withdrawals can be input to any element.
- Various other biochemical and physical processes are incorporated for BOD, sediment, algae, pathogens, and weirs and waterfalls.

QUAL2K model constituents are listed in Table 10.4. Details on algorithms can be found in Chapra et al. (2007). QUAL2K is implemented within the Microsoft Windows environment. Numerical computations are programmed in Fortran 90. Excel is used as the graphical user interface. All interface operations are programmed in the Microsoft Office macro language: Visual Basic for Applications (VBA). However, GIS can be used to generate the input files through file transfer or full integration.

The model represents a river as a series of reaches. These represent stretches of river that have constant hydraulic characteristics (e.g., slope, bottom width, etc.). As depicted in Figure 10.12, the reaches are numbered in ascending order, starting from the headwater of the river's main stem. Notice that both point and nonpoint sources and point and nonpoint withdrawals (abstractions) can be positioned anywhere along the channel's length. For systems with tributaries (Figure 10.13), the reaches are numbered in ascending order, starting at reach 1 at the headwater of the main stem. When a junction with a tributary is reached, the numbering continues at that tributary's headwater. Observe that both the headwaters and the tributaries are also numbered consecutively, following a sequencing scheme similar to that of the reaches. Note also that the major branches of the system (i.e., the main stem and each of the tributaries) are referred to as segments. This distinction has practical importance because the software provides plots of model output on a segment basis, that is, the software generates individual plots for the main stem as well as each of the tributaries.

TABLE 10.4
QUAL2K Model State Variables

Variable	Symbol	Units [a]
Conductivity	s	µmhos
Inorganic suspended solids	m_i	mg D/L
Dissolved oxygen	O	mg O_2/L
Slow-reacting CBOD	c_s	mg O_2/L
Fast-reacting CBOD	c_f	mg O_2/L
Organic nitrogen	N_o	µg N/L
Ammonia nitrogen	N_a	µg N/L
Nitrate nitrogen	N_n	µg N/L
Organic phosphorus	P_o	µg P/L
Inorganic phosphorus	P_i	µg P/L
Phytoplankton	a_p	µg A/L
Detritus	mo	mg D/L
Pathogen	X	cfu/100 mL
Alkalinity	Alk	mg $CaCO_3$/L
Total inorganic carbon	C_T	mole/L
Bottom algae biomass	a_b	mg A/m^2
Bottom algae nitrogen	IN_b	mg N/m^2
Bottom algae phosphorus	IP_b	mg P/m^2

Source: Chapra et al. 2007.

[a] mg/L = g/m^3. In addition, the terms D, C, N, P, and A refer to dry weight, carbon, nitrogen, phosphorus, and chlorophyll a, respectively. The term cfu stands for colony-forming unit, which is a measure of viable bacterial numbers.

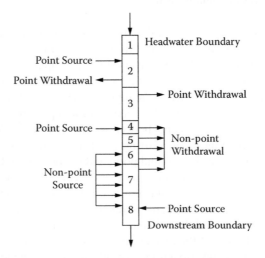

FIGURE 10.12   QUAL2K segmentation scheme for a river with no tributaries. (From Chapra et al. 2007.)

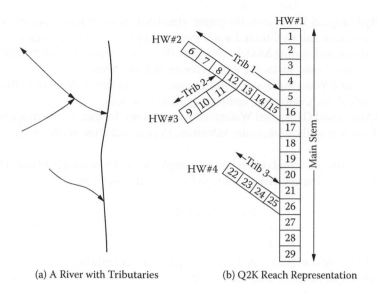

(a) A River with Tributaries   (b) Q2K Reach Representation

FIGURE 10.13   QUAL2K segmentation scheme for (a) a river with tributaries. The Q2K reach representation in (b) illustrates the reach, headwater, and tributary numbering schemes. (From Chapra et al. 2007.)

10.3.3   Nonpoint-Source Water-Quality Modeling with GIS

A study by Wong et al. (1997) illustrated the utility of an empirical rainfall-runoff model and relatively simple GIS data for identifying priorities for monitoring watersheds draining to Santa Monica Bay, California. The modeling predicted receiving-water loading on an annual basis or for single-storm events. Land-use maps detailing eight categories were used to estimate impervious percentages in establishing runoff coefficients for the runoff model. The annual total runoff was computed using local average storm rainfall (scaled for area), watershed area, and runoff coefficient; values for the total number of storms were summed to get the annual total runoff. Water-quality data for typical urban runoff pollutants were obtained from local agencies to determine mean event concentrations (ME) for each land-use type using the method of Driscoll et al. (1990). The land-use, sub-basin, and watershed themes were developed by scanning USGS maps and merging with drainage-network data. These data were integrated using the ArcInfo® GIS to form an integrated GIS model. Overlay (spatial UNION) of the three coverages and subsequent linkage with the pollutant-loading tables allowed computation of the 11 pollutant loadings for each watershed and sub-basin polygon. The GIS model was used to identify critical areas where pollutant discharges should be monitored. It was also used to assist in the development of control strategies and to evaluate the effectiveness of various best management practices (BMPs).

10.3.4   EPA BASINS

The U.S. EPA has developed a widely distributed nonpoint-source water-quality model called BASINS (Better Assessment Science Integrating Point and Nonpoint Sources) (EPA 2001a). EPA BASINS software and associated data are a free service provided by the EPA, but requires ArcView 3.x® GIS software for use. The software was developed to aid in watershed management by providing tools to model hydrological, erosion, and pollution processes, including pollutant loadings (EPA 2001b). BASINS is considered a useful tool for generating generalized scenarios for management of nonpoint sources through the use of BMPs. BASINS incorporates several models into a single suite of tools, including:

- HSPF (Hydrological Simulation Program): simulates hydrological processes on land surfaces, in in-stream and well-mixed water systems, and time-series analysis.
- PLOAD (Pollutant Loading Model): simulates hydrological processes for pollutant-loading estimation (runoff over surfaces); it provides annual averages.
- SWAT (Soil and Water Assessment Tool): assesses long-term effects of pollutant loading (mainly agricultural), sediment, and management practices.
- AGWA (Automated Geospatial Watershed Assessment): focuses on the processes involved in runoff, such as infiltration, water velocities, slopes, and erosion processes.

PLOAD calculations are based on a relatively simple method based on land use. The runoff coefficients are based on the impervious percentage for each land-use type:

$$RV_u = 0.05 + (0.009 \times I_u) \tag{10.1}$$

where
  RV_u = runoff volume coefficient for land-use type u, cm/cm rainfall
  I_u = percent impervious for land-use type u

Event mean concentrations (EMCs) are computed for each land-use type:

$$LP_u = \Sigma_u \, [P \times P_j \times RV_u \times C_u \times A_u \times (2.72/12)] \tag{10.2}$$

where
  LP_u = pollutant load for each land-use type u, lb
  P = precipitation, in./y
  P_j = ratio of storms producing runoff (default = 0.9)
  C_u = event mean concentration for land-use type u, mg/L
  A_u = area of land-use type u, acres
      (PLOAD converts square miles to acres prior to calculation)

The export-coefficient method uses single factors for each land-use type:

$$LP = \Sigma_u \, (LP_u \times A_u) \tag{10.3}$$

where
  LP = pollutant load, lb
  LP_u = pollutant loading rate for land-use type u, lb/acre/year

The export coefficients are developed from extensive studies and literature review. However, it is recommended that, to obtain the best results, these values should be as specific to the study area as possible. Export coefficients store the report's pollutants in units of lb/acre/year by land-use type (EPA 2001b).

An example application was conducted for the Bear Creek watershed near Denver, Colorado, where there is a concern with elevated nutrient (phosphorous) loading to a downstream reservoir (Heller, 2008; Figure 10.14). The minimum data required were elevation, hydrology (NHD), watershed boundaries, land use, and roads and highways. Tabular data on land use, EMCs, and impervious ratings are incorporated in the PLOAD model. Land-use data were obtained from the National Land Cover Dataset (NLCD).

The PLOAD model was run to locate the areas of high pollutant loading within the East Bear Creek sub-basin. The data loaded into the model were the delineated sub-basin watershed boundary theme, the updated NLCD, the EMCs, and the percent-impervious tables. As before, an average

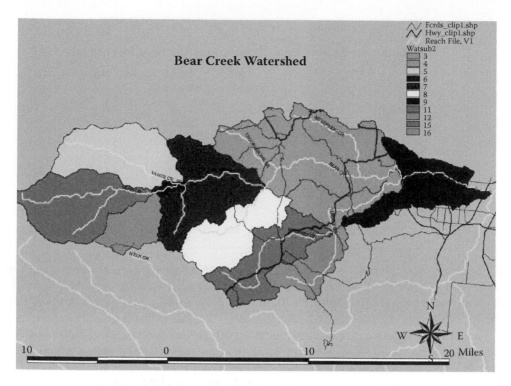

FIGURE 10.14 Bear Creek watershed sub-basins were delineated using digital elevation data. (From Heller 2008. With permission.)

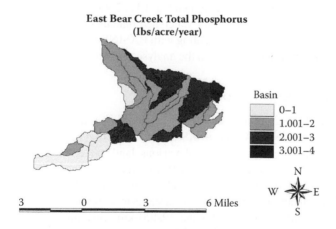

FIGURE 10.15 BASINS-simulated total phosphorous classified for East Bear Creek sub-basins.

annual rainfall of 30 in. was specified, and options to load BMPs, point sources, and any additional data were not chosen. The results were mapped by total phosphorus per acre per year (Figure 10.15). The area located in the northeastern section of the sub-basin showed the highest loading of 3.49 lb/acre/year. Sections in the northern and southern-central parts of the East Bear Creek sub-basin showed high loading rates between 2 and 3 lb/acre/year. The lowest loadings of between zero and 1 lb/acre/year were located in the southwestern region of the sub-basin, where the land use was composed of more vegetation and less transportation and urban land uses. Several simulations were conducted to assess the effectiveness of various BMPs in reducing total phosphorus loads. Table 10.5 summarizes the results.

TABLE 10.5
Summary of BASINS Total Phosphorous Loadings by BMP

BMP	Area (acres)	Total P (lb/year)	Reduction (%)
No action	0	24,703	0.0
Street sweeping (SS)	3,943	24,281	1.7
Infiltration trench +SS	4,494	23,640	4.3
Bioretention + SS	4,890	22,982	7.0
Bioretention + infiltration + SS	5,541	22,341	9.6
Detention ponds	2,285	21,610	12.5
All BMPs	78,267	19,249	22.1

Source: Heller 2008.

The accuracy of the model is dependent on the accuracy of the input data, such as EMC and impervious-percentage-value development, and on the confidence of the calculation methods used in the model (simple and export coefficient) (EPA 2001a). Lin (2004) evaluated published EMC and export-coefficient values by individual regional studies and the National Urban Runoff Program (NURP). The overall theme of the paper was that, while nationally pooled loading rates (EMCs and export coefficients) provide a reasonable estimation for large areas, region-specific values should be compiled or developed for the most accurate results. Review of case studies such as the Cottonwood Creek watershed in Idaho County, Idaho—performed with the use of BASINS—underline the importance of model calibration with measured stream flows and pollutant concentrations when using BASINS to model watershed dynamics (Baxter 2002). These calibrations were stressed less in the use of the PLOAD model; however, calibration is recommended (EPA 2006). There was one publication voicing criticism of the use of BASINS modeling for watershed management (Wittemore and Beebe 2000). They expressed concern over watershed modeling with BASINS, stating that the software takes a simplistic approach and that the data upon which the analyses are based are sometimes suspect.

10.3.5 WATERSHED ASSESSMENT MODEL

There is a trend toward integration of GIS tools into multicompartmental watershed models. The Watershed Assessment Model (WAM) is a GIS-based model that allows users to interactively simulate and assess the environmental effects of various land-use changes and associated land-use practices (Bottcher and Hiscock 2002). Called WAM*View* with an ArcView interface, the model simulates spatial water-quality loads based on land use and soils and then routes and attenuates these source cell loads through uplands, wetlands, and streams to watershed outlets. WAM was originally developed with an Arc/Info® interface for the 19,400-km² Suwannee River Water Management District (SRWMD) of northern Florida. The GIS-based processing and user interface in the WAM/WAM*View* models allow for a number of user options and features to be provided for grid sizes down to 0.1 ha. The features include:

- Source cell mapping of TSS and nutrient surface and groundwater loads
- Tabular ranking of land uses by constituent contributions
- Overland, wetland, and stream load attenuation mapped back to source cells
- Hydrodynamic stream routing of flow and constituents with annual, daily, or hourly outputs
- Optional index model for toxins, BOD, and bacteria for source-cell mapping
- Wetland indexing model for wildlife diversity impacts
- Flood proneness model
- User interface to run and edit land-use and BMP scenarios

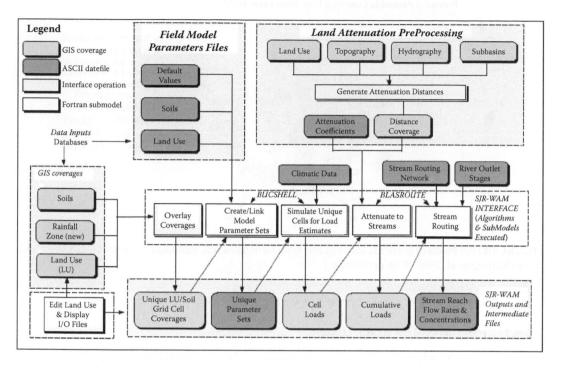

FIGURE 10.16 WAM*View* modeling process integrates GIS functions with ASCII data sets and Fortran submodels. (From Bottcher and Hiscock 2002.)

Figure 10.16 shows a flow diagram of the hydrologic contaminant transport modeling component of the overall WAM*View* model.

10.3.6 NRCS-GLEAMS

GIS databases and water-quality modeling tools provide capacity to conduct large-area and even nationwide assessments. For example, the Natural Resources Conservation Service (NRCS) has used the GLEAMS model to assess water-pollution potential from agricultural operations (Kellogg et al. 2000). The analytical framework consists of about 4700 resource polygons representing the intersection of 48 states, 280 watersheds at the six-digit hydrologic unit level, and 1400 combinations of climate and soil groups. Pesticide loss from farm fields was estimated using a process model called GLEAMS, which incorporates these complex interactions of weather, pesticide properties, and soil characteristics. GLEAMS consists of three major components: hydrology, erosion/sediment yield, and pesticides. Soil-water accounting procedures represent the principal hydrologic processes of infiltration, runoff, water application by irrigation, soil evaporation, plant transpiration, and soil-water movement within and through the root zone. A modification of the universal soil loss equation (USLE) is used to simulate storm-by-storm rill and interrill soil erosion in overland flow areas. The pesticide component of GLEAMS is designed to allow simulation of interactions among pesticide properties, climate, soils, and management, as seen in Figure 10.17. Adsorption characteristics are coupled with the hydrologic component to route pesticides within and through the root zone. A National Pesticide Loss Database was constructed using GLEAMS estimates for 243 pesticides applied to 120 generic soils for 20 years of daily weather from each of 55 climate stations. This resulted in 1,603,800 runs of 20 years each.

Potential Pesticide Leaching Loss from Farm Fields

■ Top 200 watersheds
■ Next 200 watersheds
▢ Next 400 watersheds
□ Next 400 watersheds
▢ Remaining watersheds
□ Greater than 96%
  Federal land or no
  cultivated cropland
  or CRP land or
  value equal to zero.

FIGURE 10.17   Potential pesticide leaching loss from farm fields was estimated using the GLEAMS model. (From Kellogg et al. 2000. With permission.)

10.4   GIS FOR WATER-QUALITY MANAGEMENT DECISION SUPPORT

10.4.1   TOTAL MASS-DISCHARGE LOADING

A total maximum daily load (TMDL) is a calculation of the maximum amount of a pollutant that a water body can receive and still meet water-quality standards, and an allocation of that amount to the pollutant's sources (http://www.epa.gov/owow/tmdl/). Water-quality standards are set by states, territories, and tribes. They identify the uses for each water body, e.g., drinking-water supply, contact recreation (swimming), and aquatic life support (fishing) as well as the scientific criteria to support that use. A TMDL is the sum of the allowable loads of a single pollutant from all contributing point and nonpoint sources. The calculation must include a margin of safety to ensure that the water body can be used for the purposes the state has designated. The calculation must also account for seasonal variation in water quality. The Clean Water Act, Section 303, establishes the water-quality standards and TMDL programs. GIS plays a key role in the TMDL process, given that it requires a comprehensive assessment of water quality resulting from point and nonpoint sources. The scope of modeling for TMDL includes:

- A wide range of watershed-scale loading models
- Receiving water models, including eutrophication and water-quality models, toxics models, and hydrodynamic models
- Integrated modeling systems that link watershed-scale loading with receiving water processes

In addition to modeling, the TMDL process involves extensive public-outreach activities. These too are aided by Web-based interfaces that provide maps, data, and program information.

10.4.2   ROUGE RIVER CASE STUDY

The Rouge River National Wet Weather Demonstration Project (Kluitenberg et al. 1998; http://www.rougeriver.com/) is an excellent example of the application of GIS and water-quality models for development of water-quality management plans. The Rouge River is located in the Detroit metropolitan region in Michigan. Initiated in 1992, the early focus was on combined sewer overflows (CSOs) as the priority. It was determined that CSO control alone would not provide sufficient improvements to meet water-quality standards in the watershed. This was because nonpoint stormwater runoff, failed septic tank drainage, illicit connections, wetland drainage, and land erosion all needed to be controlled. To address the problems, a holistic watershed approach was adopted that involved a comprehensive sampling and monitoring program, various water-quality and water-quantity modeling tools, data management, and GIS.

The Rouge River water-quality and ecosystem monitoring has involved an extensive effort in the collection, management, and analysis of data on rainfall, stream flow, in-stream water quality, combined sewer overflows, stormwater quality, biological communities and habitat, in-stream bottom sediment, air deposition, and aesthetic conditions. In addition, the monitoring program includes measurement of the performance of various stormwater best management practices (BMPs), including structural controls, wetlands, and nonstructural controls. Since 1993, approximately 17,000 water-quality samples have been analyzed for conventional and priority pollutants and bacteria. The 1996 biological sampling involved a comprehensive assessment of 83 sites along 200 miles of waterway. Bottom-sediment samples at over 180 locations were also analyzed for a variety of parameters. These data were collated into a water-quality database and are displayable using a GIS interface.

The Rouge River modeling effort involved linking three component models of the storm and combined sewer systems: RUNOFF, TRANSPORT, and the Water Quality Analysis Simulation Program (WASP). The RUNOFF block of the Storm Water Management Model (SWMM) simulated the flows and pollutant loads from stormwater and combined sewer subareas (Figure 10.18). The flows simulated by the RUNOFF model are input to the TRANSPORT block of the SWMM model, which simulates the hydraulic response of the Rouge River from precipitation events. Pollutant loads from the RUNOFF model were input to the Water Quality Analysis Simulation Program

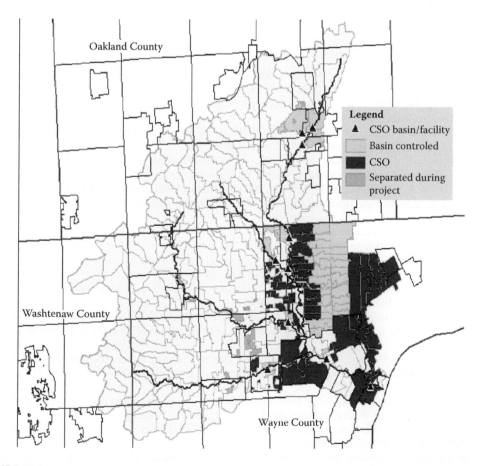

FIGURE 10.18   Rouge River basin showing locations of combined sewer areas. (From RRWWD 1995.)

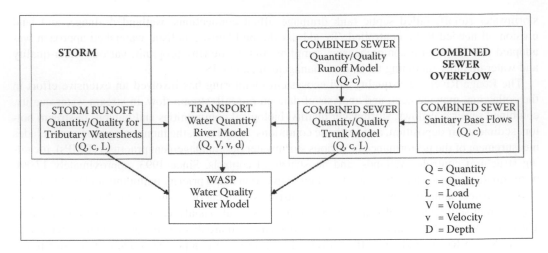

FIGURE 10.19 Linkage of storm runoff, combined sewer overflow, and river-water-quality models for Rouge River system. (From RRWWDP 1995.)

(WASP), which simulates the water-quality response of the river during wet and dry weather conditions. Integration of the models into one comprehensive model system using the Linked Watershed Model (LWM) is illustrated in Figure 10.19.

REFERENCES

Arnold, C. L., and C. J. Gibbons. 1996. Impervious surface coverage: The emergence of a key environmental indicator. *J. Am. Plann. Assoc.* 62 (2): 243–258.

Baxter, R. 2002. Modeling tools for the stormwater manager. *Stormwater* 3 (2). http://www.stormh20.com/sw_0203_modeling.html.

Benaman, J., and J. D. Mathews. 2001. The use of GIS in the development of water quality models. Paper presented at ASCE Water Resources 2000. Minneapolis, Minn.

Bottcher, D., and J. G. Hiscock. 2002. A new GIS approach to watershed assessment modeling. In *Proc. Seventh Biennial Stormwater Research and Watershed Management Conference*, 15–24.

Carleton, C. J., R. A. Dahlgrena, and K. W. Tateb. 2005. A relational database for the monitoring and analysis of watershed hydrologic functions, II: Data manipulation and retrieval programs. *Comput. Geosci.* 31 (4): 403–413.

Chapra, S. C. 1997. *Surface-water quality modeling*. New York: McGraw-Hill.

Chapra, S. C., G. J. Pelletier, and H. Tao. 2007. QUAL2K: A Modeling Framework for Simulating River and Stream Water Quality, Version 2.07: Documentation and User's Manual. U.S. Environmental Protection Agency (http://www.epa.gov/athens/wwgtsc/html/qual2k.html).

Clausen, J. C., G. Warner, D. Civco, and M. Hood. 2003. Nonpoint education for municipal officials impervious surface research: Final report. Connecticut DEP.

Dougherty, M., R. L. Dymond, S. J. Goetz, C. A. Jantz, and N. Goulet. 2004. Evaluation of impervious surface estimates in a rapidly urbanizing watershed. *Photogrammetric Eng. Remote Sensing* 70 (11): 1275–1284.

Driscoll, E. D., P. E. Shelley, and E. W. Strecker. 1990. Pollutant loading and impacts from stormwater runoff. Volume III: Analytical investigation and research report. FHWA-RD88-008. Washington, D.C.: Federal Highway Administration.

EPA (Environmental Protection Agency). 1991. Guidance for water quality-based decisions: The TMDL process. Assessment and Watershed Protection Division, U.S. EPA, Washington, D.C.

EPA (Environmental Protection Agency). 2001a. BASINS 3.0 user's manual. http://www.epa.gov/waterscience/basins/bsnsdocs.html.

EPA (Environmental Protection Agency). 2001b. PLOAD user's manual. http://www.epa.gov/waterscience/basins/bsnsdocs.html.

EPA (Environmental Protection Agency). 2004. EnviroMapper for water tutorial. http://www.epa.gov/waters/enviromapper/em_tutorial.pdf.

Frerichs, R. R. 2001. High-resolution maps of John Snow. Dept. Epidemiology, University California at Los Angeles. http://www.ph.ucla.edu/epi/snow.html.

Gibbons, D. E., G. E. Wukelic, J. P. Leighton, and M. J. Doyle. 1989. Application of Landsat Thematic Mapper data for coastal thermal plume analysis at Diablo Canyon. *Photogrammetric Eng. Remote Sensing* 55: 903–909.

Heller, E. 2007. Using EPA Basins for Best Management Practice Scenarios in the Upper South Platte Basin, Colorado. Masters Project, Dept. of Civil Eng., Univ. Colorado Denver.

Karaska, M. A., R. L. Huguenin, J. L. Beacham, M. H. Wang, J. R. Jensen, and R. S. Kaufmann. 2004. AVIRIS measurements of chlorophyll, suspended minerals, dissolved organic carbon, and turbidity in the Neuse River, North Carolina. *Photogrammetric Eng. Remote Sensing* 70 (1): 125–133.

Kauffman, G. J., M. B. Corrozi, and K. J. Vonck. 2006. Imperviousness: A performance measure of a Delaware Water Resource Protection Area ordinance. *J. Am. Water Resour. Assoc.* 42 (3): 603–615.

Kellogg, R. L., R. Nehring, A. Grube, D. W. Goss, and S. Plotkin. 2000. Environmental indicators of pesticide leaching and runoff from farm fields. In *Proc. Conference on Agricultural Productivity: Data, Methods, and Measures*. Washington, D.C. http://www.nrcs.usda.gov/TECHNICAL/land/pubs/eip_pap.html.

Kluitenberg, E. H., G. W. Mercer, and V. Kaunelis. 1998. Water quality modeling to support the Rouge River restoration. In *Proc. National Conference on Retrofit Opportunities for Water Resources Protection in Urban Environments*.

Lin, J. P. 2004. Review of published export coefficient and event mean concentration (EMC) data. WRAP Technical Notes Collection (ERDC TN-WRAP-04-3). U.S. Army Engineer Research and Development Center, Vicksburg, Miss. http://el.erdc.usace.army.mil/elpubs/pdf/tnwrap04-3.pdf.

Maidment, D. R., ed. 2002. *Arc Hydro: GIS for water resources*. Redlands, Calif.: ESRI.

McKinney, D. C., and C. Patino-Gomez. 2006. Water quality data model in GIS for the Rio Bravo/Grande basin. CRWR online report 06-13. Center for Water Resources, University of Texas at Austin.

Novotny, V. 2003. *Water quality: Diffuse pollution and watershed management*, 2nd ed. New York: John Wiley & Sons.

Oguchi, T., H. P. Jarvie, and C. Neal. 2000. River water quality in the Humber catchment: An introduction using GIS-based mapping and analysis. In *Science of the total environment*. New York: Elsevier.

Parsons, J. E., D. L. Thomas, and R. L. Huffman. 2004. Agricultural non-point source water quality models: Their use and application. Southern Cooperative Series bulletin #398. http://www3.bae.ncsu.edu/Regional-Bulletins/Modeling-Bulletin/.

Ritchie, J. C., F. R. Schiebe, and J. R. McHenry. 1976. Remote sensing of suspended sediment in surface water. *Photogrammetric Eng. Remote Sensing* 42 (2): 1539–1545.

Ritchie, J. C., P. V. Zimba, and J. H. Everitt. 2003. Remote sensing techniques to assess water quality. *Photogrammetric Eng. Remote Sensing* 69 (6): 695–704.

RRWWD. 1995. Rouge River National Wet Weather Demonstration Project (RRWWD). http://www.rougeriver.com/.

Schiebe, F. R., J. A. Harrington, Jr., and J. C. Ritchie. 1992. Remote sensing of suspended sediments: The Lake Chicot, Arkansas, project. *Int. J. Remote Sensing* 13 (8): 1487–1509.

Schueler, T. R. 1994. The importance of imperviousness. *Watershed Protection Tech.* 1 (3): 100–111.

Slonecker, T. E., D. B. Jennings, and D. Garofalo. 2001. Remote sensing of impervious surfaces: A review. *Remote Sensing Rev.* 20: 227–255.

Stefan, H., and F. R. Schiebe. 1970. Heated discharges from flumes into tanks. *J. Sanitary Eng. Div. Am. Soc. Civ. Eng.* 12: 1415–1433.

Thomann, R. V., and J. A. Mueller. 1987. *Principles of surface water quality modeling and control*. New York: Harper and Row.

Wittemore, R. C., and J. Beebe. 2000. EPA'S BASINS model: Good science or serendipitous modeling? *J. Am. Water Resour. Assoc.* 36 (3): 493–499.

Wong, K. M., E. W. Strecker, and M. K. Senstrom. 1997. GISI to estimate storm-water pollutant mass loadings. *ASCE J. Environ. Eng.* 123 (10): 737–745.

Yang, L., C. Huang, C. Homer, B. Wylie, and M. Coan. 2002. An approach for mapping large-area impervious surfaces: Synergistic use of Landsat 7 ETM+ and high spatial resolution imagery. *Can. J. Remote Sensing* 29 (2): 230–240.

11 GIS for Water Resources Monitoring and Forecasting

11.1 INTRODUCTION

Geographic information systems (GISs) are used to support water resources hazards monitoring, forecasting, and warning in a variety of ways. Central to monitoring and warning is the need for data collection in real time when the event is happening and to anticipate future conditions through forecasting. These GISs are driven by advanced data-collection systems that measure the full range of hydrological and related system parameters, including water flow and weather. Telecommunications of data by various modes, in real time, provide up-to-date databases required for timely and accurate control and analyses. GIS-based interactive displays allow water resource systems managers to quickly examine the state of the system, invoke models to forecast possible future conditions, and disseminate warnings.

The intent of this chapter is to summarize technologies for real-time data collection and transmission as well as the monitoring and forecasting systems that utilize these data. Emphasis is on flood warning systems and general concepts of flood warning/preparedness programs, flood-threat recognition, warning dissemination, emergency response, and plan management. The concepts of monitoring and forecasting apply to other resource sectors having less time-critical response factors, such as droughts. GIS concepts and tools play a central role in all aspects of data collection and archiving, support for condition assessment through display and modeling, and message dissemination.

11.2 HYDROLOGIC ASPECTS OF FLOOD WARNING PROGRAMS

A major emphasis of water resources monitoring and forecasting systems is for flood warning and preparedness. Other tasks may apply to longer time frames, such as drought monitoring. Flood warning/preparedness programs (FWPPs) have been a priority for several decades, since recognition that technologies for monitoring and forecasting of floods can support public responses for avoidance and mitigation (USACE 1996). FWPPs are a part of comprehensive flood-damage reduction programs along with other structural and nonstructural measures. These programs reduce the risk to life and, to some extent, reduce the damage potential during flood episodes. FWPPs do not prevent flood disasters, but they do enable institutional and public emergency response actions to be conducted more effectively. Many institutional arrangements required for implementing flood warning/preparedness programs are similar to those for other disasters.

Hydrologic analyses are important in evaluating the feasibility of implementing FWPPs and to ensure their validity of operation on a real-time basis during flood events. Information presented in Chapter 9 (GIS for Floodplain Management) describes how base conditions are determined on the nature of the flood hazard for a range of events (magnitude, frequency, inundation boundaries, velocities, depths, and warning times). These analyses may result in impacts to threatened properties and vital services. Enhanced condition analyses include arrangements, equipment, hardware/software, and actions that yield better responses. The analyses may include enhanced and more-reliable warnings, better warning dissemination, and response actions that can reduce the number of casualties or deaths and the threat of property damage.

The monitoring component of a modern flood-threat recognition system can range from a collection of automated precipitation and stream gauges (e.g., Automated Local Evaluation in Real Time [ALERT] systems) to more advanced systems that utilize weather radars, satellites, and sophisticated hydrometeorological numerical models of the atmosphere and watersheds. The objectives of the measurement and detection task are to monitor developing hydrometeorological and watershed conditions and project what might develop in the near term. Reliable data transmissions from the sensors to a central location for display and assessment are essential for these systems, as are transmissions of the assessments and warnings to response authorities and citizens.

Flood-threat recognition systems may include flood forecasting using technology appropriate to the situation as well as the capabilities for assessment and operational support. The value of the forecast is derived from the additional lead time made available for the response effort to reduce impacts. There are trade-offs between the accuracy and the timeliness of advanced warnings; typically, rapid assessments are less accurate, but extended analyses can lengthen the time for threat recognition and response. Warning dissemination is an additional key element in order to get messages to public safety authorities in a timely manner and in understandable formats. Advanced warnings can be used to motivate evacuations and guide deployment of assistance. With longer lead times, more aggressive actions may be accomplished, such as drawing down reservoir levels to increase capture capacities, moving vehicles and materials from low-lying areas, or initiating flood-fighting efforts.

Mathematically, warning time can be described as (USACE 1996)

$$T_W = T_{WP} - T_R \tag{11.1}$$

where

T_W = actual warning time
T_{WP} = maximum potential warning time
T_R = flood threat recognition/reaction time

The maximum potential warning time (T_{WP}) is defined as the time from the beginning of the storm event (or its anticipation) to the time that the stream reaches flood stage. The flood recognition and reaction time, T_R, includes the time required to observe or measure developing conditions. T_R is the sum of the times for several component tasks: T_{obs}, the time needed to acquire the necessary observational data; T_{data}, the time required to analyze the storm data and determine (recognize) that a flood is likely to occur; T_{fp}, the time to prepare a forecast; and the time required to disseminate the flood warning, T_{dis}. The flood recognition and reaction time can be expressed as

$$T_R = T_{obs} + T_{data} + T_{fp} + T_{dis} \tag{11.2}$$

Modern hydrometeorological observation and forecasting systems are directed to increasing the maximum potential lead time (T_{WP}) for anticipating a hazardous event. For example, the National Weather Service (NWS) weather forecast offices (WFOs) have been equipped with new-generation weather radars (NEXRAD) that provide rapid updates on storm rain-drop reflectivity and rainfall intensity. These and associated WFO observational technologies have enhanced forecasters' capabilities to observe and anticipate movements of severe rainfall events. Hydrometeorological observation and forecasting systems are continuing to advance by incorporating numerical models on storm dynamics and movement. The National Oceanic and Atmospheric Administration's (NOAA) Hydrometeorological Testbed (HMT) (http://www.esrl.noaa.gov/research/programs/) seeks to demonstrate the use of advanced observational and modeling tools on quantitative precipitation estimation (QPE) and quantitative precipitation forecasting (QPF) to improve hydrological forecasts and warnings. The HMT is described as a case study below.

During pre-event planning, GIS-based databases incorporating flood inundation maps, aerial photographs, and field surveys are used to identify locations where existing properties are threatened by various levels of flooding. Existing damageable structures should be categorized by type and number for each flood event throughout the range of the flood-frequency relationship. Frequency-discharge–elevation-damage relationships are presented in tabular form. The information is used to help in the development of warning and evacuation plans, location of mass-care centers, management of vital services, and to estimate the number of structures and people impacted by an event. It is also required to define actions and facilities for implementation of temporary flood-loss-reduction actions such as flood fighting, installation of temporary barriers, removal of or raising building contents, and relocation of potential hazardous materials. The threatened-properties analysis helps refine response measures that might be dictated due to the warning time available. The location of structures in the floodplain can determine whether more or less warning time is available.

11.3 WATER RESOURCES MONITORING SYSTEMS

11.3.1 Real-Time Data-Collection System Technologies

Methods and systems for real-time data collection have been strongly influenced by the proliferation of low-cost microprocessor data-collection platforms (DCPs). Community flash-flood warning systems, statewide water-management information systems, and river-reservoir monitoring and operations systems are increasingly common due to improved capabilities of these DCPs, real-time reporting modes, and telecommunications techniques. A variety of telecommunications modes are used, including telephone and other dedicated landlines, Internet, radio, satellite, and meteorburst (i.e., VHF radio signals that are reflected at a steep angle off the ever-present band of ionized meteorites existing from about 50 to 75 miles above the Earth). Hardware and software components of these systems are continually being refined to realize lower error and faster transmission rates.

Remote-sensing technologies play a primary role in water resources monitoring. Satellites provide visual and near-visual infrared (IR) imagery of the Earth's surface and atmospheric reflectance. These images provide the basis for tracking storm systems as well as mapping crop types and status and flood extent. Communications satellites also provide the means for transmitting DCP data from the sensors to a central receiving station. Remote sensing also includes radar technologies. The NEXRAD (next-generation radar) program of the National Weather Service comprises a nationwide network of Doppler radars that monitor storm dynamics.

Global positioning system (GPS) technology has also impacted the collection of data for water resources monitoring. GPSs have wide application for real-time monitoring where personnel and equipment can be deployed to the field and their locations established. GPS signal processing has also been used to determine the state of the atmospheric water vapor. NOAA uses existing GPS sites to receive measurements for the total amount of water vapor above a GPS antenna. Water vapor refracts radio waves, including GPS signals. When this happens, the apparent distance between a GPS satellite and a receiver on the ground is a little longer than it would be if the air were completely dry. Differences between the GPS signals at different times are correlated with the atmospheric water vapor, and a large GPS water-vapor-sensing system has been developed (http://www.esrl.noaa.gov/media/2007/gps/).

11.3.2 Automated Local Evaluation in Real Time (ALERT)

Real-time rain-gauge and stream-gauge data provide the foundation of metropolitan flood warning systems. The NWS (1997) has described automated local flood warning systems (LFWS). A popular LWFS known as ALERT (Automated Local Evaluation in Real Time) was originally developed by the National Weather Service in Sacramento, California. ALERT gauges transmit via wireless radio signals to a base station located at the agency headquarters (and other similar base stations). Through this system, real-time rainfall and stream stages are monitored remotely for an entire metro region. In addition to real-time data, the base station software also archives all data so that

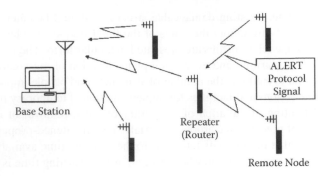

FIGURE 11.1   ALERT-2 network. (After Salo 2007. With permission.)

historic records can be analyzed. Precipitation is observed in a standpipe tower by a 12-in.-diameter funnel. All weather stations measure precipitation, air temperature, relative humidity, and wind speed/direction, and some stations also measure barometric (station) pressure. Stream flow and water height are observed by submersible pressure transducers and digital shaft encoders.

ALERT systems are one-way, event-based environmental sensing networks (Figure 11.1). DIAD Corp. (2000) provided a summary description. Each data-collection platform (DCP) is programmed to transmit a brief data burst when triggered by environmental changes (for example, after receiving 1 mm of rainfall or recording a 1-mm change in stream depth). A typical ALERT "event" is the tip of a tipping bucket signaling the accumulation of 1 mm of rain. The DCP sends a 4-byte message using frequency-shift keying modulation at 300 baud. Software at the receiving site identifies the transmitted ID and decodes the data value into appropriate engineering units using stored information about the sensor. Some limitations of ALERT stem from its event-driven nature. Two DCPs may transmit at nearly the same time on the same radio channel such that the two transmissions partly or completely overlap. This can result in the loss of one or both data transmissions. For rain, this loss of data is tolerable, because each transmission encodes an accumulator value, and the base station compares this value to the last successfully received value. Therefore, a missed rain report usually causes no inaccuracy in rainfall totals, but only a loss of information about its temporal distribution.

GIS tools were implemented to enhance the Denver region Urban Drainage and Flood Control District (UDFCD) ALERT system (http://alert.udfcd.org/). The UDFCD Flood Detection Network contains 149 ALERT gauges that provide 135 real-time rain measurements, 68 real-time stream-water or reservoir-water level measurements, and 17 real-time weather stations that measure atmospheric variables. Precipitation is measured using standard 1-mm ALERT tipping buckets. USGS digital line graph (DLG) files were obtained for roads, lakes, and streams for the Denver metro area. These were used as the basis for the background displays for the rainfall and stream-flow data. The individual rain- and stream-flow gauges were located by map coordinates and generated as a point coverage in ArcInfo®. Attributes were added to the point coverages for additional information, such as ID number and name and location of the gauges. Finally, drainage basins were digitized as polygons from drainage maps for later analysis.

A Web deployment was originally developed using ArcView, with customization with Avenue® scripts (Rindahl 1996). A script was developed to query the ALERT base station computer for real-time (or historic) data. Time steps (1 hour, 1 day, etc.) are saved as global variables from a menu item in ArcView. A button then executes a script to request the selected information from the ALERT base station via a remote shell command. The rainfall data are returned in the form of an ASCI text table to the machine running display. The script then joins the text table to the rainfall point coverage using the ID field (included in the ALERT table). ArcView then displays the rainfall on the active View using the Auto-Label feature. This entire process takes approximately 20 seconds, with most of the time taken by the ALERT base station to process the request. Historic data can be displayed in an identical fashion. More recently, the Web deployment interface is being migrated to open-source

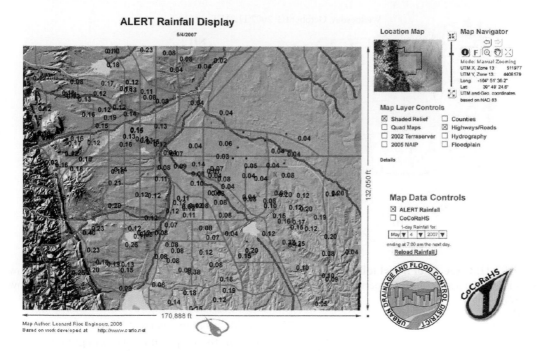

FIGURE 11.2 ALERT rain-gauge data for Urban Drainage and Flood Control District in Denver, CO. (*Source:* http://alert.udfcd.org/.)

coding using scalable vector graphics (SVG) (http://www.carto.net/). Figure 11.2 illustrates the Web interface for the UDFCD ALERT rain-gauge data. SVG allows representation of all graphical elements producible by graphics and cartographic software along with high interaction possibilities and animation, all based on open and standardized file formats and programming languages.

11.3.3 RAINFALL MONITORING

Considerable efforts have been directed to adjusting the radar-rainfall estimates provided by the NEXRAD radars. In general, the procedures developed involve some correlation of the reflectivity-based estimates with ground-based rain gauges. This topic was discussed in more detail in Chapter 5 (GIS for Surface-Water Hydrology). Seo et al. (1999) summarized the background research and described the National Weather Service (NWS) procedures that estimate mean field bias in real time. To reduce systematic errors in radar-rainfall data due to lack of radar calibration and inaccurate $Z–R$ relationships (i.e., the multiplicative constant), precipitation estimation by the NWS (Fulton et al. 1998) uses procedures that estimate radar umbrella-wide biases in radar-rainfall data in real time (Seo and Breidenbach 2002). The estimated "mean field bias" is then applied to the entire radar umbrella to produce "bias-adjusted" radar-rainfall data.

11.3.4 USGS HYDROLOGICAL MONITORING

The U.S. Geological Survey (USGS) Water Resources Division maintains a nationwide network of hydrological monitoring stations that collect data on stream flows, water quality, and other data. The online National Water Information System (NWIS) provides the public with access to more than 100 years of water data collected by the USGS. The Web site (http://waterdata.usgs.gov/nwis/) allows users to access several hundred million pieces of historical and real-time data. For example, Figure 11.3 shows a typical retrieval; the colored dots on the map depict current stream-flow conditions as a percentile of the period of record for the current day. A percentile between 25 and 75 is

FIGURE 11.3 (See color insert following page 136.)   NWIS Web retrieval portrays river-gauge daily flow conditions as a percentile of the period of record flows for the selected day. (*Source:* http://waterdata.usgs.gov/nwis/.)

considered normal, and a value greater than 75 is considered above normal. Only stations with at least 30 years of record are used. For the example shown, there appears to be drought conditions along the east coast of the United States.

The NWIS Automated Data Processing System (ADAPS) contains more than 850,000 station years of time-series data that describe stream-water levels, stream flow (discharge), reservoir water levels, surface-water quality, groundwater levels, and rainfall. ADAPS consists of a collection of computer programs and databases. The water data stored in ADAPS result from the processing of data collected by automated recorders and by observations and manual measurements at field installations around the nation. The data from these sites are transported by field personnel or are relayed through telephones or satellites to offices where USGS personnel, using ADAPS procedures, process the data. The data relayed through the Geostationary Operational Environmental Satellite (GOES) system are processed automatically in near-real time, and in many cases are available within minutes on the USGS Web pages.

11.3.5   PRISM

PRISM (Parameter-elevation Regressions on Independent Slopes Model) is an expert system that uses point data and a digital elevation model (DEM) to generate gridded estimates of climate parameters (Daly et al 1994; http://www.prism.oregonstate.edu/). Although PRISM was originally developed for precipitation mapping, it was quickly recognized that the model philosophy, i.e., that the topographic facet is an important climatic unit and that elevation is a primary driver of climate patterns, could be extended to other climate parameters. PRISM has since been used to map temperature, snowfall, weather generator statistics, and others. PRISM was developed to overcome the deficiencies of standard spatial interpolation methods, where orographic effects strongly influence weather patterns.

In operation, PRISM gives the user the option to use the DEM to estimate the elevations of precipitation stations at the proper orographic scale, and to use the DEM and a windowing technique to group stations onto individual topographic facets. For each DEM grid cell, PRISM develops a weighted precipitation/elevation (P/E) regression function from nearby stations, and predicts precipitation at the cell's DEM elevation with this function. In the regression, greater weight is given to stations with locations, elevations, and topographic positionings similar to that of the grid cell. Whenever possible, PRISM calculates a prediction interval for the estimate, which is an approximation of the uncertainty involved.

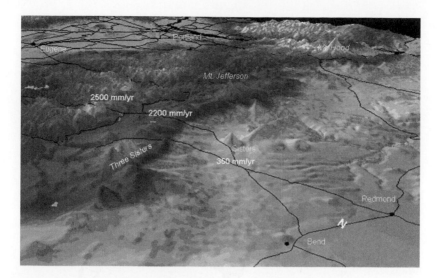

FIGURE 11.4 (See color insert following page 136.) Three-dimensional view of the Oregon Cascades rain shadow. Mean annual precipitation drops from 2200 mm/year at the crest of the Cascades to only 350 mm/year just down the hill to the east. (From Daly et al. 1994. With permission.)

PRISM results have been compared with results for kriging, detrended kriging, and cokriging in the Willamette River basin, Oregon (Daly et al. 1994). In a jackknife cross-validation exercise, PRISM exhibited lower overall bias and mean absolute error. PRISM was also applied to northern Oregon and to the entire western United States. Detrended kriging and cokriging could not be used in these regions because there was no overall relationship between elevation and precipitation. PRISM's cross-validation bias and absolute error in northern Oregon increased a small to moderate amount compared with those in the Willamette River basin; errors in the western United States showed little further increase. PRISM has since been applied to the entire United States with excellent results, even in regions where orographic processes do not dominate precipitation patterns. By relying on many localized, facet-specific P/E relationships rather than a single domain-wide relationship, PRISM continually adjusts its frame of reference to accommodate local and regional changes in orographic regime with minimal loss of predictive capability. Figure 11.4 shows a mapping of the rain shadow effect in Oregon.

Applications of PRISM include regional isohyetal analysis and hydrologic modeling and forecasting. For isohyetal analysis, the ability of PRISM to maintain predictive accuracy over large areas while still responding to local P/E relationships makes it extremely useful for developing isohyetal maps for states or regions. PRISM is currently being used to develop state-of-the-art isohyetal maps of monthly and annual precipitation for all 50 states as part of a National Resources Conservation Service project. For hydrological modeling, PRISM can provide time series of spatially distributed precipitation for a variety of watersheds and drainages at time intervals ranging from years to single events, serving as high-quality input to water-supply forecast models and decision-making activities involving flood control. The Oregon Department of Water Resources has recently abandoned its previous precipitation fields in favor of PRISM estimates for water-supply forecasting. PRISM is also a useful framework for examining the effects of gauge undercatch on runoff prediction. Also, the NWS River Forecast Centers use PRISM (or its alternative name, Mountain Mapper) to check rain-gauge readings and to determine mean areal precipitation (MAP) for watershed models (Daly 2004, 2007).

11.3.6 DROUGHT MONITORING

Drought is a recurring, slow-onset natural phenomenon that is unlike other natural hazards such as hurricanes, floods, and tornadoes. There are various activities to accomplish drought monitoring in the United States and worldwide. One program, the National Integrated Drought Information Systems

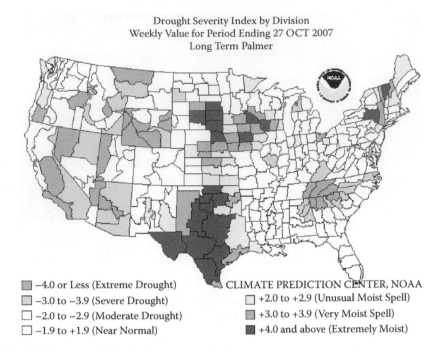

FIGURE 11.5 Palmer Drought Severity Index Map. (*Source:* http://www.cpc.ncep.noaa.gov/products/monitoring_and_data/drought.shtml.)

(NIDIS; http://www.drought.gov/), is being implemented to help proactively manage drought-related risks by providing those affected with the best available information and tools to assess the potential impacts of drought, and to better prepare for and mitigate the effects of drought. The NIDIS Web portal provides access to various drought-monitoring and forecast products, including the U.S. Drought Monitor. Indicators based on climatology enable people to describe drought in a consistent way across different times and places. These indicators combine variables such as precipitation and temperature.

The Climate Prediction Center of NOAA provides a similar functionality for monitoring drought conditions (http://www.cpc.noaa.gov/products/forecasts/). The Climate Prediction Center (CPC) is responsible for issuing seasonal climate outlook maps for 1 to 13 months in the future. In addition, the CPC issues extended-range outlook maps for 6–10 and 8–14 days as well as several special outlooks, such as degree day, drought and soil moisture, and a forecast for daily ultraviolet (UV) radiation index. Many of the outlook maps have an accompanying technical discussion. The CPC's outlook and forecast products complement the short-range weather forecasts issued by other components of the National Weather Service (e.g., local weather forecast offices and National Centers for Environmental Prediction).

As an example, the Palmer Drought Severity Index (PDSI) (Figure 11.5) and Crop Moisture Index (CMI) are indices of the relative dryness or wetness affecting water-sensitive economies. The Palmer Index uses temperature and rainfall information in a formula to determine dryness. It has become the semiofficial drought index. The Palmer Index is most effective in determining long-term drought—a matter of several months—and is not as good with short-term forecasts (a matter of weeks). It uses a 0 as normal, and drought is shown in terms of minus numbers; for example, –2 is moderate drought, –3 is severe drought, and –4 is extreme drought. Tools used in the Drought Outlook (Figure 11.6) include the official CPC long-lead precipitation outlook for November–January 2007/08, the four-month drought termination and amelioration probabilities, various medium- and short-range forecasts, and models such as the 6–10-day and 8–14-day forecasts, the soil-moisture tools based on the Global Forecast System model and the Constructed Analogue on Soil Moisture, and the Climate Forecast System (CFS) monthly precipitation forecasts.

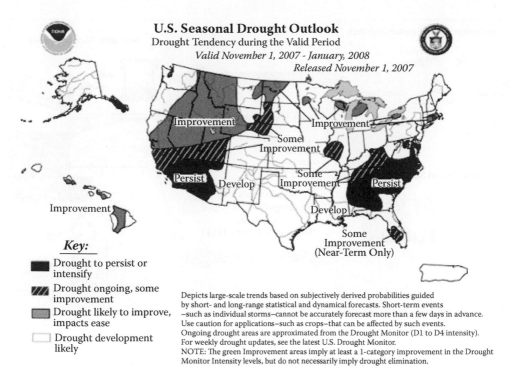

U.S. Seasonal Drought Outlook
Drought Tendency during the Valid Period
Valid November 1, 2007 - January, 2008
Released November 1, 2007

Key:

- Drought to persist or intensify
- Drought ongoing, some improvement
- Drought likely to improve, impacts ease
- Drought development likely

Depicts large-scale trends based on subjectively derived probabilities guided by short- and long-range statistical and dynamical forecasts. Short-term events –such as individual storms–cannot be accurately forecast more than a few days in advance. Use caution for applications–such as crops–that can be affected by such events. Ongoing drought areas are approximated from the Drought Monitor (D1 to D4 intensity). For weekly drought updates, see the latest U.S. Drought Monitor.
NOTE: The green Improvement areas imply at least a 1-category improvement in the Drought Monitor Intensity levels, but do not necessarily imply drought elimination.

FIGURE 11.6 Drought outlook is generated from global climate models and other sources. (*Source:* http://www.cpc.noaa.gov/products/expert_assessment/seasonal_drought.html.)

11.3.7 SENSOR NETWORKS

Recent advances in the technology of sensor networks are expected to lead to new forms of monitoring networks. A sensor network is a computer-accessible network of many spatially distributed devices using sensors to monitor conditions at different locations, such as temperature, hydrostatic pressure, or pollutants. A *sensor Web* refers to Web-accessible sensor networks and archived sensor data that can be discovered and accessed using standard protocols and application program interfaces (APIs) (Botts et al. 2006). The Open Geospatial Consortium, Inc. (http://www.opengeospatial.org) has an initiative called Sensory Web Enablement (SWE) that will provide a framework of open standards for exploiting Web-connected sensors and sensor systems of all types: flood gauges, air pollution monitors, Webcams, satellite-borne Earth-imaging devices, and countless other sensors and sensor systems. SWE presents many opportunities for adding a real-time sensor dimension to the Internet and the Web. In much the same way that hypertext markup language (HTML) and hypertext transfer protocol (HTTP) standards enabled the exchange of any type of information on the Web, the OGC's SWE initiative is focused on developing standards to enable the discovery, exchange, and processing of sensor observations, as well as the tasking of sensor systems. The functionality that OGC has targeted within a sensor Web includes:

- Discovery of sensor systems, observations, and observation processes that meet an application's or a user's immediate needs
- Determination of a sensor's capabilities and quality of measurements
- Access to sensor parameters that automatically allow software to process and geolocate observations
- Retrieval of real-time or time-series observations and coverages in standard encodings
- Tasking of sensors to acquire observations of interest
- Subscription to and publishing of alerts to be issued by sensors or sensor services based upon certain criteria

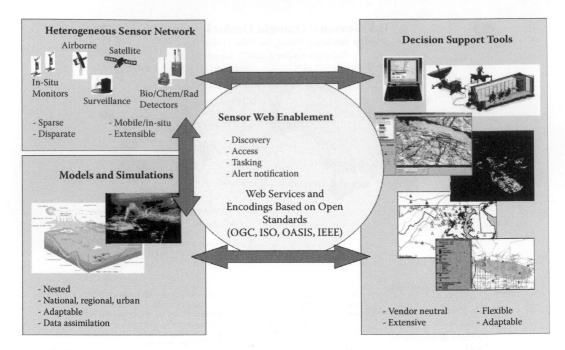

FIGURE 11.7   Sensor Web enablement concept. (From Botts et al. 2006. With permission.)

The goal of SWE is to enable all types of Web- and/or Internet-accessible sensors, instruments, and imaging devices to be accessible and, where applicable, controllable via the Web (Figure 11.7). The vision is to define and approve the standards foundation for "plug-and-play" Web-based sensor networks. Advances in digital technology are making it practical to enable virtually any type of sensor or locally networked sensor system with wired or wireless connections. Such connections support remote access to the devices' control inputs and data outputs as well as their identification and location information. For both fixed and mobile sensors, sensor location is often a vital sensor parameter. A variety of location technologies such as GPS and Cell-ID with triangulation make mobile sensing devices capable of reporting their geographic location along with their sensor-collected data. When the network connection is layered with Internet and Web protocols, eXtensible Markup Language (XML) schemas can be used to publish formal descriptions of the sensor's capabilities, location, and interfaces. Then Web brokers, clients, and servers can parse and interpret the XML data, enabling automated Web-based discovery of the existence of sensors and evaluation of their characteristics based on their published descriptions. The information provided also enables applications to geolocate and process sensor data without requiring a priori knowledge of the sensor system.

11.4 HYDROLOGICAL FORECASTING SYSTEMS

11.4.1 Hydrological Forecasting

Forecasting of stream flows and hydrometeorological conditions leading to floods incorporates real-time monitoring data, archival databases, and linked models. Models used for river forecasting include the types described in Chapter 5 (GIS for Surface-Water Hydrology). GIS tools are used to develop and manage the watershed spatial data supportive of the hydrology models (e.g., DEMs, hydrography, gauge locations and readings). GIS spatial data-management and -modeling tools are used extensively in processing and displaying hydrometeorological data as well. Several case study examples follow that illustrate these capabilities.

11.4.2 NWS River Forecast Centers

The U.S. National Weather Service (NWS) is charged with providing weather forecasts for the nation. For hydrologic forecasts and related services, the NWS is organized by major river basins through its River Forecast Centers (RFCs). There is significant use of GIS at each of the RFCs. The Arkansas-Red Basin RFC (ABRFC) located in the central United States provides an excellent example of the use of GIS for river basin monitoring and flood forecasting (http://www.srh.noaa.gov/abrfc/).

The ABRFC's primary operational duties include hydrologic forecasting, hydrometeorologic analysis and support (HAS) functions, and the monitoring/quality control of associated data sets that are input to and output from operational computer models. The hydrologic forecaster is responsible for the daily production of river forecasts, flash-flood guidance (FFG), data summary products, execution of the river forecast computer model, and coordination of river forecasts as required. The HAS forecaster is responsible for preparing the observed and forecast precipitation and temperature input for the river forecast model, the hydrometeorologic-discussion product, and the coordination with and support of the hydrologic forecaster and the NWS Forecast Offices (NWSFOs). Other operational functions are performed on a seasonal or as-needed basis. These functions include production of water supply forecasts, flood outlooks, and drought summaries.

Each morning after 12Z (6 a.m. or 7 a.m.), the ABRFC issues river-stage forecasts for approximately two dozen locations along the main-stem Arkansas and Red Rivers and their tributaries. These forecasts are used as guidance by NWS weather forecast offices to support their hydrologic mission. The ABRFC also issues river-stage forecasts for other locations, depending upon expected or ongoing flooding. The hydrometeorologist on duty issues a hydrometeorological discussion (HMD) concerning the precipitation that has occurred and is forecast to occur, the state of the rivers and streams in the ABRFC, and river-flood and flash-flood potential outlooks for the next 24 hours. These forecasts and discussions may be updated throughout the day as conditions warrant.

When various points along the rivers within the ABRFC are forecast to rise near or exceed the defined flood stage, the ABRFC issues forecast guidance for these points (Figure 11.8). These forecasts include a brief description of when the river will rise above flood stage, when and what the crest will be, and when the river is forecast to fall below flood stage. At the ABRFC Web site, the current and forecast flow conditions can be determined by clicking on the site of interest.

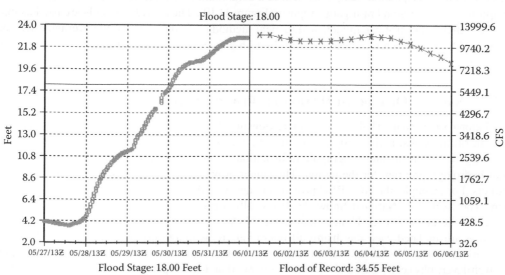

FIGURE 11.8 Reported and forecast flood stage product. (*Source:* http://www.srh.noaa.gov/abrfc/.)

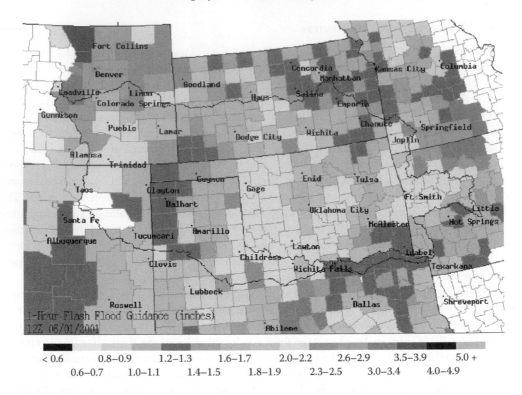

1-Hour Flash Flood Guidance (inches)
12Z 06/01/2001

| < 0.6 | 0.8–0.9 | 1.2–1.3 | 1.6–1.7 | 2.0–2.2 | 2.6–2.9 | 3.5–3.9 | 5.0 + |
| 0.6–0.7 | 1.0–1.1 | 1.4–1.5 | 1.8–1.9 | 2.3–2.5 | 3.0–3.4 | 4.0–4.9 | |

FIGURE 11.9 (See color insert following page 136.) Flash-flood guidance product. (*Source:* http://www. srh.noaa.gov/abrfc/.)

The ABRFC routinely issues a flash-flood guidance (Figure 11.9) three times daily (00UTC, 12UTC, and 18UTC). These values are estimated precipitation amounts and durations that could cause flash flooding for each county. The NWS weather forecast offices use this guidance when issuing flash-flood warnings to the public.

Every hour (usually between half past and the top of the hour), the HAS forecaster creates a gridded (4 × 4 km) precipitation field. This field is a combination of both WSR-88D NEXRAD radar precipitation estimates and rain-gauge reports (Figure 11.10). These gridded fields are used as input into the ABRFC's hydrologic model. A statistical method is used to remove the average difference (mean field bias) between radar estimates at the rain-gauge locations and the corresponding gauge rainfall amounts.

11.4.3 National Operational Hydrologic Remote-Sensing Center

The National Operational Hydrologic Remote-Sensing Center (NOHRSC) (http://www.nohrsc.nws. gov/), in Minneapolis, is the National Weather Service (NWS) center of expertise in satellite and airborne remote-sensing and geographic information systems (GIS) used to support NWS's operational hydrology program for the nation. The center provides remotely sensed and GIS hydrology products. For example, the NOHRSC provides imagery of the extent of snow cover in the ABRFC basin. The NOHRSC National Snow Analyses (NSA) provide daily comprehensive snow information for the conterminous United States. The NSA are based on modeled snowpack characteristics that are updated each day using all operationally available ground, airborne, and satellite observations of snow water equivalent, snow depth, and snow cover. The NOHRSC snow model is a multilayer, physically based snow model operated at 1-km² spatial resolution and hourly temporal resolution for the nation. Snow data used to update the model include observations from the NOHRSC's Airborne Snow Survey Program, NWS and FAA field offices, NWS cooperative

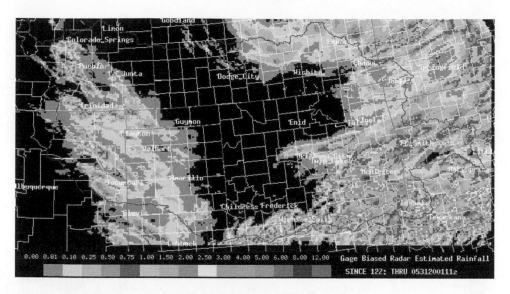

FIGURE 11.10 (See color insert following page 136.) NEXRAD radar-rainfall display. (*Source:* http://www.srh.noaa.gov/abrfc/.)

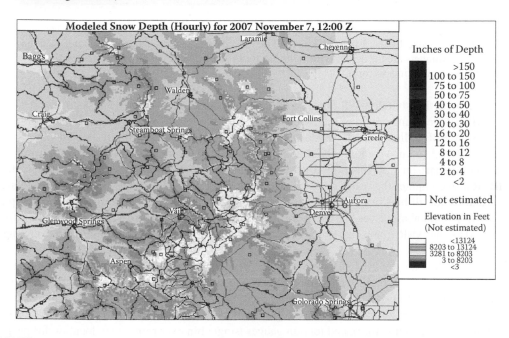

FIGURE 11.11 Snow-depth product from National Operational Hydrologic Remote-Sensing Center. (*Source:* http://www.nohrsc.noaa.gov/.)

observers, the NRCS SNOTEL and snow course networks, the California Department of Water Resources snow pillow networks, and snow cover observations from NOAA's GOES and AVHRR satellites. Figure 11.11 shows a snow-depth product from NOHRSC.

11.4.4 NWS Areal Mean Basin Effective Rainfall

The National Weather Service has developed various methods for assessing the threat of flash floods for local forecast regions. The AMBER (areal mean basin effective rainfall) algorithm was

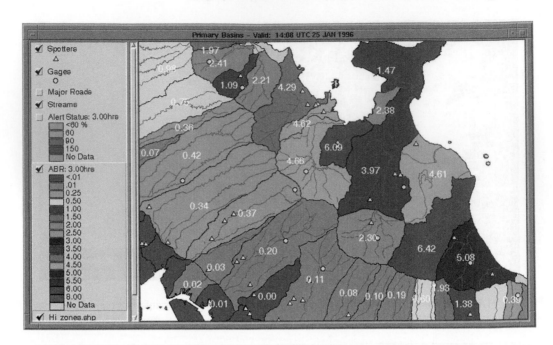

FIGURE 11.12 (See color insert following page 136.) Sample ArcView GIS AMBER basin display. Basins are color coded by the 3-h ABR value for the basin. Streams and gauge and spotter locations also are displayed. Numbers are the actual ABR in inches for the basin. (*Source:* http://www.erh.noaa.gov/er/rnk/amber/amberov/ambrerindex.html.)

developed at the Pittsburgh NWS Forecast Office in the early 1990s (Davis and Jendrowski 1996). The AMBER program provides the field forecaster direct guidance for issuance of flash-flood warnings. AMBER directly links WSR-88D radar-rainfall estimates with all defined watersheds, down to a 2-mi.^{2} (3 km^2) area. AMBER computes average basin rainfall (ABR) in each watershed for direct comparison with flash-flood guidance (FFG) or other thresholds set by the forecasters. AMBER also computes the "rate" or intensity of each watershed ABR every 5–6 min. Scan-to-scan accumulation for all bins in a basin is summed using area of bin as a weighting factor to compute the ABR. The database saves the ABR for each basin for each scan-to-scan accumulation time period. Also, the ABR from scan-to-scan periods is summed to produce accumulation over longer time periods.

Basins delineations for AMBER were derived nationwide using GIS terrain-processing procedures (see Chapter 5). A basin was defined for each segment of a stream network, and basins were defined for various thresholds of basin scale, including: (a) headwaters (e.g., less than 50 mi.^{2}), (b) streams (less than 200 mi.^{2}), (c) rivers (greater than 200 mi.^{2}), (d) urban areas (i.e., any area known to be prone to flooding), and (e) rain gauges (single bin over rain-gauge location for gauge–radar comparison). Figure 11.12 shows an example AMBER display. As implemented for flash-flood monitoring, the interface provides access to the various levels of basin scale. It can be difficult to monitor 5,000 to 10,000 small basins, so only those basins that exceed rainfall and rainfall thresholds can be displayed. Also, a database on results for all basins can be sorted, and the ABR, FFG, alert status, and basin rate of accumulation (BRA) output for each basin can be obtained for each alert time period.

11.4.5 NEXRAD FLOOD WARNING

The NEXRAD rainfall products provide opportunity for input to hydrologic models to assess the threat of flooding. The AMBER case study described above is one example of this, although a higher level of sophistication of the watershed modeling is being demonstrated (e.g., Smith et al. 2004). Given the relatively rapid update cycle of the radar (≈5–10 min), flash-flood threats can be assessed; these are events that occur in short time frames, on the order of less than 1 h.

One example of interfacing a distributed hydrologic model with the NEXRAD rainfall products was demonstrated by Skahill and Johnson (2000) for the Buffalo Creek watershed near Denver, Colorado. The two-dimensional (2-D) flood-runoff model, F2D, was applied to simulate a flash flood that occurred in 1995, killing two people. F2D is an event-based, kinematic, infiltration-excess, distributed rainfall-runoff model developed to account for spatial variability of watershed rainfall, abstractions, and runoff processes. The F2D rainfall-runoff model operates on a square grid of specified spatial resolution; typically 200–800 m. The main model outputs include a volume summary, discharge hydrographs for interior locations as well as for the main basin outlet, and raster maps of diverse variables, such as cumulative infiltration and water depth throughout the basin. Figure 11.13 illustrates the types of output from the model.

Radar-rainfall remote-sensing imagery was the primary dynamic input to the AMBER and F2D models. The Mile-High Radar (MHR) was the source for the rainfall estimates; MHR was a NEXRAD prototype radar with 0.95° beam width and 225-m gate spacing. The temporal resolution of the radar reflectivity data was approximately 6 min. The Z–R power-law relationship: $Z = 500R^{1.3}$, where Z and R represent the radar reflectivity (in mm^6/m^3) and rainfall rate (in mm/h), respectively, was used to transform reflectivity into rainfall rate. This Z–R relation has been used with reasonable success for the summertime climate of the Colorado Front Range (Smith and Lipschutz 1990). A reflectivity threshold of 53 dBZ was applied before converting to a rainfall rate. A bias-correction

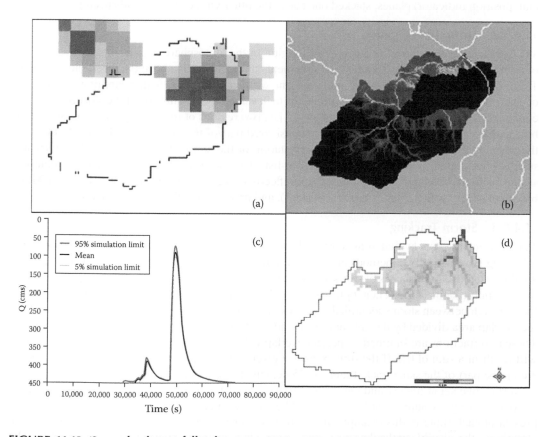

FIGURE 11.13 (See color insert following page 136.) F2D model for Buffalo Creek: (a) radar-rainfall input, (b) cumulative infiltration, (c) runoff hydrograph, and (d) runoff flow distribution. (From Skahill and Johnson 2000.)

factor was computed for the storm event based on the ratio of the precipitation measured by the rain gauges to the precipitation measured by the radar. Rain-gauge-adjusted radar-rainfall estimates were subsequently determined by multiplying the original radar-rainfall estimates by the correction factor. Bias corrections similar to those found by Fulton (1999) were found; 63% to 25% reductions are possible with bias correction (generally overestimates).

11.4.6   NCAR TITAN

The rainfall products from the NEXRAD generation weather radars also provide the basis for forecasting storm movement and growth. The Thunderstorm Identification, Tracking and Analysis (TITAN) system, developed at the National Center for Atmospheric Research (NCAR), identifies storms in three-dimensional or two-dimensional (3-D or 2-D) Cartesian radar data. Dixon and Weiner (1993) provided the initial detailed description of the TITAN procedures. Since that time, TITAN has been implemented at a number of sites and applications. TITAN is public domain and is available for download and installation by interested users (http://www.ral.ucar.edu/projects/titan/).

11.4.6.1   Storm Radar Data

The data are stored in radar "volumes," where a volume refers to one 3-D scan of the sky surrounding the radar. Radars scan in modified polar coordinates, therefore the data are converted into Cartesian coordinates before the storm identification step is performed. These Cartesian volumes can be thought of as a series of horizontal slices through the storms, or CAPPI (constant-altitude plan position indicator) planes, stacked one above the other with equal vertical spacing.

11.4.6.2   Storm Identification

TITAN identifies a "storm" as a contiguous region in the atmosphere with a reflectivity in excess of a set threshold. First, horizontal (E–W) segments in the data grid above the threshold are found. Then, overlaps between a segment and adjacent segments are identified. Searching for overlaps occurs in both the N–S and up–down directions to develop the 3-D analysis of the storms. Although configurable, the overlap is generally set to "1," i.e., two regions of reflectivity only need to overlap by a single grid cell in order for both to be considered part of the same storm. Figure 11.14 shows the storm-cell ellipses as a simplified representation of the storm shape; an alternative format is storms represented as polygons. An extensive list of properties is computed for each identified storm, including: volumetric centroid (x,y,z), reflectivity-weighted centroid (x,y,z), top (km MSL), base (km MSL), volume (km^3), mean area (km^2), and precipitation flux (m^3/s), to name a few.

11.4.6.3   Storm Tracking

TITAN tracks storms by matching storms identified at one scan time with storms identified at the following scan time. Two methods are applied in order. First is an attempt to match storms using areal overlap. If storm shapes at two successive times overlap significantly, these shapes are likely to be from the same storm. Then, optimal centroid matching is applied, which determines the most likely match between storms identified at successive scans. Two overlap fractions are computed: (1) the overlap area divided by the area at time 1, and (2) the overlap area divided by the area at time 2. These two fractions are summed. A perfect overlap will result in a sum of 2.0, and no overlap at all will result in a sum of 0.0. If the sum exceeds a given value (normally 0.6), the storms are considered to be part of the same track. Figure 11.15 depicts the centroids and projected areas of two sets of storms, one set at time t1 and the other at time t2, with the difference, Dt, being the time taken to collect a single volume scan (≈5–10 min). There are not necessarily the same number of storms present at each time; in this example, there are four storms at t1 and five storms at t2. The figure also shows the possible paths the storms may have taken during the period between t1 and t2. The problem is to match the t1 storms with their t2 counterparts, or, equivalently, to decide which set of logically possible paths most likely is the true one.

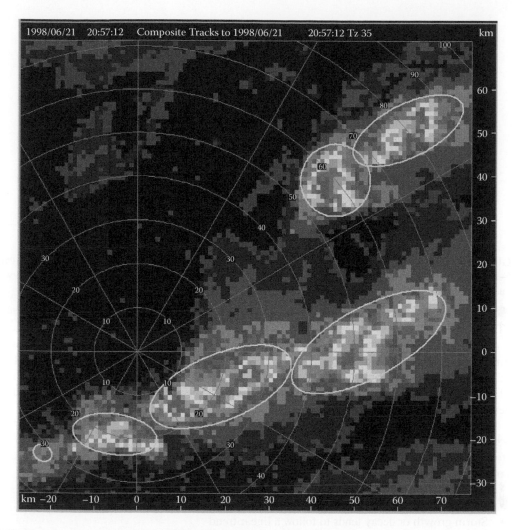

FIGURE 11.14 (See color insert following page 136.) TITAN identifies storms as ellipses (or polygons) where radar reflectivity exceeds some threshold level. (*Source:* http://www.ral.ucar.edu/projects/titan/home/index.php.)

For the purposes of setting up the matching problem as one of optimization, the following assumptions are made about the correct set of matching storms:

- The correct set will include paths that are shorter rather than longer. This is true for thunderstorms that are observed frequently (Dt ≈ 5 min), because the ratio of the size of the storm (≈2–20-km diam.) to the distance moved in Dt (≈1–10 km) is such that it is unlikely that a storm will move well away such that its former position (or one close to it) is occupied by a different storm. Therefore, given a set of possible alternatives as shown in Figure 11.15, the shorter the path, the more likely it is to be a true one.
- The correct set will join storms of similar characteristics (size, shape, etc.). For example, a 2000-km³ storm at t1 is more likely to match a 1500-km³ storm than a 100-km³ storm at t2.
- There is an upper bound to the distance a storm will move in Dt, governed by the maximum expected speed of storm movement (advection plus lateral development). In the figure, the paths that exceed this upper bound are drawn as faint lines.

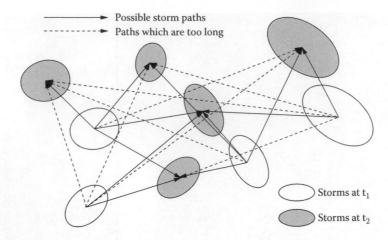

FIGURE 11.15 TITAN storm-tracking algorithm seeks to identify storm cells at successive volume scans (≈5–10 min) and to resolve their origin-destination using logical rules and optimization. (*Source:* http://www. ral.ucar.edu/projects/titan/.)

The problem of determining the true set of storm paths may be posed and solved as one of optimization. There is a search for the optimal set of paths, where the problem is solved as an optimal assignment using techniques from the field of combinatorial optimization. The TITAN model also has algorithms for handling mergers and splits. For mergers, a maximum of one track will be extended, and the remainder will be terminated. For a split, a maximum of one track will be extended, and a new track will be generated for the unmatched storms.

11.4.6.4 Forecast

In formulating the storm-forecast algorithm, the following simplifying assumptions are made based on observations of typical storm behavior:

- A storm tends to move along a straight line
- Storm growth or decay tends to follow a linear trend
- Random departures from the above behaviors occur

As a result, TITAN relies purely on extrapolation for forecasting. A weighted linear fit is performed to the time-history of the storm property for which a forecast is required. Figure 11.16 shows a 30-min forecast of storm location, presented as a series of forecasts at 6-min intervals. The storm speed is also shown in km/h. Growing storms have increasing forecast areas, while decaying storms have decreasing forecast areas.

An application of TITAN to Denver's Urban Drainage and Flood Control District (UDFCD) demonstrated the usefulness of TITAN data in support of the UDFCD flood-threat monitoring and dissemination mission. Radar data were combined to produce watershed-specific estimates and 30-min forecasts of rainfall accumulation over GIS-specified UDFCD watersheds. These data were transformed into XML format, placed on an ftp server for broadcast, and then picked up by Web-based and Microsoft Excel® tools operated by the UDFCD for visualization of storm movement and basin discharge estimates and forecasts. Sharif et al. (2006) applied NCAR Research Applications Program (RAP) Auto-Nowcast (ANC) system to flood forecasting. The ANC system combines a host of remotely sensed data, including gridded radar fields, surface mesonet data, sounding data, and satellite data to produce 0–60-min quantitative precipitation "nowcasts" through deterministic and fuzzy-logic algorithms. The potential benefits of ANC precipitation nowcasts were demonstrated by forcing them through a physically based, distributed hydrologic model to predict flash

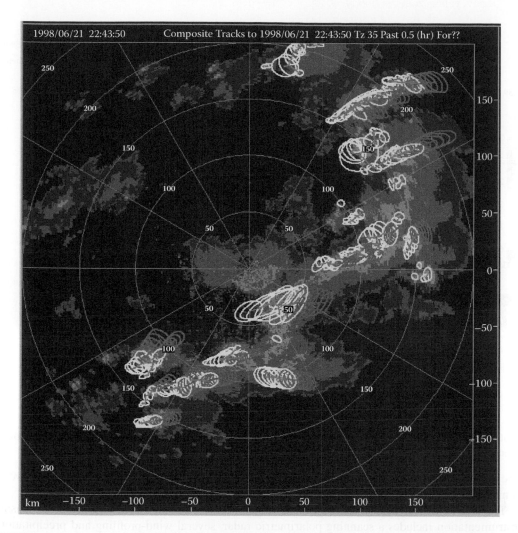

FIGURE 11.16 (See color insert following page 136.) TITAN forecast display shows the current storm location in cyan, the recent (30-min) past in yellow, and the 30-min forecast in red. (*Source:* http://www.ral. ucar.edu/projects/titan/.)

floods in a small, highly urbanized catchment in Denver, Colorado. Two rainfall events on July 5 and 8, 2001, in the Harvard Gulch watershed, corresponding to times during which the ANC system operated, were analyzed. In addition to analyzing the nowcasts for these events, the July 8 case was used to evaluate the hydrologic model's ability to accurately reproduce the watershed's response to the precipitation based on either rain-gauge or high-resolution radar estimates.

11.4.7   NOAA's Hydrometeorological Testbed

NOAA's Hydrometeorological Testbed (HMT) is a demonstration program that focuses the use of advanced observational and modeling tools on quantitative precipitation estimation (QPE) and quantitative precipitation forecasting (QPF) for the purpose of improving hydrologic forecasts and warnings. Full-scale deployments of HMT were focused on the North Fork of the American River basin located between Sacramento and Reno on the western slopes of the Sierra Nevada (Figure 11.17). Water from the American River Basin is a critical resource for California's economy and natural ecosystems, and the threat of flooding poses an extremely serious concern for the heavily populated downstream area.

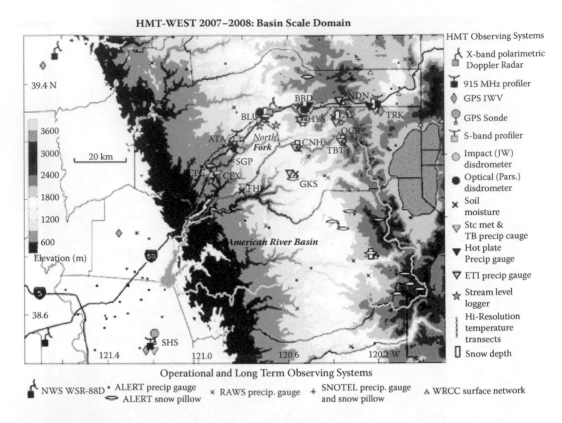

FIGURE 11.17 HMT instrumentation for the American River basin represents advanced technologies intended for full deployment. (*Source:* http://www.esrl.noaa.gov/psd/programs/2008/hmt/.)

The frequent impact of prolonged, heavy winter precipitation from concentrated "atmospheric rivers" of moisture, originating in the tropical Pacific, underscores the area's flood vulnerability.

The HMT involves deployment of a wide range of remote and in situ sensors. Remote-sensing instrumentation includes a scanning polarimetric radar, several wind-profiling and precipitation-profiling radars, and a network of GPS receivers for measuring precipitable water vapor. Precipitation gauges and disdrometers (used to measure the drop size distribution and velocity of falling hydrometeors), surface meteorological stations, soil moisture/temperature probes, snow-depth sensors, and stream-level loggers are among the in situ sensors that are deployed. In addition, rawinsondes are released serially immediately upwind of the area during storm episodes. Data from HMT and operational sources are merged by algorithms to produce multisensor (e.g., radar + gauge) estimates of precipitation. These blended estimates overcome deficiencies inherent in estimates based on single instrument types.

Demonstrations and evaluations of precipitation-forecast improvements were conducted for the American River basin using high-resolution numerical prediction models. The Local Analysis and Prediction System (LAPS) integrates data from virtually every meteorological observation system into a high-resolution 3-D gridded framework centered on a forecast office's domain of responsibility (Figure 11.18). Thus, the data from local mesonetworks of surface observing systems, Doppler radars, satellites, wind and temperature profilers, radiometric profilers, as well as aircraft are incorporated every hour into a three-dimensional grid covering a 1040 × 1240-km area. LAPS has analysis and prediction components. The prediction component is configured using the RAMS, MM5, WRF, and ETA models. Any or all of these models, usually being initialized with LAPS analyses, are run to provide short-term forecasts. Ensemble forecasts are produced using multiple models and initialization methods, with verification.

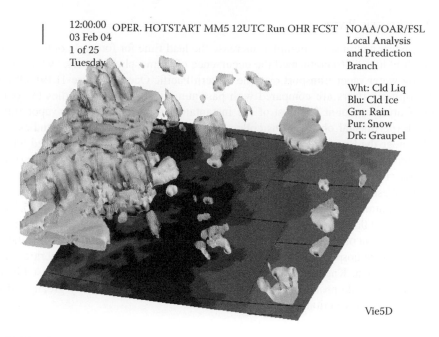

FIGURE 11.18 Cloud-cover forecast from LAPS using MM5 as the forecast model. (*Source:* http://laps.fsl. noaa.gov/.)

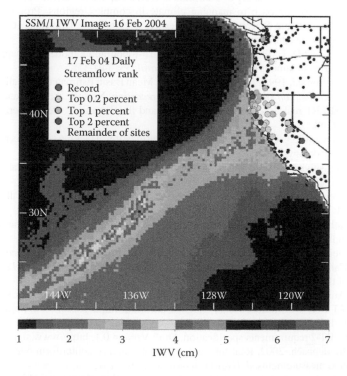

FIGURE 11.19 Atmospheric river phenomenon channels water vapor from tropical regions onto the coast of California, resulting in record rainfalls and river flood flows. (From Ralph et al. 2006. With permission.)

11.4.8 ATMOSPHERIC RIVERS

Weather forecasters continue to attempt to increase the lead time for forecasts of heavy rainfall. A study by Ralph et al. (2006) confirmed the occurrence of "atmospheric rivers," which are narrow bands of strong water-vapor transport over the eastern Pacific Ocean (Figure 11.19). Observations of these atmospheric rivers are compared with past numerical modeling studies to confirm that such narrow features account for most of the instantaneous meridianal water-vapor transport at midlatitudes. Experimental observations collected during meteorological field studies conducted by the NOAA near the Russian River of coastal northern California were combined with Special Sensor Microwave/Imager (SSM/I) satellite observations offshore to examine the role of land-falling atmospheric rivers in the creation of flooding. The study documented their impact when they strike the U.S. West Coast. One such atmospheric river produced more than 100 cm (40 in.) of rainfall in the mountains of southern California in only four days in early January 2005. That storm produced widespread flooding and caused a massive mudslide in La Conchita, California, that took 10 lives. An assessment of all seven floods on the Russian River since October 1997 indicated that, for all seven floods, atmospheric river conditions were present and caused heavy rainfall through orographic precipitation. Knowledge of the atmospheric river phenomenon and establishment of means for forecasting hold promise for increasing forecast lead time, thereby greatly aiding flood preparedness and response actions.

REFERENCES

Botts, M., M. Botts, G. Percivall, C. Reed, and J. Davidson. 2006. OGC® sensor Web enablement: Overview and high level architecture. OGC 06-050r2, Version: 2.0. Open Geospatial Consortium, Inc. http://www.opengeospatial.org/pressroom/papers.

Daly, C. 2004. PRISM: A knowledge-based approach to mapping precipitation in complex regions. *Proc. American River Watershed Institute*, University of California at Davis. http://www.arwi.us/precip/2004.php.

Daly, C. 2007. The PRISM approach to mapping climate in complex regions. Presentation to Northwest Alliance for Computational Science and Engineering, Oregon State University, Corvallis, Ore. http://www.prism.oregonstate.edu/docs/.

Daly, C., R. P. Neilson, and D. L. Phillips. 1994. A statistical-topographic model for mapping climatological precipitation over mountainous terrain. *J. Appl. Meteorol.* 33: 140–158.

Davis, R. D., and P. Jendrowski. 1996. The operational Areal Mean Basin Estimated Rainfall (AMBER) module. In *Preprints 165th Conf. on Weather Analysis and Forecasting*, Norfolk. Am. Meteor. Soc., 332–335.

DIAD Corp. 2000. ALERT real-time weather monitoring and flood warning. September. http://www.alert-2.com/resources.html.

Dixon, M., and G. Wiener. 1993. TITAN: Thunderstorm Identification, Tracking, Analysis and Nowcasting—A radar-based methodology. *J. Atmos. Ocean. Technol.* 10: 785–797.

Fulton, R. A. 1999. Sensitivity of WSR-88D rainfall estimates to the rain-rate threshold and rain gauge adjustment: A flash flood case study. *Weather Forecasting* 14: 604–624.

Fulton, R. A., J. P. Breidenbach, D.-J. Seo, and D. A. Miller. 1998. The WSR-88D rainfall algorithm. *Weather Forecasting* 13: 377–395.

NWS (National Weather Service). 1997. Automated local flood warning systems handbook. Weather Service hydrology handbook No. 2. http://www.nws.noaa.gov/oh/docs/alfws-handbook/.

Ralph, F. M., P. J. Neiman, G. A. Wick, S. I. Gutman, M. D. Dettinger, D. R. Cayan, and A. B. White. 2006. Flooding on California's Russian River: Role of atmospheric rivers. *Geophys. Res. Lett.* 33: L13801. U.S. Department of the Interior. doi:10.1029/2006GL026689.

Rindahl, B. 1996. Analysis of real-time raingage and streamgage flood data using ArcView 3.0. Paper presented at ESRI User's Conference. San Diego, California. http://gis.esri.com/library/userconf/proc96/TO50/PAP040/P40.HTM.

Salo, T. J. 2007. ALERT-2 requirements specification. Draft Version 0.1. http://www.alert-2.com.

Seo, D. J., and J. Breidenbach. 2002. Real-time correction of spatially nonuniform bias in radar rainfall data using rain gauge measurements. *J. Hydrometeorol.* 3 (2): 93–111.

Seo, D.-J., J. Breidenbach, and E. Johnson. 1999. Real-time estimation of mean field bias in radar rainfall data. *J. Hydrol.* 223: 131–147.

Sharif, H., D. Yates, R. Roberts, and C. Mueller. 2006. The use of an automated nowcasting system to forecast flash floods in an urban watershed. *J. Hydrometeorol.* 7 (1): 190–202.

Skahill, B., and L. Johnson. 2000. F2D: A kinematic distributed watershed rainfall-runoff. NOAA technical memorandum ERL FSL-24. Boulder, Colo.: Forecast Systems Lab.

Smith, J. K., and R. C. Lipschutz. 1990. Performance of the NEXRAD precipitation algorithms in Colorado during 1989. Paper presented at Eighth Conference on Hydrometeorology, American Meteorological Society. Kananaskis Park, Alberta, Canada, October 22–26.

Smith, M., V. Koren, Z. Zhang, S. Reed, D. Seo, F. Moreda, V. Kuzmin, Z. Cui, and R. Anderson. 2004. NOAA NWS distributed hydrologic modeling research and development. NOAA technical report NWS 45.

USACE (U.S. Army Corps of Engineers). 1996. Hydrologic aspects of flood warning preparedness programs. Technical letter No. 1110-2-540 (ETL 1110-2-540).

12 GIS for River Basin Planning and Management

12.1 OVERVIEW

This chapter addresses river basin planning and management models and their integration with GIS. River basin models include those that simulate and optimize the operation of reservoirs and associated facilities to serve multiple purposes. River basin models and supportive GIS databases include representations for surface-water and groundwater hydrology to varying levels of sophistication. Many of these topics are addressed in other chapters. This chapter emphasizes the purposes and structures of GIS-based river basin models and their application as spatial decision-support systems.

12.2 RIVER BASIN PLANNING AND MANAGEMENT

12.2.1 RIVER BASIN SYSTEMS

The natural "region" for water resources planning and development is often the river basin. River basins provide a geographic and functional context for many water-related purposes. Most water projects, particularly reservoirs, can serve more than one of the basic purposes—water supply, irrigation, hydroelectric energy, navigation, flood mitigation, recreation, pollution control, and wildlife conservation. Linsley et al. (1992) summarized the various purposes for which water stored in reservoirs can be used, as well as the nature of reservoir operations required to achieve multiple purposes (Table 12.1).

12.2.2 RIVER BASIN PLANNING AND MANAGEMENT

Water resource planning is directed toward the achievement of certain objectives while avoiding serious negative impacts external to a given project. Each specific action in water management is likely to have consequences downstream, and sometimes upstream. To measure the feasibility of a project or program, the purposes or objectives must be specified. In the United States, national objectives for water planning are (a) enhancement of national economic development and (b) enhancement of the quality of the environment (U.S. Water Resources Council 1983). At the river basin level, these objectives can be stated in more specific terms.

Typically, there are multiple objectives involved in the development of river basin plans. For example, the national objectives for regional economic development and environmental enhancement constitute multiple objectives. These objectives may be competitive in that maximizing one will result in reductions of achievement in the other. Sometimes objectives may be complementary, whereby each can be increased. For any given river basin planning situation, formulation and evaluation of alternative plans will require that trade-offs between objectives be defined. This involves specifying the degree of achievement of each of the objectives for the set of feasible alternatives. It is then the responsibility of the decision makers, involved stakeholders, and the political processes to determine which set of alternatives is preferred.

As an example, Hayes et al. (1998) applied an optimal control model to the daily operation of the Cumberland River basin system with explicit integration of both water quantity and water quality

TABLE 12.1
Reservoir Multiple Purposes

Purpose	Description
Water supply	Reservoir conservation storage is mainly designated for municipal and industrial (M&I) and irrigation water supply requirements. M&I demands are more nearly constant throughout the year than are irrigation requirements, although lawn irrigation causes a summer maximum. Planning for growth is a major ongoing requirement. Storage of an adequate reserve supply is determined to avoid shortages during "design" or "most-adverse" drought scenarios.
Irrigation	Water requirements for irrigation are typically seasonal, with a maximum during the summer dry season and little or no demand during the winter. Irrigation is a highly consumptive use of water, as one-half or more of the applied water can be lost to evapotranspiration.
Hydroelectric energy	Energy demand has a marked seasonal and daily variation in response to peak summer air-conditioning loads. Hydropower facilities have the advantage that they can be turned on to maximum power output in a short time. Also, water passed through turbines to produce power flows downstream is available for other uses.
Recreation	Substantial benefits are usually realized from recreational uses, regardless of the primary purposes. The ideal recreational reservoir is one where water levels are held nearly constant during the recreation season to permit boating, fishing, and swimming and other water sports. In contrast, downstream recreation uses, such as rafting and kayaking, can be dependent on reservoir releases.
Fish and wildlife	Reservoirs may serve to enhance fish and wildlife habitats in general, but construction of a reservoir blocks migratory fish from traveling upstream unless fish ladders are implemented. Large and rapid fluctuations in water levels are harmful to fisheries, particularly during the spawning season. Reservoir releases can sustain downstream fisheries during low-flow drought periods in support of the Endangered Species Act.
Flood mitigation	Flood mitigation requires empty storage space to capture flood runoff. The amount of storage required is determined through hydrologic analyses of watershed rainfall-runoff using a design-storm approach.
Navigation	Reservoir releases are sometimes made to sustain navigation flows. Run-of-the-river reservoirs developed to maintain navigation water levels are of limited height because of the need for ship locks, and hence their storage capacity is limited.
Pollution control	Low-flow releases may be considered for dilution of wastewater. In the United States, however, reservoirs are not supposed to be designed for this, since wastewater dischargers are required to treat their discharges adequately. Reservoir releases may actually degrade water quality, because releases from the lower levels of a thermally stratified reservoir can be cold and anoxic.
Multiple purposes	Reservoirs often have potential for multiple purposes, as benefits may be increased without a proportionate increase in costs. Irrigation, navigation, and water supply all require a volume of water that cannot be jointly used; thus a reservoir project combining these functions must provide a designation of storage space to each as well as rules for operation.

Source: Adapted from Linsley et al. (1992).

components. The optimal control model optimizes hydropower production revenues while allowing direct control over water characteristics of temperature and dissolved oxygen through alternative operational schemes. Figure 12.1 displays the trade-offs between loss in hydropower revenues and improving water quality in the critical reaches of the Cumberland River near the City of Nashville, Tennessee.

The planning process involves specification of objectives, identification of constraints, formulation of alternatives, and evaluation of alternatives as per the objectives and constraints. Usually, the process is incremental, beginning with a reconnaissance study, proceeding to increased levels of detail in a feasibility study, and ultimately to a design study. Since planning is always for the future, forecasts of future climatic, water supply, water demand, and socioeconomic conditions over time frames up to 50 years create a high degree of uncertainty, relegating the planning problem to one of minimizing the risks of a wrong decision as a result of poor forecasts. The various project alternatives can be tested to see how they perform under a range of possible future scenarios. An

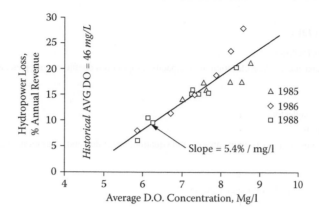

FIGURE 12.1 Trade-off between loss of hydropower revenues and water-quality enhancement for the Cumberland River basin system. (From Hayes et al. 1998. With permission.)

alternative that is robust enough to perform well regardless of assumed future conditions is likely to be the best alternative.

12.2.3 RIVER BASIN SYSTEMS ANALYSIS

Systems analysis techniques have found wide application for identification of feasible sets of alternatives as well as the "best" of all possible alternatives. A river basin system comprises an extensive array of components with complex spatial, temporal, and regulatory interactions. The strong spatial and topological characteristics of river basin systems particularly lend themselves to application of GIS. The systems analysis techniques of simulation and optimization are used to examine how all the components interact, and how they can be combined into efficient systems for maximizing the prescribed objectives. River basin systems include combinations of reservoirs; hydropower generation facilities; navigation lock and dam structures; transbasin diversion projects; treatment works; municipal, industrial, and agricultural demands; pump stations; distribution networks; groundwater basins with complex stream–aquifer interactions; and ecologically sensitive river reaches providing habitats for endangered species. Perhaps the most distinguishing aspect of a river basin systems analysis is its focus on the joint function of components of the system under various conditions that the system may be subject to. These circumstances may be associated with wet or dry periods, and they may represent short or long periods of time. In addition, institutional aspects of the systems being dealt with must be considered. For example, laws or regulations (e.g., water rights priorities, endangered species) may act as constraints on technical options to water allocation problems.

GIS procedures are used extensively in river basin modeling for data capture and archiving, database integration, and development of user-friendly interfaces for model pre- and postprocessing. Table 12.2 lists river basin data conducive to spatial representation as GIS layers, along with the important attribute data associated with these features.

Water resources systems are analyzed using computerized mathematical models. However, river basin systems are complex, involving both physical and human dimensions, and thus may not be described exactly by mathematical methods. Also, there may be inadequate data or scientific understanding to properly support the modeling of some systems. However, computerized mathematical models have proven advantageous in river basin planning and are now considered the state of the art. Also, the conceptual process of building a model provides for understanding the interactions of river system elements; such understanding helps pinpoint data needs and ultimately leads to better models. GIS tools and databases have been widely integrated with mathematical models to form spatial decision-support systems (SDSS) supportive of river basin planning and management.

TABLE 12.2
River Basin Data

Water resources facilities

 Reservoirs (capacities, zones, elevation-area-capacity, evaporation coefficient, seepage, release policies)

 Demand sites

 Diversions

 Canals (source, collector)

 Pumping stations

 Hydroelectric power plants (output capacity and equation, target demand, operational policies)

 Treatment plants

Land

 Urban lands (residential, commercial, etc.)

 Irrigated lands

 Open space, parks, and habitats

Transportation

 Roads

 Railroads

 Other utilities

 Transmission lines

Administrative boundaries

 Political boundaries (countries, states, counties, municipalities, etc.)

 Water districts (water supply, irrigation, etc.)

Hydrography and hydrology

 Basins and sub-basin boundaries

 Stream network

 Main stem, tributaries, river reaches

 Capacity, slope, roughness

 Stream gauges and records

 Precipitation and weather stations and records

 Water-quality stations and records

Aquifers

 Type and extent

 Wells

 Safe yield and transmissivity

The discussion of simulation and optimization models in Chapter 4 (GIS Analysis Functions and Operations) is particularly relevant for SDSS.

12.2.4 RIVER BASIN RESERVOIR SIMULATION SYSTEM

An example of a river basin simulation system is the popular HEC-ResSim (HEC-Reservoir Simulation or HEC-ResSim) model developed by the U.S. Army Corps of Engineers Hydrologic Engineering Center (Klipsch and Hurst 2007). HEC-ResSim has been developed to aid in predicting the behavior of reservoirs and to help reservoir operators plan releases in real time during day-to-day and emergency operations. HEC-ResSim is composed of a Windows® graphical user interface (GUI), a computational program to simulate river basin reservoir operations, and graphics and reporting utilities. The HEC Data Storage System (HEC-DSS) is used for storage and retrieval of input and output time-series data. HEC-ResSim includes three sets of modules with which to set up and conduct a river basin simulation: (a) Watershed Setup, (b) Reservoir Network, and (c) Simulation. Each module has a set of functions accessible through menus, toolbars, and schematic elements.

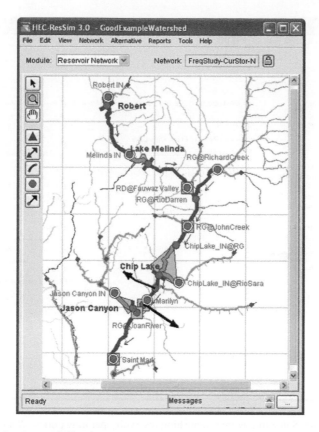

FIGURE 12.2 (See color insert following page 136.) HEC-ResSim is a general-purpose river basin simulation package. (*Source:* Klipsch and Hurst 2007; http://www.hec.usace.army.mil/software/hec-ressim/.)

HEC-ResSim provides a realistic view of the physical river/reservoir system using a map-based schematic with a set of element drawing tools. The program's user interface allows the user to draw the network schematic either as a stick figure or an overlay on one or more georeferenced maps of the watershed (Figure 12.2). HEC-ResSim represents a system of reservoirs as a network composed of four types of physical elements: junctions, routing reaches, diversions, and reservoirs. By combining these elements, the HEC-ResSim modeler is able to build a network capable of representing anything from a single reservoir on a single stream to a highly developed and interconnected system rules like that of California's Central Valley. A reservoir is the most complex element of the reservoir network and is composed of a pool, a dam, and the associated operating rules.

12.2.4.1 Hierarchical Outlet Structure

The dam is the root of an outlet hierarchy or "tree" that allows the user to describe the different outlets of the reservoir in as much detail as is deemed necessary. There are two basic and two advanced outlet types. The basic outlet types are controlled and uncontrolled. An uncontrolled outlet can be used to represent an outlet of the reservoir, such as an overflow spillway, that has no control structure to regulate flow. Controlled outlets can be used to represent any outlet capable of regulating flow, such as a gate or valve. The advanced outlet types are power plant and pump, both of which are controlled outlets with additional features to represent their special purposes. The power plant has the ability to compute energy production. The pump is even more specialized because its flow direction is opposite that of the other outlet types, and it can draw water up into the reservoir from the pool of another reservoir (i.e., pumped storage operations). The pump outlet type was added to enable the user to model pump-back operation in hydropower systems, although hydropower is not required for its operation.

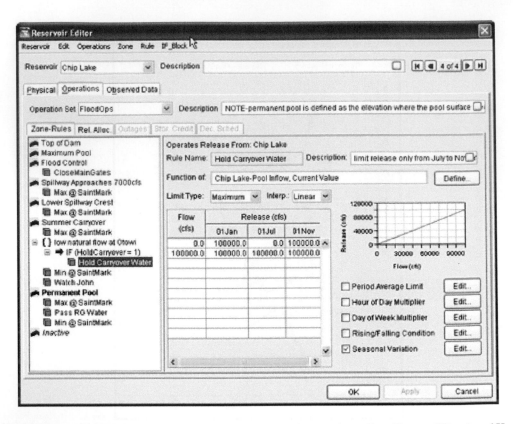

FIGURE 12.3 HEC-ResSim window for establishing reservoir operating policy. (*Source:* Klipsch and Hurst 2007; http://www.hec.usace.army.mil/software/hec-ressim/.)

12.2.4.2 Rule-Based Operations

Most reservoirs are constructed for one or more of the following purposes: flood control, power generation, navigation, water supply, recreation, and environmental quality. These purposes typically define the goals and constraints that describe the reservoir's release objectives. Other factors that may influence these objectives include time of year, hydrologic conditions, water temperature, current pool elevation (or zone), and simultaneous operations by other reservoirs in a system. HEC-ResSim attempts to reproduce the decision-making process that human reservoir operators must use to set releases. It uses an original rule-based description of the operational goals and constraints that reservoir operators must consider when making release decisions. As HEC-ResSim has developed, advanced features such as outlet prioritization, scripted state variables, and conditional logic have made it possible to model more-complex systems and operational requirements. Figure 12.3 shows the GUI for establishing the operating policy.

12.3 SPATIAL DECISION-SUPPORT SYSTEMS IN RIVER BASIN MANAGEMENT

Concepts for a spatial decision-support system (SDSS) were introduced in Chapter 4 (GIS Analysis Functions and Operations). SDSSs are finding wide application for river basin systems due to the complexities of the physical water resources component interactions and the multiple jurisdictions involved. The SDSS provides the essential means for integration of databases and models, along with interactive display capabilities to support exploration and learning about resource allocation choices and trade-offs.

River basin models are often based on a link–node structure forming a network of system components and the topological relationships between them. For this modeling approach, links represent river reaches, canals, or pipelines serving to convey water from one node to another. Other links may be more abstract to account for flows to a particular user, even though that water flows along the main river channel. Nodes represent river junctions, diversion points, reservoirs, pumping stations, and demand locations. The collection of nodes and links form a network that can be efficiently analyzed with a variety of powerful analysis tools available in geographic information systems.

The link–node structure provides a logical construct whereby each component is characterized as an "object" and the topological relationships are paramount. An object represents a physical or abstract feature, and its placement in the network establishes its relationship to other objects. The actual distances between objects can be derived from the GIS, which is important, since the travel time of flow through a channel is fundamentally related to distance. With an object-oriented approach, distance is an attribute of the link in the database. Also, it is important that the network structure relates visually to the actual physical structure of the river system.

The object orientation supports the coupling of databases and models in a geographic information system. The attributes of an object are stored in the database, thus gaining the advantages that a database offers for handling alphanumeric and other data. For a given object, these attributes include its descriptive data, which define its capacity, etc., as well as prescriptive data on how it would be operated. For example, descriptive data for a reservoir would include its capacity, area–elevation–volume relation, and outlet capacity, to name a few. The reservoir prescriptive data could include maximum and minimum allowed pool levels, release schedules according to priority, and rule curves.

A powerful array of tools is provided in ArcGIS® for converting existing geospatial data into the geodatabase framework. In addition to the convenience of storage, access, and management of geospatial data in the geodatabase framework, additional capabilities include representation of rules customized by the user governing how river basin network objects are logically interrelated. The geodatabase framework in ArcGIS is implemented in a relational database-management system (DBMS) as a set of joined and related tables.

A geometric network is a specialized type of geodatabase designed specifically for analysis and modeling of the networks. The terminology used for geometric networks defines edges as network links and junctions as network nodes. As illustrated in Figure 12.4, a particular advantage of geometric networks is that a collection of different feature classes can be incorporated into a feature data set. The geometric network integrates a geographic network composed of edges and junctions, with a logical network for setting connectivity rules that bind together the feature classes that make up the network. The geometric and logical elements of the geometric network are always synchronized, even during editing of the geometric network. This allows consistent interconnection of all important elements of a river basin network, with logical rules governing which types of network objects can be connected.

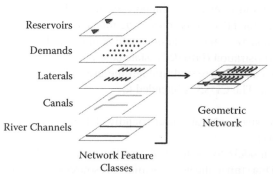

FIGURE 12.4   Integration of river basin network feature classes into a geometric network.

In the following sections of this chapter, there are descriptions of various river basin modeling and spatial decision-support systems.

12.4 COLORADO'S DECISION-SUPPORT SYSTEMS

12.4.1 OVERVIEW OF COLORADO WATER MANAGEMENT

River compacts, international treaties, Supreme Court decrees, and numerous federal and state laws play an important role for water management in the arid western United States. By virtue of geography, the state of Colorado contains the headwaters of a number of major river basins, including the Colorado, Arkansas, Rio Grande, and South Platte. The state is constantly evaluating management of its water resources in response to population and demand for growth, droughts, and reductions in federal water programs funding. Management and more efficient use of existing supplies are receiving greater attention by both water users and state agencies.

Water management issues include those concerning:

- Interstate compacts: In Colorado, demands for water by downstream states pose management challenges and place additional burdens on already overappropriated systems. A severe sustained drought could cause curtailment of Colorado uses in order to meet compact delivery obligations. River and reservoir operating policy modifications can be evaluated using systems models to minimize local impacts and reduce interstate compact conflicts.
- Water resources planning: It is of great interest to be able to assess the impacts of water-rights transfers or water resource developments. Water rights in Colorado can be viewed as private property, whereby an individual owns the rights to beneficial use of the water. These rights can be changed from one type of use to another or from one place to another, but only if no other water right is injured. Such changes require water court approval. Because changes of agricultural ditch and reservoir rights to municipal use are fairly common, concerns arise over whether a transfer of water will affect the supply available to owners of other decreed water rights. There are also significant concerns with maintaining flow conditions to protect and recover endangered fish species.
- Water resources administration: State of Colorado water allocations are based on the Doctrine of Prior Appropriation or the first-in-time, first-in-right doctrine. Colorado Division of Water Resources officials have established procedures for allocating water to users according to decreed priorities. This is a challenging task made complicated by changing flows and reservoir levels, and increased uses of the state's waters.

12.4.2 COLORADO'S DECISION-SUPPORT SYSTEMS

The principal goal of Colorado's Decision-Support System (CDSS) is to provide credible information on which to base informed decisions concerning management of the state's water resources per the three categories summarized above: (a) interstate compact policy, (b) water resource planning, and (c) water-rights administration (http://cdss.state.co.us/). It is felt that the water supply available to the different interests can be increased through the use of comprehensive data, planning tools, and real-time information. The CDSS is intended to:

- Provide comprehensive, accurate, user-friendly databases compatible with the CDWR HydroBase database
- Provide data and models to evaluate alternative water development and administration strategies that can maximize the use of available resources in all types of hydrologic conditions and the development of sound water resources management strategies
- Provide a functional, integrated system that can be maintained and upgraded by the state

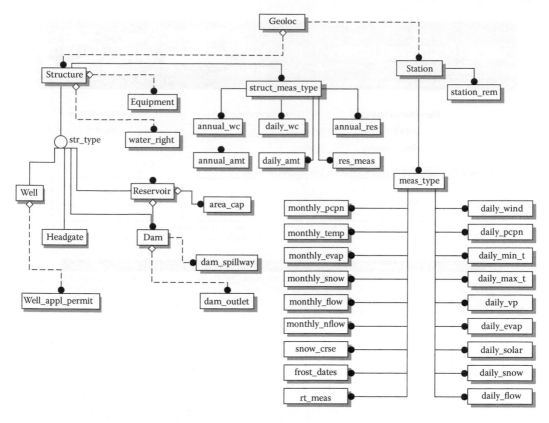

FIGURE 12.5 CRDSS database schema. (*Source:* http://cdss.state.co.us/.)

- Have the capability to accurately represent current and potential federal and state administrative and operational policies and laws
- Promote information sharing among government agencies and water users

The CDSS is a data-centered system integrating analysis tools and models that help water users and water managers make decisions. CDSS uses can be divided into three main classes: (a) planning DSS, (b) administrative DSS, (c) integrated planning and administrative DSS. The planning DSS is typically used where water resources development and protection are the central issues. Such studies typically involve studying long periods of data in order to validate a decision based on the long-term behavior of a basin. The DSS devoted to administrative tasks relies more on real-time information to help make daily decisions for reservoir releases and diversions. The integrated planning and administrative DSS provides tools for planning studies and water resources administration. For example, an administrative tool may rely on displays of historical data and the planning model results to identify reasonable bounds of a real-time decision. The database schema is shown in Figure 12.5; it has been deployed with a Web-based interface to make access available to users (Figure 12.6). Key components of the CDSS include various GIS and attribute databases, models, and associated graphical user interface (GUI) and data management interface (DMI) tools (Table 12.3).

12.4.3 CDSS Application Scenarios

12.4.3.1 Scenario Using Spatial Database: Determine Irrigated Acreage

The spatial database functions of CDSS could be used to review one or more of the data sets incorporated into the CDSS. For example, a state legislator might request that the DWR determine the irrigated acreage in his or her district. The DWR engineer can use the Spatial Data Browser to

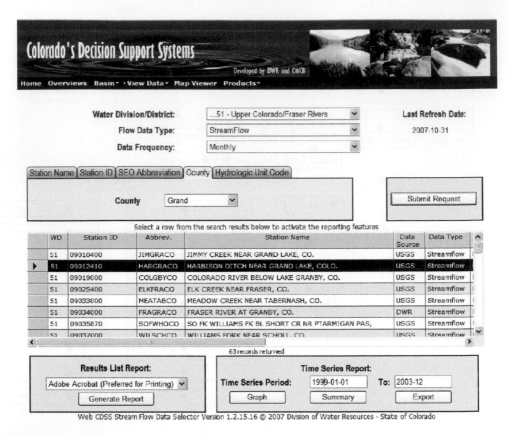

FIGURE 12.6 CRDSS databases are accessible through a Web-based interface. (*Source:* http://cdss.state. co.us/.)

display a map of irrigated acreage at any location in the Colorado River basin in Colorado, as illustrated in Figure 12.7. The engineer could display the regional map, select a subpart of the region pertaining to the area of interest, and display the irrigated lands for that area. Total irrigated acreage for the selected area can then be computed and a table of all irrigated acreage and crop types created.

12.4.3.2 Scenario on Water Resource Planning: Evaluate Basin Development Proposal

The water resource planning model of CDSS is used by CWCB personnel to evaluate a proposed water project on a river on the western slope. The CWCB personnel posit a level of project development that would not adversely impact decreed water rights and that establishes the project's feasibility on a technical and economic basis. Information development actions can be performed using data-management and modeling functions accessed through the CDSS interface (Figure 12.8). The Map Utility can be used to identify possible locations for development. A development plan can be defined using a spreadsheet utility to compile and edit water-rights priorities, amount, and timing. These data can then be merged with the other basin water rights and baseline conditions to establish input to the Water Resource Planning Model. An overplot of the time-series output for the baseline or comparison case can be obtained to assess the impact of the proposed project on other water resources.

12.5 RIVERWARE®

12.5.1 OVERVIEW

RiverWare® is a general-purpose river basin modeling system for planning and operations (Zagona et al. 2001). RiverWare is capable of modeling hydrologic processes of reservoirs, hydropower, river

TABLE 12.3
Colorado River Decision-Support System (CRDSS) Components

Component	Description

Centralized relational database (HydroBase)[a]	HydroBase includes data on:
	    Historic and real-time stream flows and water-rights calls, water rights, diversions, climate, and well permits
	    Geographic locations of stream gauges, water rights, and climate gauges

Centralized spatial database	GIS data layers (as ESRI shapefiles) include current and historic land use and irrigated vegetation, water source and service area data (location of structures, water service areas, irrigated parcels, wells, gauges, etc.), aquifer coverages, highways, PLSS (public land survey system), soils, and wetlands.
Colorado Consumptive-Use Model (StateCU)	StateCU was developed to estimate/report both crop and noncrop consumptive use within the state. The primary crop consumptive-use method employed in the program and the user interface is a modified Blaney-Criddle consumptive-use method with calculations on a monthly basis. Other crop consumptive-use methods include the Penman-Monteith and modified Hargreaves methods.
Colorado Water Resources Planning Model (StateMod)	StateMod is the water resources planning model used to assess past, present, and future water management policies in Colorado basins. StateMod is a water allocation and accounting model that computes natural flow and then allocates flow to water rights (via diversions, storage, and in-stream flow requirements) in strict conformance with the prior appropriation doctrine. The model operates on a monthly and daily time step.
Colorado Water Budget Model (StateWB)	StateWB was developed to perform a water balance for a wide range of basins or sub-basin combinations. StateWB is a Visual Basic computer program that calculates the mass balance of a defined area's surface-water and groundwater inflows, outflows, and changes in storage on a monthly, annual, or average annual time step.

Groundwater model	Groundwater modeling (e.g., ModFlow for the Rio Grande DSS) has the capability to:
	    Provide information on the location and timing of groundwater return flows to stream systems
	    Characterize aquifer yields
	    Characterize groundwater flow at critical locations within the basin
	    Provide maps of current and predicted groundwater level data under alternative management scenarios.

Colorado Water Rights Administration Tool (CWRAT)	CWRAT provides a spreadsheet that performs computations for daily administration by the Colorado State Engineer's office. This tool:
	    Allows import of real-time stream gauge data and gauged diversions
	    Allows import of external data
	    Includes features to compute gain/loss
	    Allows for dry river reaches
	Point flow, natural flow, and delivery flows are tracked from upstream to downstream. CWRAT has features to synchronize data between the central database and a local copy of the database. CWRAT displays both real-time and historic data.

Interstate compact-analysis models	The Colorado River Basin Simulation Model (Big River Model) and the Colorado Annual Operating Plan Model, developed by the USBR, are used to evaluate river and reservoir operations throughout the Colorado River system, allowing examination of present and future interstate compact policies and operating criteria (http://www.usbr.gov/lc/riverops.html).

[a] Figure 12.5 illustrates a generalized database schema.

reaches, diversions, distribution canals, consumptive uses, and surface–groundwater conjunctive use. It also can model water rights and water-accounting transactions.

A central concept of RiverWare is that river basin system features are represented as "objects" that encapsulate data and process information. The RiverWare interface provides icons that can be

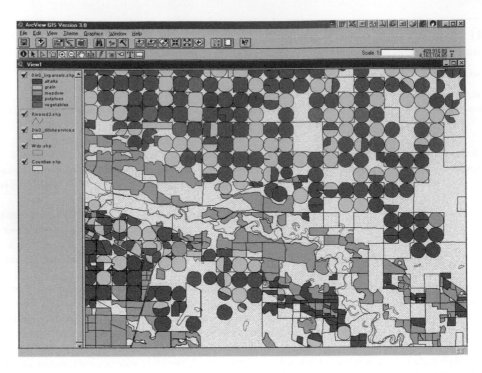

FIGURE 12.7 (See color insert following page 136.) Display of irrigated acreage obtained through interactive query. (*Source:* http://cdss.state.co.us.)

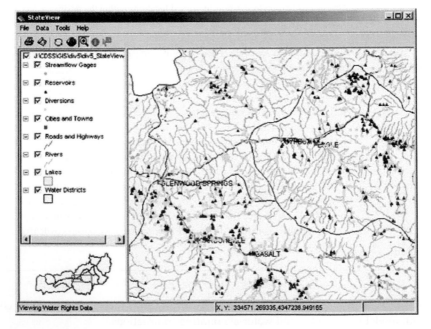

FIGURE 12.8 CRDSS data retrieval for stream-flow gauges. (*Source:* http://www.riverside.com/projects/marketing_flyers/crdss.asp.)

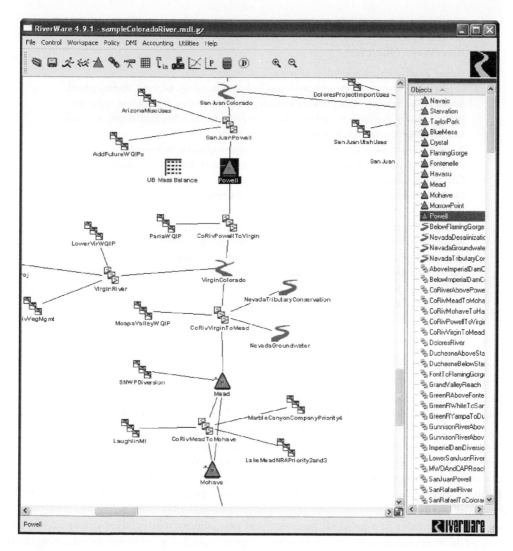

FIGURE 12.9 RiverWare model schema for the Lower Colorado River.

dragged into position relative to other objects when the system model is being constructed. Each object is given a name and can be opened to reveal a list of "slots," which are the variables associated with the physical model equations for that feature. Reservoir objects, for example, have slots for inflow, outflow, storage, and pool elevation, among others. Links between objects are established interactively to form the topology of the river system; this establishes links between the slots of the objects. When simulating, the solution processes involve propagation of information among objects through the links. Figure 12.9 shows a model schema for the Lower Colorado River basin. Table 12.4 lists the types of objects and model processes in RiverWare.

RiverWare object slots are where the required data are entered through direct manual entry or through file importation integrated with the database. Three types of data are entered: time series, tabular, and scalar. Time-series slots include river inflows, diversions, reservoirs, and the like. Table slots contain parameter data needed for model equations and functional relationship data such as reservoir area–elevation–volume tables. Scalar slots are single values. Methods for each object define the physical processes and equations for the feature. Table 12.5 lists selected examples of methods for a few objects. These methods are encapsulated with the object definition. There are options for users to specify their own methods as might be required by a given agency.

TABLE 12.4
RiverWare Objects

Storage Reservoir: mass balance, evaporation, bank storage, spill

Level Power Reservoir: Storage Reservoir plus hydropower, energy, tailwater, operating head

Sloped Power Reservoir: Level Power Reservoir plus wedge storage for very long reservoirs

Pumped Storage Reservoir: Level Power Reservoir plus pumped inflow from another reservoir

Reach: routing in a river reach, diversion and return flows

Aggregate Reach: many Reach objects aggregated to save space on the workspace

Confluence: brings together two inflows to a single outflow, as in a river confluence

Canal: bidirectional flow in a canal between two reservoirs

Diversion: diversion structure with gravity or pumped diversion

Water User: depletion and return flow from a user of water

Aggregate Water User: multiple Water Users supplied by a diversion from a reach or reservoir

Aggregate Delivery Canal: generates demands and models supplies to off-line water users

Water Quality: dissolved solids, temperature, and dissolved oxygen

Groundwater Storage Object: stores water from return flows

River Gauge: specified flows imposed at a river node

Thermal Object: economics of thermal power system and value of hydropower

Data Object: user-specified data; expression slots or data for policy statements

Source: Zagona et al. (1998).

There are three types of RiverWare solution procedures: simple simulation, rule-based simulation, and optimization. Simple simulation allows straightforward scenario runs in which user-supplied inputs drive the solution. The solution is based on an object-oriented modeling paradigm: each object waits until it has enough information to solve, and then it executes its method. The method solves for the unknown slots on the object, and information is propagated across links to other objects. Too much (conflicting) information results in an error state and termination of the run. Insufficient information results in parts of the model being left unsolved. RiverWare allows some flexibility in specifying models where the solution is not propagating from upstream to downstream and forward in time. For example, river reaches with time lags may solve for inflow given outflow, setting the inflow value at a previous time step and propagating that value upstream. Also, target operations on reservoirs may be specified, where a future target storage is met by adjusting the reservoir's outflow over a specified time frame.

In the other two RiverWare solution techniques, operational policies drive the solution. For rule-based simulation, there is not enough information on the objects to solve the system. The additional information is added by prioritized policy statements (rules) that (a) are specified by the user and interpreted by the rule processor and (b) set slot values on the objects based on the state of the system. The rules themselves are basically if–then constructs that examine the state of the system (functions of values of slots on the objects) in the antecedent (if) clause and then set slot values depending on that state. RiverWare's optimization utilizes preemptive goal programming, using linear programming (LP) as an engine to optimize each of the prioritized goals input by the user. The optimal solution of a higher priority goal is not sacrificed to optimize a lower priority goal. The goals are input by the user through the graphical Constraint Editor tool. Each goal can be either a simple objective or a set of constraints that is turned into an objective to minimize the deviations from the constraints. RiverWare provides automatic linearizing of nonlinear variables.

12.5.2 RiverWare Application

RiverWare was applied to assess alternative policies on the Colorado River (Wheeler et al. 2002). The Colorado River model included the existing policy, frequently called the "Law of the River."

TABLE 12.5
Selected Methods in RiverWare

Object Type	User Method Category	User Methods
Reservoirs	Evaporation & Precipitation	No Evaporation Pan and Ice Evaporation
		Daily Evaporation
		Input Evaporation
		CRSS Evaporation
	Spill	Unregulated Spill
		Regulated Spill
		Unregulated Plus Regulated
		Regulated Plus Bypass
		Unregulated Plus Regulated Plus Bypass
Power Reservoirs	Power	Plant Power
		Unit Generator Power
		Peak Base Power
		LCR Power
	Tailwater	Tailwater Base Value Only
		Tailwater Base Value Plus Lookup Table
		Tailwater Stage Flow Lookup Table
		Tailwater Compare
		Hoover Tailwater
Reaches	Routing	No Routing
		Time Lag Routing
		Variable Time Lag Routing
		SSARR
		Muskingum
		Kinematic Wave
		Muskingum-Cunge
		MacCormack
Water User (on AggDiversion)	Return Flow	Fraction Return Flow
		Proportional Shortage
		Variable Efficiency

Source: Zagona et al. (1998).

The Law of the River includes international treaties, interstate compacts, court decisions and decrees, state and federal statutes, and operating criteria. A variety of policy scenarios were examined, including: (a) Interim Surplus Guidelines Study, (b) Secretarial Implementation Agreement, (c) Multi-Species Conservation Program Study, (d) restoration of the Colorado River Delta, and (e) operation of Flaming Gorge Dam. For these studies, a Graphical Policy Analysis Tool (GPAT) was developed that can compare the output from several RiverWare runs that differ only in terms of policy and hydrologic scenarios.

The Interim Surplus Guidelines Study was to evaluate the potential environmental effects of alternative proposed management scenarios, which establish specific criteria for the declaration of surplus conditions for the lower basin of the Colorado River through 2016. Inherent to this purpose was the reduction of California's dependency upon surplus Colorado River water by this date. This dependency has developed for the past several years due to frequent unused apportionment water causing frequent diversions in excess of the allotted 4.4 maf (million acre-ft) as declared by U.S. Supreme Court decree. The intention of the Interim Surplus period was to provide a "soft landing" for California. The Interim Surplus Guidelines analysis was a landmark study for reclamation by allowing multiple Colorado River stakeholders to have their policy alternatives analyzed within the modeling framework and participating in the analysis through interactive graphical representations of the modeled results. The RiverWare model schematic is presented in Figure 12.9.

Various scenarios were developed by the participating parties, including:

- Baseline condition: represents current operational conditions against which to compare alternative management policies
- Flood-control alternative: surplus conditions are determined to exist only when flood-control releases from Lake Mead are occurring
- Shortage protection: maintain an amount of water in Lake Mead necessary to provide a normal annual supply of 7.5 maf for the lower-basin states and 1.5 maf for Mexico, while also ensuring, with an 80% probability, that Lake Mead's elevation would stay above the "80P-1083" line through 2050
- California alternative: three-tiered approach to declaration of surplus water based on Lake Mead water levels
- Six-states alternative: similar to the California alternative but required higher Lake Mead levels before surplus conditions could be declared
- Basin states alternative: same as the six-states alternative but adjusted to better meet California's projected demand (selected as preferred alternative)
- Environmental alternative: guarantee of 32,000 af (acre-ft) of base flow water to reach the Colorado River Delta during years when Lake Mead's elevation exceeds 1120.4 ft and an additional flood pulse of 260,000 af to the delta when Lake Mead's elevation exceeds the 70% flood-control avoidance (70A1) elevation

Figure 12.10 shows the probabilistic projected reservoir elevations at the end of each year and demonstrates potential comparative impacts on the lower basin. At the probabilistic 10th, 50th, and 90th percentiles of the scenarios, clear differences can be seen with the baseline and flood-control alternatives maintaining higher reservoir elevations, and the California and Shortage Protection alternatives resulting in the largest storage declines. While the policy differences between the alternatives end in 2016, meaningful differences in Lake Mead's elevation persist for years until converging in roughly 2040.

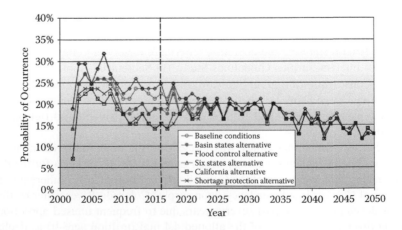

FIGURE 12.10   Lake Mead elevation interim surplus criteria alternatives. (From Wheeler et al. 2002. With permission.)

12.6 GEO-MODSIM RIVER BASIN NETWORK MODEL

12.6.1 GEO-MODSIM OVERVIEW

MODSIM 8.0 (Labadie 2006) is a generalized river basin network flow model designed to aid stakeholders in developing a shared vision of planning and management goals while gaining a better understanding of the need for coordinated operations in complex river basin systems that may impact multiple jurisdictional entities. MODSIM provides for integrated evaluation of hydrologic, environmental, and institutional/legal impacts as related to alternative development and management scenarios, including conjunctive use of surface-water and groundwater resources. Although the MODSIM graphical user interface (GUI) allows displays of background maps for network creation, it lacks capabilities for georeferencing network objects to real-world coordinate systems and incorporating GIS capabilities. As described by Triana and Labadie (2006), Geo-MODSIM has been developed as an extension of MODSIM 8.0 for creating and analyzing georeferenced river basin networks inside ArcGIS 9. Geo-MODSIM allows full utilization of the available spatial data processing, display, and analysis tools in ArcGIS in conjunction with the powerful MODSIM model functionality.

The custom Geo-MODSIM data model is applied to developing georeferenced MODSIM networks by creation of an ESRI geometric network in ArcGIS using imported feature classes such as the National Hydrologic Dataset (NHD) stream and canal layers, reservoirs, gauging stations, diversion structures, and wells. Various tools programmed under the MS .NET Framework utilize ESRI ArcObjects® to develop a MODSIM-compatible georeferenced network of river basin features from the geometric network, and then perform all data import, processing, network simulation, and display of scenario results entirely within the ArcMap® interface to ArcGIS, which essentially serves as the GUI for MODSIM. This allows full exploitation of all spatial data processing and display functionality in ArcGIS for development of MODSIM data sets, and establishes Geo-MODSIM as a spatial decision-support system for river basin management.

In the following subsections, the features and capabilities of Geo-MODSIM are described, followed by a presentation of two case studies demonstrating these capabilities. The first case study applies Geo-MODSIM to the irrigated stream-aquifer system of the Lower Arkansas River basin in Colorado for development of salinity management strategies under complex water rights and interstate compact agreements. For the second case study, Geo-MODSIM is used to assess water conservation programs for the large and complex Imperial Irrigation District (IID) water delivery system in support of the IID/SDCWA Water Transfer Agreement that could provide significant additional water supplies to San Diego County from the Colorado River.

12.6.2 RIVER BASIN NETWORKS BUILT FROM GEOMETRIC NETWORKS

The topology and infrastructure of a river basin network are represented using ArcGIS geometric networks. The geometric network contains the geometry and location of edges and junctions, along with connectivity information between edges and junctions and rules of behavior (e.g., which edge classes can be connected to a particular junction class). The geometric network is assembled from a custom Geo-MODSIM data model designed to accommodate MODSIM features in a geographic environment. Features include: ArkQuestions, Modsim_Canals, Modsim_Demands, Modsim_Gauges, Modsim_Gauges_Inactive, Modsim_Network_Net, Modsim_Network_Net_Junctions, Modsim_ReservoirNodes, Modsim_Sinks, and Modsim_Streams.

System edges in the geometric network are based on the USGS National Hydrography Dataset (NHD), which is a comprehensive set of digital spatial data containing information on surface-water features such as lakes, ponds, streams, rivers, canals, drains, springs, and wells. The NHD data are filtered to an adequate degree of detail, with streams then separated from canals and drainages and imported into the Modsim_Streams and Modsim_Canals elements of the Geo-MODSIM data model. Reservoir nodes are created into the Modsim_ReservoirNodes feature class based

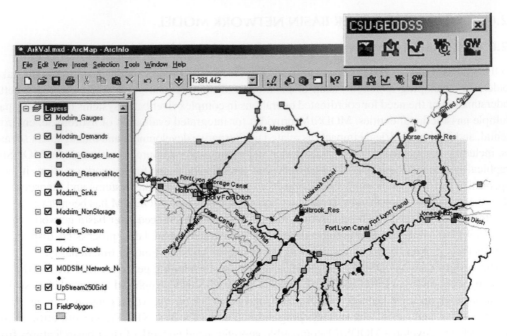

FIGURE 12.11   Example geometric network based on the Geo-MODSIM data model. Inserted CSU-GeoDSS toolbar provides access to network objects using the ArcMap GIS. Area shown is the Lower Arkansas River in Colorado. (From Labadie 2006. With permission.)

on locations of water bodies in the NHD layer and linked to the appropriate streams and canals. Demand nodes are created in the Modsim_Demands feature class and connected to canal edges close to the diversion points. The NHD digitized direction provides a reasonable initial estimate of flow direction in the geometric network, although manual editing is sometimes required in the drainage system. Figure 12.11 shows a geometric network based on the Geo-MODSIM data model as displayed in the ArcMap interface.

Once the geometric network is loaded into ArcMap, the toolbar for the Geo-MODSIM extension is accessed from the Tools > Customize > Toolbars tab by selecting the CSU-GeoDSS toolbar. Clicking the GeoDSS-MODSIM dialog icon in the toolbar opens the Geo-MODSIM GUI in ArcMap. All the functionality for connecting ArcGIS and MODSIM is available in this custom dialog, which docks into the ArcMap work space, providing a convenient means of capturing the objects and characteristics of the geometric network and transforming them into a MODSIM network flow model (Figure 12.12). New MODSIM networks can be generated from this dialog, or existing networks updated based on changes in the geometric network. The network is saved in the MODSIM *.xy file format to accommodate storage of MODSIM-specific data. Synchronization between the geometric network and MODSIM networks is automatically maintained, while also providing access to general MODSIM user dialogs, run options, and customized tools for spatial/temporal data preprocessing.

12.6.3   GEO-MODSIM FUNCTIONALITY

As an extension of MODSIM, Geo-MODSIM networks include all the essential functionality for comprehensive water resources modeling and river basin management. Figure 12.13 summarizes the important river basin features and infrastructure that can be integrated into Geo-MODSIM networks, including dynamic, time-lagged stream–aquifer interaction for conjunctive use modeling. Complex management criteria incorporated into Geo-MODSIM include water-rights decrees, costs and benefits of water use, reservoir operation rules, shortage rules for water allocation during

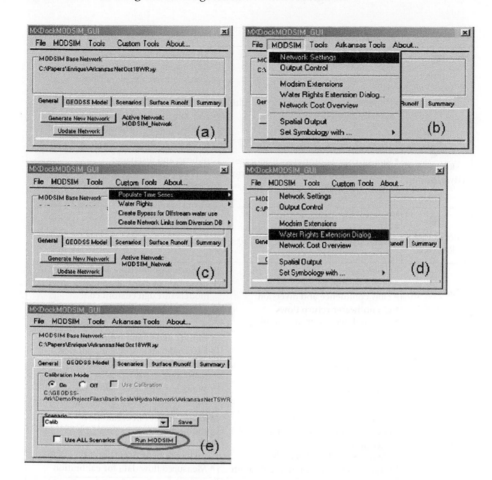

FIGURE 12.12 Geo-MODSIM GUI displayed in ArcMap: (a) start menu, (b) network settings, (c) time-series data import, (d) water-rights extension, (e) run MODSIM. (From Labadie 2006. With permission.)

drought conditions, energy production targets from hydropower, storage ownership accounting, in-stream flow requirements for environmental/ecological protection, flood-control rules, and dynamic stream-flow routing. MODSIM simulates water allocation mechanisms in a river basin through sequential solution of a minimum-cost network-flow optimization problem. The objective function and constraints are automatically formulated in the Geo-MODSIM interface and solved with the highly efficient Lagrangian relaxation algorithm RELAX-IV (Bertsekas and Tseng 1994), which is up to two orders of magnitude faster than the revised simplex method of linear programming. Although Geo-MODSIM is primarily a simulation model, the network flow optimization provides an efficient means of ensuring allocation of flows in a river basin in accordance with specified water rights and other priority rankings, including economic valuation.

12.6.4   GEO-MODSIM CUSTOMIZATION

Several river basin management models that incorporate GIS to some degree, such as MIKE BASIN (DHI Water & Environment 2006), offer internal rainfall-runoff, water quality, and groundwater flow models within their software packages. In addition to these modules, WEAP (Stockholm Environmental Institute-Boston; Yates et al. 2005) includes consumptive use, demand forecasting, and economic valuation modules. These modules are generally simplified for ease of use, whereas many users will often have their own, more accurate models already calibrated to their system. Rather than including simplified, internal modules, Geo-MODSIM is designed to provide customization

Icon	Functional Elements	Data Requirements
Reservoir (Operations)	▪ Main-stem and offstream reservoir operations ▪ Flood control, conservation pools; dead storage ▪ Zones for storage balancing in multi-reservoir systems	▪ Elevation-area-capacity tables ▪ Maximum, minimum, initial storage ▪ Reservoir storage guidecurves ▪ Reservoir balance tables ▪ Hydraulic outlet capacity tables ▪ Net evaporation loss; seepage ▪ Inflow forecasts (if available)
Reservoir (Hydropower)	▪ High-head hydropower ▪ Run-of-river hydropower (0 storage) ▪ On-peak, secondary and firm energy ▪ Pumped storage	▪ Nonlinear efficiency tables as functions of head and discharge ▪ Tailwater-discharge tables ▪ Powerplant capacity ▪ Load factors for pumped storage
StorageRight Reservoir	▪ Storage right accounts ▪ Storage ownership maintenance ▪ Water banking and service contracts	▪ Storage right users ▪ Group ownerships
NonStorage	▪ Watershed runoff ▪ Tributary inflow ▪ Flow confluence and diversion ▪ Groundwater return flows ▪ Stream depletion from pumping	▪ Imported inflow time series data ▪ Execution of external rainfall-runoff models through custom code
Demand	▪ Consumptive demand ▪ Groundwater pumping ▪ Stream-aquifer modeling with glover model or USGS stream depletion factor (sdf) method	▪ Import of demand time series data ▪ External consumptive use models ▪ Demands/priorities conditioned on hydrologic state ▪ Water use efficiency (time variable) ▪ Aquifer parameters; pumping capacity
Flowthru	▪ Instream flow requirements environmental, ecological or navigation purposes ▪ Nonconsumptive demands ▪ Gaging station for model calibration	▪ Time series of instream flow requirements ▪ Flow-through demands and priorities vary with hydrologic conditions ▪ Measured flow data for calibration
NetworkSink	▪ River basin outlet (multiple outlets for several basins allowed)	
Link	▪ Channel losses ▪ Maximum and minimum flow	▪ Time series of maximum capacities ▪ Link costs and benefits
MultiLink	▪ Represent nonlinear discharge-channel loss functions ▪ Nonlinear cost-discharge functions ▪ Multiple water sources and rights	▪ Time series of maximum capacities ▪ Link costs and benefits
RoutingLink	▪ Streamflow and channel routing	▪ Muskingum method coefficients ▪ User defined lag coefficients

FIGURE 12.13   Geo-MODSIM functional objects and features. (From Labadie 2006. With permission.)

capabilities allowing users to attach their own preferred modules. The Custom Code Editor shown in Figure 12.14 is accessed in the MODSIM GUI for preparation of customized code in the Visual Basic.NET or C#.NET languages that are compiled with MODSIM in the Microsoft .NET framework. The .NET CLR produces executable code, as opposed to other applications requiring scripts to be prepared in an interpreted language such as PERL or JAVASCRIPT with poorer runtime performance. The Custom Code Editor guides users in the preparation of customized code, allowing interfacing with MODSIM at any desired strategic locations, including data input, execution at the beginning of any time step, processing at intermediate iterations, and model output. Access to all key variables and object classes in MODSIM is provided, allowing customization for any

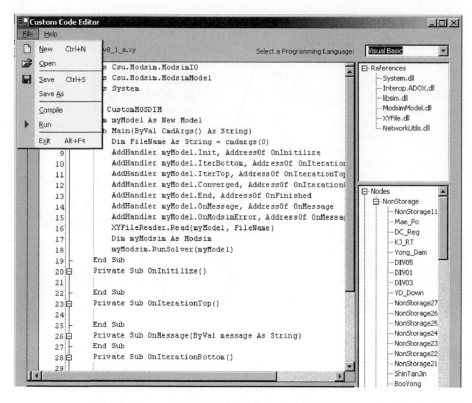

FIGURE 12.14   Custom Code Editor in MODSIM. (From Labadie 2006. With permission.)

complex river basin operational and modeling constructs without reprogramming and recompiling the MODSIM source code.

12.6.5   Assigning River Basin Feature Attributes in ArcMap

Once the MODSIM network is generated, clicking the MODSIM tab in the Geo-MODSIM GUI opens forms for setting MODSIM network properties, selecting variables for output display, activating various MODSIM extensions such as the Water Rights Extension and the Storage Rights Extension, importing water-rights data, and specifying cost and priority structures governing water allocation in the basin (Figure 12.12b).

MODSIM user dialogs can be opened in the ArcMap interface using Geo-MODSIM for import and editing of necessary data for the network objects. ESRI ArcObjects provides access to events triggered in ArcMap that allow association of the MODSIM object-oriented database with the georeferenced features. The Select Features button in the Geo-MODSIM toolbar opens access to the MODSIM dialog for any network object. Users can click on any network feature in the ArcMap display to open a dialog that permits entry and editing of all pertinent data for that feature (Figure 12.15). The GUI then saves the data in MODSIM format. By incorporating MODSIM dialogs in this way, Geo-MODSIM affords the same functionality as the stand-alone MODSIM interface, but with the added capabilities of a GIS.

12.6.6   Time-Series Data

The Populate Time Series Tool (Figure 12.12c) opens the Import TimeSeries dialog for importing time-series data from database-management systems. Supported DBMS software includes MS Access, MS Excel, and *.CSV ASCII files. This tool processes measured data from various agencies, including diversion records, pumping records, and reservoir contents, with the time-series data

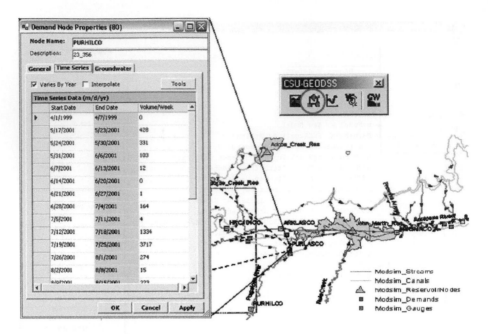

FIGURE 12.15 Georeferenced data input via MODSIM forms in ArcMap. (From Labadie 2006. With permission.)

processed in the correct model time step and units. The processed time series are automatically incorporated into the corresponding MODSIM objects in the ArcMap display.

12.6.7 WATER-RIGHTS DATABASE

The Water Rights Extension Dialog menu item in the Geo-MODSIM GUI (Figure 12.12d) opens the Water Rights–Priorities utility for processing large numbers of water rights and transactions and locating them in a MODSIM network at georeferenced sites (Figure 12.16). Most states record original water rights, as well as modifications in decrees, in a database-management system. Water rights can be abandoned or changed (diversion location, amounts, original appropriation dates). State administrative numbers are used to logically analyze all water rights-records and assign them to links connected to the corresponding demand node. The utility is easily modified to accommodate various state water-rights database formats.

12.6.8 MODSIM EXECUTION FROM ARCMAP

MODSIM networks can be directly executed from the Geo-MODSIM GUI in ArcMap in either Calibration Mode or Management Mode. Calibration Mode automatically calculates unknown river reach gains and losses, such that calculated flows at gauging-station locations match the imported times series of historical measured flows at that location and measured reservoir storage levels agree with historical data. These estimated gains and losses can then be utilized for executing MODSIM networks in Management Mode, where the impacts of various water allocation priorities and reservoir operation policies can be simulated.

12.6.9 GEO-MODSIM OUTPUT DISPLAY AND SCENARIOS ANALYSIS

MODSIM model output is also georeferenced with access available by simply clicking the Display MODSIM Output icon on the Geo-MODSIM toolbar in ArcMap and selecting any desired network feature in the ArcMap display (Figure 12.17). The results display contains a comprehensive

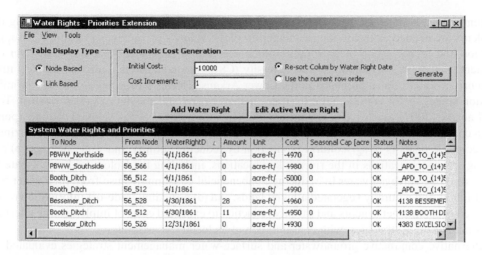

FIGURE 12.16 MODSIM Water Rights Extension user dialog. (From Labadie 2006. With permission.)

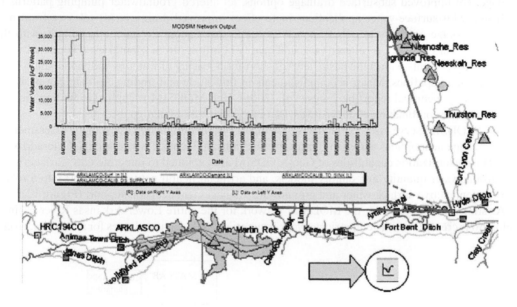

FIGURE 12.17 Georeferenced MODSIM output results displayed in ArcMap. (From Labadie 2006. With permission.)

summary of the variables modeled in each time step, including flow, link losses, routed network flows, water demands and shortages, groundwater variables, reservoir storage, storage-rights accounting, and many others. Output options include extensive statistical analysis and plots of flow duration or exceedance probability curves from Monte Carlo analysis. Scenario analysis tools allow comparison of the performance of several management scenarios for any selected output variable. A tool has been developed in ESRI ArcObjects where users may play an animated movie of MODSIM simulation results in the ArcMap display, with dynamically varying sizes and colors of MODSIM nodes and links reflecting flow and storage magnitudes occurring during the simulation.

12.6.10 MODSIM Application to the Lower Arkansas River Basin, Colorado

The Lower Arkansas River Valley in Colorado is currently experiencing the damaging effects of waterlogging from shallow water tables, excessive salt buildup, and high selenium (Se) concentrations, both on the land and in the river ecosystem. Innovative methods for solving these problems are needed to ensure the sustainability of the valley's productive agricultural base, preservation and revitalization of its rural communities, and enhancement of the overall river environment. Since 1999, Colorado State University (CSU) has conducted studies to develop insight into current water-related problems in the Lower Arkansas River Valley and to identify promising solution strategies for consideration by water managers and stakeholders. As described in Triana and Labadie (2007), extensive field data and modeling tools have been developed and incorporated into a decision-making framework based on Geo-MODSIM that is focused on: (a) maximizing the net economic benefits of agricultural production by reducing salinity and waterlogging; (b) minimizing salt concentrations in the river at key locations, including the Colorado–Kansas state line; (c) maximizing salvaged water by reducing nonbeneficial consumptive use from high water tables under fallow and naturally vegetated alluvial lands; and (d) providing trade-off information relating expected economic net benefits and salinity reduction to expected costs of implementing alternative water management strategies, with consideration of budgetary constraints.

Alternative conjunctive groundwater and surface-water management strategies evaluated with Geo-MODSIM include: (a) reduction of recharge from field irrigation and reduction in canal seepage, (b) improved subsurface drainage options, (c) altered groundwater pumping patterns in exchange with surface-water applications or to effect improved drainage, (d) releases from new reservoir storage accounts to offset impacts of reduced irrigation diversions to comply with the Arkansas River Compact, (e) altered rates and quality of inflows from tributaries, (f) optional water exchange agreements within the basin, and (g) short-term leasing of water by individual ditch companies to municipalities. Major accomplishments in this study and findings to date can be found in a summary technical report by Gates et al. (2006). Geo-MODSIM is applied to finding the best alternatives, or combination of alternatives, that can achieve the stated goals.

Geo-MODSIM serves as a spatial decision-support system designed to assist in the assessment of water management options across the entire river basin from Pueblo Reservoir to the Colorado State line. The customization capabilities of Geo-MODSIM are exploited to integrate GIS, surface-water and groundwater quantity and quality models, and artificial neural networks (ANN) into a robust tool for conjunctive surface-water and groundwater management decision support (Figure 12.18).

As shown in Figure 12.11, the MODSIM network for the entire Lower Arkansas River basin in Colorado is created from a geometric network integrating NHD feature classes for reservoirs, rivers,

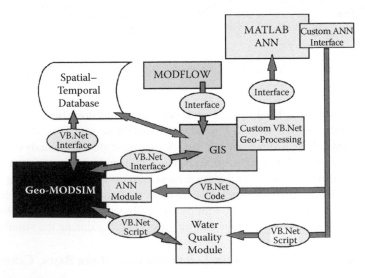

FIGURE 12.18 Spatial decision-support system integrating Geo-MODSIM with groundwater module, water-quality module, and ANN for stream–aquifer modeling. (From Triana and Labadie 2007. With permission.)

tributaries, canals, diversions, demands, and gauging stations by applying the Geo-MODSIM data model. Using the aforementioned tools, attribute data for all river basin feature classes are imported in ArcMap, along with tools for importing time-series data on tributary inflows, flow gauging-station records, meteorological data, and recorded reservoir levels that automatically assign the time-series data to the appropriate river basin feature. The Water Rights Extension not only imports all relevant water-rights data, but creates physical links conveying flows for each water right to the appropriate decree-holder. Various calibration tools are applied to estimate unmeasured river-reach gains and losses using historical gauging-station records.

An innovative methodology developed by Triana et al. (2003) trains an artificial neural network (ANN) using data sets generated from the Groundwater Modeling System (GMS) (Brigham Young University 1999), which links the MODFLOW finite-difference groundwater flow model (Harbaugh et al. 2000) and the MT3DMS (Zheng and Wang 1999) contaminant transport model. Stream–aquifer response relationships from irrigation return flows and stream-flow depletion from groundwater pumping are captured by the ANN and linked to Geo-MODSIM using the customization capabilities available through the MS .NET Framework. Geo-MODSIM can also be directly integrated with the MODFLOW numerical groundwater flow model using a new GIS-based tool that provides a seamless linkage of the surface-water and groundwater flow models. Shown in Figure 12.19 is the Geo-MODFLOW tool for overlaying a Geo-MODSIM surface-water network on MODFLOW grids in the ArcMap interface for calculating return flows to selected Geo-MODSIM links.

Custom tools in ArcMap have been developed for importing and processing data from both intermittent and periodic water-quality sampling in a river basin. Specific conductance data are imported as total dissolved solids (TDS) using a user-selected conversion equation and visualized in the ArcMap environment through user dialogs activated by the water-quality modeling tool (WQM) in the Geo-MODSIM toolbar (Figure 12.20). A variety of regression equations can be selected for filling in missing or sporadic TDS concentration data by estimating flow vs. concentration relationships

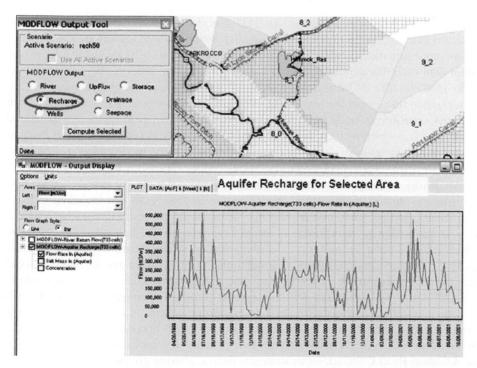

FIGURE 12.19 (See color insert following page 136.) Geo-MODFLOW tool in ArcMap for integrating MODFLOW output with the Geo-MODSIM surface-water network. (From Gates et al. 2006. With permission.)

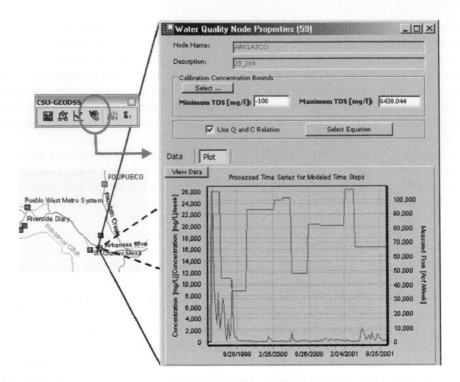

FIGURE 12.20   Access to water-quality tools from Geo-MODSIM toolbar. (From Gates et al. 2006. With permission.)

(Q and C relations). Minimum and maximum TDS concentration ranges can be selected by users or automatically calculated for nodes with water-quality data in the surrounding area. A semiautomatic calibration procedure then allows adjustment of concentrations at upstream nodes with missing concentration data such that measured calculations at downstream nodes in the river are matched as closely as possible. The WQM features an efficient georeferenced network-tracing algorithm that allows navigation throughout the network from upstream to downstream for computing changes in water-quality constituents while having access to user data, MODSIM flows, and geospatial data. Geo-MODSIM is coupled with the WQM and the ANN module at run time to provide conjunctive surface-water and groundwater salt-mass routing throughout the modeled basin. Combined with flow results, the user is capable of monitoring solute concentrations throughout the river system.

Geo-MODSIM, as integrated with the groundwater and water-quality modules, is being applied to investigate the impacts of alternative conjunctive management strategies, including increased irrigation efficiency and reduction in canal seepage. These strategies result in reductions in soil-water salinity and waterlogging, with consequent increases in crop yield. Salt loadings and Selenium concentrations in return flows are markedly reduced under these strategies, thereby enhancing river water quality. However, associated changes in rates and timing of canal diversions and return flows alter the patterns of river flows available for downstream diversion, in-stream use, and discharge into Kansas. Geo-MODSIM and associated tools are being applied to the development of strategies for offsetting these impacts by (a) establishing new accounts in existing on-stream and off-stream reservoirs to store water volumes resulting from reduced canal diversions and (b) optimizing the timing of releases that would adequately preserve historical river flow patterns in compliance with Colorado water law and the Arkansas River Compact. Other strategies include altering groundwater pumping patterns to facilitate efficient irrigation practices in exchange for use of surface-water rights, altered rates and quality of inflows from Fountain Creek (primarily composed of drainage and stormwater runoff from the city of Colorado Springs) upstream of the study area, optional water-exchange agreements within the basin, and short-term leasing of water by individual ditch companies or collective entities such as the recently proposed *super ditch* (http://www.waterinfo.org/arkansas-river-basin).

A Scenario Manager is implemented in the Geo-MODSIM GUI for compiling utilities for scenario management and analysis. The Scenario Manager endows the user with tools to simulate the implementation of these management options at the basin scale. Scenario preferences are loaded and stored for ease of access and repeatability. The tools allow graphical comparison of quantity and quality modeling results from multiple scenarios within the ArcMap interface.

12.6.11 APPLICATION TO IMPERIAL IRRIGATION DISTRICT WATER TRANSFER AGREEMENT

As described in Triana and Labadie (2007), Geo-MODSIM is being applied to analysis of the water-transfer agreement between the Imperial Irrigation District (IID) and the San Diego County Water Authority in Southern California (Miller et al. 2005). Over 3.7 billion m^3 of Colorado River flow are diverted annually to the world's largest irrigation canal, the All American Canal, which supplies the IID. Since IID farmers receive the great majority of these flows, Geo-MODSIM is applied to identifying opportunities for water conservation in the IID that can generate up to 370 million m^3 per year of transferable flow to the large urban areas of Southern California. The IID includes over 2000 km^2 of irrigated farmland served by 2736 km of canals and laterals distributing flow to 5300 farm delivery gates. The high efficiency of the Geo-MODSIM solver allows fully integrated modeling of the IID network, comprising over 10,000 nodes and links. In addition to conservation planning, Geo-MODSIM is also being configured to provide daily, and possibly hourly, real-time regulation guidance for automated control of gates in the IID system.

As illustrated in Figure 12.21, modeling the complex, large-scale water delivery system of the Imperial Irrigation District begins with the design and implementation of an ArcGIS data model. Georeferenced objects are created that embrace the data required for application of Geo-MODSIM within the ArcMap interface. A key feature of this application is that all MODSIM-related data are stored in the ArcGIS geodatabase, allowing all data editing to be directly performed in ArcMap.

A three-pass computational scheme has been designed and implemented to accurately simulate important structural and operational elements of the IID network (Figure 12.22). The three-pass scheme is performed sequentially, with each subsequent network building on computations from the previous pass. The network is automatically transformed and loaded with data for each of the three passes used to simulate IID system operations.

Several obstacles needed to be overcome for accurate modeling of the IID water delivery system in Geo-MODSIM. Computer hardware and software resources were exhausted when attempting to simulate the large-scale IID system in daily time steps over several years. Run times improved significantly with computer upgrades to dual-core Xeon processors and 2 GB of DRAM memory. Software for postprocessing results from Geo-MODSIM runs were designed to take advantage of multithreading technologies that improve computational efficiency. The use of carefully designed object types in the GIS data model (i.e., canal interfaces, terminal interfaces, reservoir interfaces, reservoir-bypass links, etc.) facilitated the implementation of network transformations and changing priority structures in the three-pass computational scheme. Computer memory leakage, created by combining Microsoft COM (Component Object Model) technology as used in ArcGIS with the MS.NET-based framework employed in Geo-MODSIM, was discovered and researched, with methods implemented to avoid or at least minimize these problems. Restrictions in the size of the MODSIM database output files (2 GB per database) were found to be a serious limitation. Automatic output control methods were implemented to reduce the enormous output files produced from daily network simulations over several years by including only selected output variables.

The challenges of calibrating Geo-MODSIM to measured flow and storage data throughout the IID network required the development of customized tools in ArcMap for facilitating adjustment in model parameters and providing both time-series displays and statistical results, as shown in Figure 12.23. Results analysis shows reasonable matching of flows at control points throughout the

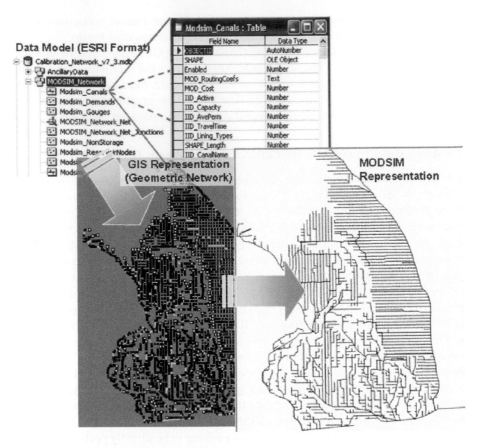

FIGURE 12.21 (See color insert following page 136.) Generation of MODSIM network from Geometric Network developed from Geo-MODSIM data model. (From Triana and Labadie 2007. With permission.)

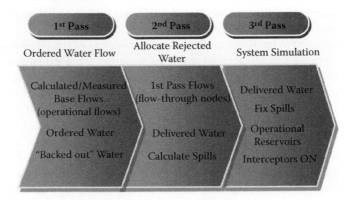

FIGURE 12.22 Three-pass computational scheme for accurate simulation of required water deliveries in the IID network.

system, with most discrepancies and differences in the results identified and explained. The calibrated network model is applied to determine minimum deliveries to IID from the All American Canal that satisfy all IID irrigation requirements, thereby determining the amount of transferable flow available. Dynamic, spatial visualization tools similar to those developed for Geo-MODSIM to the Arkansas Valley salinity-management project provide animated simulations of flow and storage conditions throughout the network under optimum water distribution schemes.

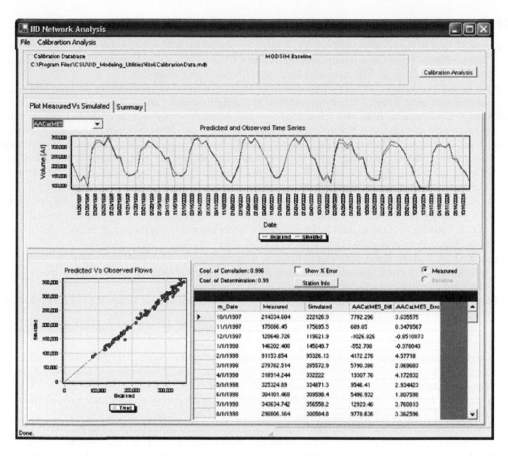

FIGURE 12.23 Customized calibration tools for analysis of the IID network. (From Triana and Labadie 2007. With permission.)

12.6.12 MODSIM CONCLUSIONS

A georeferenced version of MODSIM (Geo-MODSIM) is implemented to incorporate GIS functionality in the powerful modeling tools of MODSIM. Geo-MODSIM is developed as an extension in ArcGIS, thereby providing full access to GIS tools for creating network topology, importing and processing temporal-spatial data, and spatially preprocessing required model data. A spatial decision-support system that incorporates Geo-MODSIM is applied for developing and evaluating salinity-management strategies in the Lower Arkansas River Valley, Colorado. Geo-MODSIM, as integrated with water-quality and groundwater modules, makes full use of the MODSIM customization capability to accommodate specific features dealing with the Arkansas River basin modeling, but without requiring any modification or updating of the original MODSIM source code. Relational data in the spatial-temporal database allow access to and processing of time-series data, water-rights priorities, and storage-water activities, and inserting them directly into Geo-MODSIM. Geo-MODSIM is also applied to the large-scale water delivery system of the Imperial Irrigation District in support of the IID/SDCWA Water Transfer Agreement. The ability to automatically generate the large-scale, complex MODSIM network for the IID system from GIS layers already developed for IID proved to be essential to the success of this project. Again, the unique customization capabilities embodied in Geo-MODSIM, along with the speed and efficiency of the MODSIM network optimization solver, allowed complex operational schemes to be effectively modeled in the IID network.

REFERENCES

Bertsekas, D., and P. Tseng. 1994. RELAX-IV: A faster version of the RELAX code for solving minimum cost flow problems. Completion report under NSF Grant CCR-9103804. Department of Electrical Engineering and Computer Science, Massachusetts Institute of Technology, Cambridge.

Brigham Young University. 1999. The Department of Defense Groundwater Modeling System: GMS v. 3.0 reference manual. Provo, Utah: Environmental Modeling Research Laboratory.

DHI Water & Environment. 2006. MIKE BASIN: A versatile decision support tool for integrated water resources management and planning. Hørshelm, Denmark. http://www.dhisoftware.com/mikebasin/index.htm.

Gates, T., L. Garcia, and J. Labadie. 2006. Toward optimal water management in Colorado's Lower Arkansas River Valley: Monitoring and modeling to enhance agriculture and environment. Completion report No. 205. Colorado Water Resources Research Institute and Colorado Agricultural Experiment Station, Colorado State University, Fort Collins.

Harbaugh, A., E. Banta, M. Hill, and M. McDonald. 2000. MODFLOW-2000, the U.S. Geological Survey modular ground-water model: User guide. Open-file report 00-92. Washington D.C.: U.S. Geological Survey.

Hayes, D., J. Labadie, T. Sanders, and J. Brown. 1998. Enhancing water quality in hydropower system operations. *Water Resour. Res.* 34 (3): 471–483.

Klipsch, J. D., and M. B. Hurst. 2007. HEC-ResSim, reservoir system simulation user's manual, Version 3.0. U.S. Army Corps of Engineers, Hydrologic Engineering Center (HEC), Davis, Calif. http://www.hec. usace.army.mil/software/hec-ressim/.

Labadie, J. 2006. MODSIM: River basin management decision support system. Chapter 23 in *Watershed models*. Boca Raton, Fla.: CRC Press.

Linsley, R. K., J. B. Franzini, D. L. Freyberg, and G. Tchobanoglous. 1992. *Water resources engineering*. New York: McGraw-Hill.

Miller, D., J. Eckhardt, and A. Keller. 2005. The Imperial Irrigation decision support system: Evolution from project planning to operations. In *Proc. World Water and Environmental Resources Congress*. Environmental and Water Resources Institute, ASCE, Anchorage, Alaska.

Triana, E., and J. Labadie. 2007. GEO-MODSIM: Spatial decision support system for river basin management. In *Proc. 2007 ESRI International User's Conference*. San Diego, Calif.

U.S. Water Resources Council (WRC). 1983. Economic and environmental principles and guidelines for water and related land resources implementation studies. Washington, D.C.: U.S. Government Printing Office.

Wheeler, K., T. Magee, T. Fulp, and E. Zagona. 2002. Alternative polices on the Colorado River. In *Proc. NRLC Allocating and Managing Water for a Sustainable Future: Lessons from around the World*, Boulder, Colo.

Yates, D., J. Sieber, D. Purkey, and A. Huber-Lee. 2005. WEAP21: A demand-, priority-, and preference-driven water planning model: Part 1, model characteristics. *Water Int.* 30: 487–500. http://www.weap21.org/.

Zagona, E. A., T. F. Fulp, H. M. Goranflo, and R. M. Shane. 1998. RiverWare: A general river and reservoir modeling environment. In *Proc. First Federal Interagency Hydrologic Modeling Conference*, Las Vegas, Nev., pp. 5-113–120.

Zagona, E. A., T. J. Fulp, R. Shane, T. Magee, and H. M. Goranflo. 2001. RiverWare: A generalized tool for complex reservoir systems modeling, *J. Am. Water Resour. Assoc.* 37: 913–929.

Zheng, C., and P. Wang. 1999. A modular three-dimensional multispecies transport model for simulation of advection, dispersion, and chemical reactions of contaminants in groundwater systems: Documentation and user's guide. Contract report No. SERED-99-1. U.S. Army Engineer Research and Development Center, Vicksburg, Miss.

Index

Printed and bound by CPI Group (UK) Ltd, Croydon, CR0 4YY

24/10/2024

01778288-0010